Describing Species

The Far Side by Gary Larson

"I'm one of those species they describe as 'awkward on land.'"

Gary Larson cartoon celebrates species description, an important and still unfinished aspect of taxonomy.

DESCRIBING SPECIES

Practical Taxonomic Procedure for Biologists

Judith E. Winston

COLUMBIA UNIVERSITY PRESS
NEW YORK

Columbia University Press
Publishers Since 1893
New York Chichester, West Sussex

Library of Congress Cataloging-in-Publication Data
♾

Winston, Judith E.
Describing species : practical taxonomic procedure for biologists / Judith E. Winston.
p. cm.
Includes bibliographical references and index.
ISBN 0–231–06824–7 (alk. paper)—0-231-06825–5 (pbk.: alk. paper)
1. Biology—Classification. 2. Species. I. Title.
QH83.W57 1999
570'.1'2—dc21 99–14019

Casebound editions of Columbia University Press books are printed on permanent and durable acid-free paper.
Printed in the United States of America
c 10 9 8 7 6 5 4 3 2 1
p 10 9 8 7 6 5 4 3 2 1

For my daughter, Eliza, who has grown up (and put up) with this book

Contents

List of Illustrations

List of Tables

Preface

This book is intended to introduce students and professional scientists to basic taxonomic procedure and enable them to carry out whatever taxonomic writing they need in their studies or careers. It is intended not as a systematics textbook, but as a supplement to a systematics course and a desk reference and guide to nomenclatural procedure and taxonomic writing. Because it covers both the botanical and zoological codes of nomenclature, it should be useful to workers in most fields of biology and paleontology. It will be most valuable to ecologists, field biologists, and others who encounter new taxa in the course of their research, but it will be useful to anyone who has to write any document in which taxonomic descriptions figure: reports, checklists, floras, faunal surveys, revisions, monographs, or guides. Systematics is a global and multicultural enterprise, as the examples in this book show. Although the book was written for an English-speaking audience, it should be useful anywhere Taxonomy is spoken.

Acknowledgments

This book could not have been written without the assistance of many colleagues. Service on the Scientific Publication Committee at the American Museum of Natural History during 12 years as a curator there gave me the experience to begin writing it. However, a work so dependent on examples of taxonomic writing for many groups of organisms would have been impossible to complete without the generous responses by many colleagues to requests for assistance in finding good and bad examples of taxonomic writing or for discussions of practice and procedure in their specialties. My thanks to Patricia Bergquist, Dale Calder, Jim Carlton, Paisley Cato, Alan Cheetham, Fu-Shiang Chia, Liz Chornesky, Kerry Clark, Richard Cloney, Jay Cole, Masahiro Dojiri, Bill Emerson, Kristian Fauchald, Nick Fraser, Jane Fromont, Carl Gans, Glenys Gibson, Ivan Goodbody, Peter Hayward, Gordon Hendler, Lee Herman, Bob Higgins, Richard Hoffman, Ron Karlson, Nancy Knowlton, Alan Kohn, Mary Lecroy, John Lee, Linda Mantel, Les Marcus, Shunsuke Mawatari, Malcolm McKenna, Ken McKinney, John McNeill, Nancy Moncrief, Guy Musser, Chuck Myers, Jon Norenburg, Dave Pawson, John Pearse, Robert Raven, Mary Rice, Pamela Roe, Jerry Rosen, Ed Ruppert, Mike Salmon, Ralph Smith, Dick Strathman, Megumi Strathman, Dick Tedford, Charlie Wahle, Buck Ward, Mary Wicksten, and Phil Yund. Thanks to Brenda Jones, AMNH, for sharing her editorial expertise, and to Kris Metzger, HBOI, and Pat Christenbury, VMNH, for bibliographic and interlibrary loan assistance. I am especially gratefully to those who reviewed or read all or parts of drafts of the manuscript: Gordon Hendler, Richard Hoffman, Ken McKinney, and Dennis Stevenson. Thanks for your improvements–I take full responsibility for any mistakes that remain. Thanks to these publishers and journals for generously giving permission for the use of quotations or figures: American Museum of Natural History, *The Beagle, Biochemical Ecology,* Biological Society of Washington, *Bulletin of the National Science Museum Tokyo, Bulletin of Zoological Nomenclature,* California Academy of Science, *Canadian Journal of Botany,* Cincinnati Museum Center, Columbia University Press, Elsevier Science, *Fishery Bulletin, Gulf Research Reports, Huntia,* International Association for Plant Taxonomy, *Marine Biology* (Springer-Verlag), Missouri Botanical Garden, Muséum National d'Histoire Naturelle, Paris, the Queensland Museum, Southern California Academy of Science, Zoological Journal of the Linnaean Society (Academic Press), and *Zoologische Anzeiger* (Fischer-Verlag).

Finally, thanks to taxonomists everywhere, for their adamantine courage in pursuing their discipline regardless of fashions in science or funding; it is their devotion that will make informed planet stewardship possible.

JUDITH E. WINSTON
Director of Research
Virginia Museum of Natural History
Martinsville, Virginia, 1999

Describing Species

Part One

INTRODUCTION

Chapter 1

Introduction

The first step in wisdom is to know the things themselves; this notion consists in having the true idea of the object; objects are distinguished and known by their methodical classification and appropriate naming; therefore Classification and Naming will be the foundation of our Science.

–LINNAEUS (1735), QUOTED IN STEVENS (1994:201)

The laws of biology are written in the language of diversity.

–E. O. WILSON (1989:243)

Describing the Living World

When Carl Linnaeus published the first edition of his *Systema Naturae* in 1735, he probably thought he was close to fulfilling his ambition of classifying all the plants and animals in the world. But by 1749 he was writing desperately to his friend Abraham Bäck, "Am I to work myself to death, am I never to see or taste the world? What do I gain by it?" (Lindroth 1983:31). The job was already too much for any one person. Linnaeus persevered with his immense project, however, compiling 12 editions of the *Systema Naturae* before his death in 1777. He was succeeded by his students and by other systematists, who were explorers themselves or who had available to them the treasure trove of specimens collected by the great eighteenth- and nineteenth-century expeditions into the exotic, unknown regions of the planet.

Today, more than two hundred years and at least eight generations of systematists later, most of those unexplored jungles and forests have been populated, developed, and connected to the rest of the world by airplane, modem, and fax machine. The amount of wilderness remaining has shrunk to about one-third of Earth's land surface (McCloskey and Spalding 1989), much of that third in the Arctic or Antarctic. The number of known species of eukaryotic organisms is currently estimated to be about 1.4 million (Lean et al. 1990; Systematics Agenda 2000 1994). It seems reasonable to suppose that most of Earth's species have been described.

But this is far from true. There are millions of species still undescribed and an increasing amount of concern over whether we can discover them before they become extinct (Roberts 1991; Vane-Wright et al. 1991). Taxonomists themselves are considered by many to be an endangered species (Wilson 1989; Anonymous 1991; Harvey 1991; Feldman and Manning 1992; Scheltema 1996). This book was written to help bring about a positive solution to this biodiversity crisis by providing a basic practical guide to the process of describing a new species. This chapter explains why species description is still necessary, who finds and describes new species, how this book is organized, and what it covers.

Why Is Species Description Necessary?

Of the 1.4 million species known to science, about 300,000 are higher plants (bryophytes and tracheophytes), 70,000 are fungi, and 87,000 are protistans (algae and protozoans). That leaves more than a million species of animals, with insects, arachnids, mollusks, crustaceans, and vertebrates the largest groups (Goto 1982; Barnes 1989; Hawksworth 1991; May 1991; Stevenson, personal communication 1997).

Does that mean we are done with species description? Not by any means. Estimates of numbers of undescribed species range from 5 or 10 million (Office of Technology Assessment 1987) to 30 or even 50 million, when small and host- or habitat-specific organisms such as mites, beetles, and nematodes are taken into account (Erwin 1982, 1988; Marcus 1993). On the basis of even the 5 million figure, only 30 percent of existing organisms have been described. If the highest figure is correct, then less than 2 percent of Earth's inhabitants are known.

In many cases new collecting techniques and new technology have led to a dramatic increase in the number of species described for a group of organisms. For example, the combination of modern mountain climbing techniques and insecticide fogging at various levels in the previously unexplored rainforest canopy of Brazil led Terry Erwin to project that there might be as many as 30 million species of arachnids and insects on Earth (Erwin 1982, 1988).

Similar projections have been made for fungi. David Hawksworth (1991) surveyed the relationship between the number of species of vascular plants and number of species of fungi in well-studied geographic areas such as the British Isles, coming up with a ratio of 1:6. Projecting this ratio using the known number of vascular plant species gives an estimate of 1.8 million species of fungi. So as much as 96 percent of all fungi could still be undescribed. Considering their value as genetic resources for the production of new medicines, as well as their major importance as decomposers, it does seem urgent to learn more.

It is unlikely that many nonarthropod invertebrate groups are this greatly underestimated. However, exploration of a new region (where a group has never been studied) or a new habitat for almost any group of organisms can still result in a large number of new species being discovered and described. Our knowledge of the taxonomy of marine organisms, in particular, is far from complete, perhaps because most of them are hidden from our sight in an environment inhospitable to human bodies (Earle 1991). In the last few decades new techniques, new kinds of dredges (figure 1.1), scuba diving, and manned and unmanned submersibles have changed our idea of marine biodiversity and made us aware of entirely new habitats such as whale carcasses, hydrothermal vents, and human-made debris (Norse 1993; Committee on Biological Diversity in Marine Systems 1995). About 250,000 marine species have been described so far (Barnes and Hughes 1982; Winston 1992). Most of them are from shallow (intertidal to continental shelf) depths and well-studied coasts. In other areas 50 to 80 percent of the inhabitants may be unknown. Estimates of the number of species still to be discovered in the ocean range from a minimum of 50,000 (for the continental shelves and coastal margins) to a maximum of 10 million, if the deep sea is included (Grassle 1991; Winston 1992).

Even in well-studied marine areas such as the seas around the British Isles, that have had scientific attention since Linnaeus's day, new species are still being discovered. A plot of the number of new marine species described per year between 1757 and 1992 showed that curves for a number of small-bodied groups (e.g., polychaetes, copepods, and nematodes) still haven't leveled off, indicating that a large number of species still remain to be described (Costello et al. 1996). At the same time, many large marine

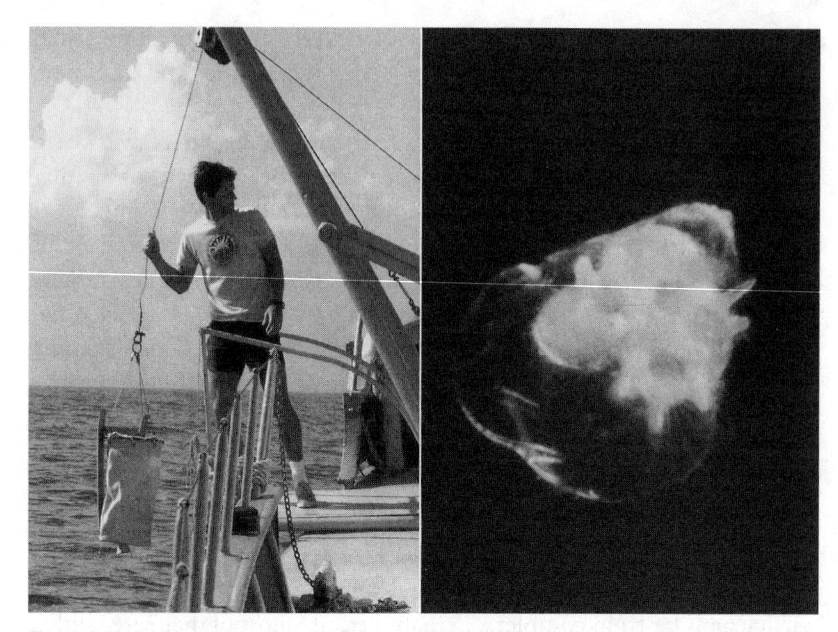

Figure 1.1 New equipment and new habitats yield new species. **Left,** A small sled dredge being lowered over the side of a research vessel to sample biota and sediments on a Florida shoal. This recently invented dredge skims the sediment–water interface and top layers of sediment as it slides along the bottom. Larger versions have been used in deep-sea sediments, helping revolutionize our ideas of the biodiversity of the deep ocean floor. **Right,** In 1984, while using such a dredge to sample sandy sediments at Capron Shoal, Florida, for epibenthic cupuladriid bryozoans, biologists discovered a previously unknown habitat for encrusting bryozoans: living interstitially on grains of sand. On this one shoal, nine new bryozoan species were discovered, including *Trematooecia psammophila,* pictured here on a grain of quartz sand (Winston and Håkansson 1986). *–Photos: J. E. Winston.*

animals (such as the Mediterranean monk seal, the coelacanth, and the Hong Kong pink dolphin) are approaching the brink of extinction.

We tend to think that unknown species must be microscopic in size or cryptic in habitat, but this is not necessarily true. Small mammals are still discovered fairly often; they are mostly rodents such as the Panay cloudrunner, *Crateromys heaneyi* (Gonzales and Kennedy 1996), pictured in figure 1.2, but 16 large mammals have been described since 1937 (Pine 1994). For example, in 1991 a new species of beaked whale was described from the eastern tropical Pacific (Reyes et al. 1991). In 1993, a new bovid, the Vu Quang ox, was described from Vietnam. Molecular and morphological studies showed that it was not only a new species, but a new genus in the family to which cattle, goats, and antelope belong (Dung et al. 1993). We might also expect that we would at least know all of our closest relatives, the primates, by now, yet in 1990 a new species of primate, the black-faced lion tamarin, was discovered in a densely populated area near São Paulo, Brazil (Lorini and Persson 1990). As one primatologist put it, this was "almost like finding a major new species in the suburbs of Los Angeles" (R. Mittermeier, quoted in Anonymous 1990:20). Birds are probably the most intensively studied vertebrate group, but about three new species are found per year (Blackburn and Gaston 1995). Most of them come from the few remaining unexplored areas of the globe, such as the most recent, a spectacularly colored new barbet (a small toucanlike bird) from Peru (Stap 1997).

Some very large trees have recently been discovered also. In 1995 the Wollemi pine, a member of a group of plants thought to have been extinct

Figure 1.2 According to some estimates, more than half of all Earth's species will be lost by 2100 (National Research Council 1980). The Panay cloudrunner, *Crateromys heaneyi,* was not described until 1996. This nocturnal tree-dwelling rodent comes from a mountainous region of western Panay in the Philippines. It is probably still common in the remaining forests there, but, like many other organisms, its continued existence is threatened by habitat destruction (Gonzales and Kennedy 1996). *–Photo: Dr. Robert S. Kennedy, © Cleveland Museum Center.*

since the Jurassic, was found in Wollemi National Park in Australia's Blue Mountains (only 125 miles from Sydney), and some trees in a remote Colombian rainforest were described as *Pseudomonotes tropenbosii,* a new species of dipterocarp, a kind of tree previously known to occur only in Asia and Africa (Anonymous 1995). It *is* rare to find a new tree, primate, or cetacean species these days, of course, but clearly not impossible, and millions of invertebrates and plants remain to be described. Some of them can even be found in well-explored regions, as in the case of the Virginia Beach bug (figure 1.3).

How New Species Are Described

This book is about taxonomy, particularly taxonomic procedure. There has been a considerable discussion in the literature about the difference

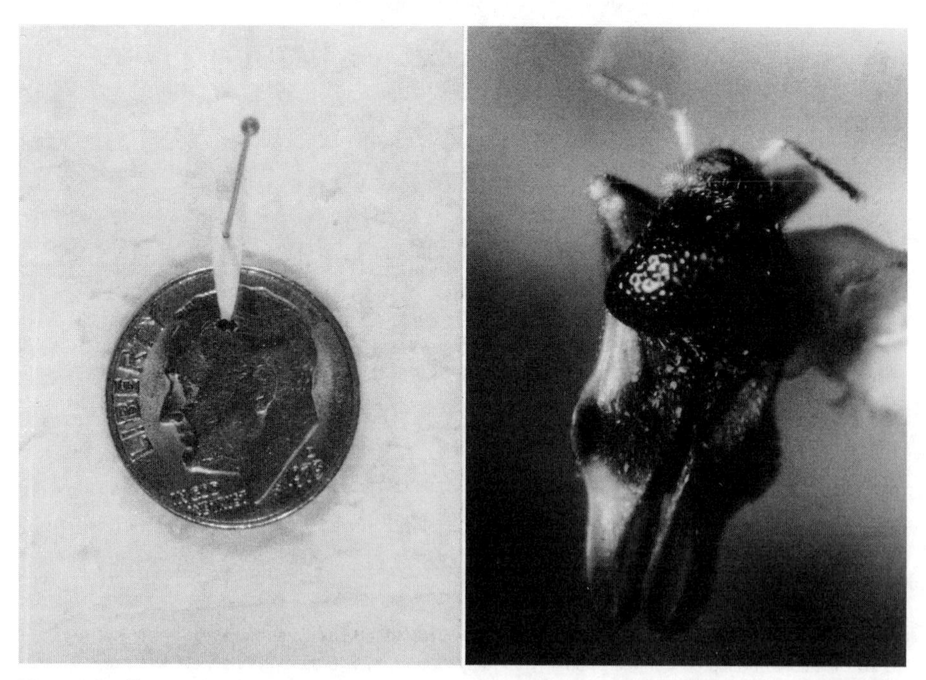

Figure 1.3 New species, both small and large, may come from parks as well as wilderness. This specimen of *Pycnoderiella virginiana,* the Virginia Beach bug was discovered in Seashore State Park, Virginia Beach. It is a new genus of a subfamily of plant bugs mostly found in the Neotropics (Henry 1993). **Left,** Pinned type specimen, shown against a dime for scale. **Right,** Closeup of same specimen. *–Photo: Bonnie Helms, VMNH.*

between taxonomy and systematics–which term was used first, which is most comprehensive, and so on–and some practitioners still say there's no distinction between the terms. However, one generally accepted definition is that *systematics* is the study of biological diversity and of the evolutionary relationships among organisms (Simpson 1961; Mayr 1969; Wilson 1985). By this definition *taxonomy* is a subdivision of systematics, consisting of three associated activities: *identification* (referring a specimen to a previously classified and named group), *classification* (ordering organisms into groups based on perceived similarities or differences), and *nomenclature* (naming groups of organisms according to rules developed for the process). Systematics also includes the study of the process of evolution and phylogeny. *Taxonomic procedure* is the practical process of identifying, recognizing, researching, or redescribing a taxon for scientific publication according to the current rules of biological nomenclature.

Biological nomenclature, the system of scientific naming of organisms, was developed to ensure that every organism can have a name that is unambiguous (refers only to that particular kind of organism) and globally understood. A species becomes known in the scientific sense when a Latin *binomial,* a name consisting of two parts (a genus term and a species term), and a description are published in the scientific literature, according to the rules of botanical (The International Code of Botanical Nomenclature, ICBN) or zoological (The International Code of Zoological Nomenclature, ICZN) nomenclature. Although they are written in legal language, there is no agency to enforce these codes; there is only the consensus of biologists to observe and accept them (Jeffrey 1989).

Describing new species is still an important part of taxonomy. About a third of all taxonomic papers published over the last 28 years contain a description of at least one new species. A survey of the BIOSIS© database (Winston and Metzger 1998) showed that the number of papers in which at least one new species is described increased every year, both in numerical terms and in proportion to the rest of the taxonomic literature up until 1988, and remains at a high level (figure 1.4).

The majority of new species are found and described by professional systematic biologists and paleontologists in the course of their research, as they study specimens generated by their own field work or by expeditions or surveys made by others and preserved in museum collections. Their work is augmented by the efforts of graduate students working on systematic thesis or dissertation topics. However, a large number of new species are discovered by other scientists: ecologists, paleoecologists,

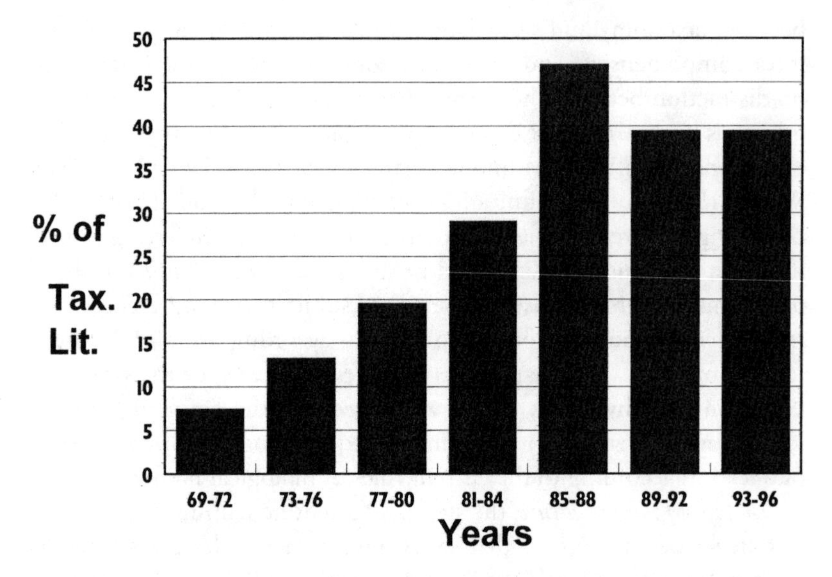

Figure 1.4 Species description is still a vital part of taxonomy. The bar graph shows taxonomic papers containing one or more new species published as a percentage of all taxonomic papers published during each time period. Greatest rate of production was between 1985 and 1988, but on average new species have appeared in more than 35 percent of all taxonomic papers during the last 28 years.

and stratigraphers in the course of field studies, physiologists carrying out comparative research, molecular biologists or biochemists seeking new compounds or processes, and so on.

If you find yourself working on what you think may be an undescribed species, your first choice, of course, should be to consult a professional systematist, and chapter 5 of this book gives advice on how to find and maximize the potential for cooperation with such people. But that course of action may not be feasible or even possible for two reasons. In the first place, despite common opinion (even among other scientists), professional systematists do not spend most of their time identifying specimens for other people, nor do they usually write descriptions of new species on request. They have their own research programs to carry out, to which they are probably already overcommitted. They may describe new species as part of their efforts to clarify the relationships of the groups in which they specialize, but they are unlikely to drop their own work to write a description of a single species for you, unless it is a very interesting species (Winston 1988).

In the second place, for many groups of organisms there simply may not be a systematist available. Despite many reports on the plight of systematics over the last 25 years (Steere et al. 1971; Lee et al. 1978; Stuessy and Thompson 1981; Edwards et al. 1987; Scudder 1987), the situation has continued to worsen. The average age of systematists (like that in the academic professions in general) is rising (Claridge and Ingrouille 1992). Many systematists now practicing are already retired from careers at universities or museums. Some work on systematics only part time (and support themselves by teaching or by some other kind of scientific research). Others specialize in only a few taxa or in the fauna of a limited geographic area (Winston 1988). In addition, the numbers and specialties of existing taxonomists do not correspond with the groups in which the most work remains to be done, nor with the areas of the world in which new species are most likely to be found (Gaston and May 1992). For example, for the 150,000 or so marine species in the groups most commonly collected for use in natural products chemistry, there are fewer than 200 specialists in the whole world, even including those who are retired or work only part time on systematics projects (Winston 1988).

In some invertebrate groups (such as medically important arthropod disease vectors) the situation has reached crisis proportions (Oliver 1988; Harvey 1991). Furthermore, the closing of a number of major natural history collections and cutbacks in personnel and funding at others have increased the work load for remaining taxonomists (Culotta 1990, 1992, 1993).

In summary, most biologists who find a new species, whether living or fossil, must describe it themselves. This book aims to explain the procedure by which scientists who find a new species in the course of their research can perform the necessary background studies and write a publishable description of that species. It is intended to serve as a guide to the common taxonomic problems the working biologist may encounter and to suggest steps for resolving those problems. The book is also intended to introduce students of biology and paleontology to the practical basics of taxonomy. They are the ones who will inherit the issues of biodiversity preservation with which the world's scientific community is just beginning to grapple. Several recent studies have corroborated biologists' intuitive perception that species-rich ecosystems should be more productive and stable than impoverished ones (Culotta 1996; Bengtsson et al. 1997; Tilman et al. 1997; McGrady-Steed et al. 1997; Naeem and Li 1997). The variety of genetic potential available to develop new drugs, new crops, and new biological products of all kinds has hardly been tapped (Wilson 1994). There is no way that Earth's biological resources can be preserved or properly managed without knowledge of both the organisms inhabiting a locality and their position (in terms of relationships and functioning) in that community. Paleontological studies documenting past trends in biodiversity are also essential to conservation because they help put long-term trends, including extinction, into perspective (Lee 1997).

Scope and Organization of This Book

Although new philosophies and techniques of systematics have generated a good deal of excitement in recent years, the basics of taxonomic procedure and the mechanics of species description are hardly taught anymore. This essential task of systematics–dealing with the rules of scientific nomenclature for the formal description of new species and other taxa–is considered old-fashioned and unworthy of a place in the modern science curriculum. Even books on systematics give species description short shrift; for instance, Ernst Mayr (in *Principles of Systematic Zoology,* 1969:260) dismissed the whole problem (except when part of a professional taxonomist's revisionary work) in a few words: "Such isolated descriptions are justified only in connection with biological or economic work." The problem is that species descriptions are increasingly needed for exactly those reasons.

I began this book at the request of some ecologist colleagues who were finding undescribed species in the course of their research and wanted to know what steps they should take to get those species named (it is a little embarrassing to have to publish 6 years of data on the ecology of "species X"). When my friends first brought the problem to my attention I went to the library to look for a book they could use. After a little research, I discovered that, although most of what they would need to know had been covered previously, many of the relevant books (e.g., Schenk and McMasters 1936, 1948, 1956; Mayr et al. 1953; Simpson 1961; Savory 1962; Blackwelder 1967; Mayr 1969) were out of print or based on outdated version of the codes of nomenclature, leaving no one source the beginner could readily consult to learn the basic practical aspects of taxonomic procedure.

This book is an attempt to solve that problem. It is divided into four parts. Part I (chapters 1–2) is an introduction to the history of species descriptions and the naming of organisms. Those who dislike history may be tempted to skip chapter 2. Please do not, because the procedures biologists and paleontologists follow today make better sense once you understand how they evolved. Part II (chapters 3–5) covers the background research necessary before a species can be described, the recognition of species in different groups of organisms, and the library and museum research necessary for identification and description. Part III (chapters 6–16) covers the section-by-section writing of species descriptions and their publication. Part IV (chapters 17–22) goes beyond the basic species description to address subspecies recognition and description, redescription and revision, key construction, description of higher taxa, common problems that may be encountered, and more advanced studies in systematics. The book is not intended as a general systematics textbook. There are already good books available for that purpose (see this chapter's *Sources* list for titles). Nor does it go into detail on systematics philosophies and methodologies (although chapter 22 is intended to point those interested toward sources for further study).

Although I have aimed to make this guide as broadly applicable as possible, I am a zoologist, so more of my examples deal with animals and with the International Code of Zoological Nomenclature (ICZN). But, realizing that people in fields such as ecology might be dealing with animals in the course of one project and plants in the next, I have tried to trace the history of the International Code of Botanical Nomenclature (ICBN) and, wherever possible, point out differences in procedure in describing organisms from these two groups. Luckily, there are a number of excellent books on plant taxonomy and nomenclature (e.g., Hawksworth 1974; Stace 1989; Stuessy 1990, and others in this chapter's *Sources* list). There is a separate version of the botanical code for those who deal with cultivated plants (Trehane et al. 1995).

Much presented here will also be useful for those working with bacteria. Descriptions of bacteria are handled according to the International Code of Nomenclature for Bacteria (ICNB), latest version, Sneath (1992). This code is explained well by Jeffrey (1989), as are the special nomenclatural problems of cultivated plants and domesticated animals.

Virologists are still working on establishing a universally acceptable system for their nomenclature (see Mathews 1981, 1985; Murant 1985). Anyone attempting to deal with viral taxonomy should see the publications of the International Committee on Taxonomy of Viruses in the journal *Intervirology*, as well as Francki et al. (1991), *Classification and Nomenclature of Viruses.*

The descriptive and nomenclatural problems a biologist faces will vary with the groups of organisms involved. Those who work on birds or mammals may never have to describe a new species, but they may well have to resolve a synonymy, designate a neotype, or grapple with subspecific variation. Those who work on invertebrates, fungi, algae, or protistans may have to deal with not one but many new taxa, for which not only must species be described, but higher-level taxonomic placement assigned.

The Pleasures of Systematics

I anticipate grumbling from a few of my colleagues who seem to feel that systematics should remain an esoteric pursuit. They are convinced that if "amateurs" acquire the information they need to publish taxonomic papers, the result will be a horrendous proliferation of poorly delimited species and a great increase in confusion in the literature. It must be noted that with certain groups (such as mollusks), there has been some justification for this viewpoint.

However, although following established taxonomic procedure may be time-consuming and occasionally tedious, it is not especially difficult. If you follow the steps outlined in this manual, you will be able to make a positive contribution to the systematics of the group you are studying. You will gain a greater appreciation of the field of systematics as a whole and you may develop a new avocation or subspecialty, for systematics is a pursuit to enjoy. In a sense, systematics is not a specialty at all; at its best it encompasses *all* of biology, addressing most basic questions regarding life itself. A good systematist must be able to learn a new field of biology or a new technology, or even create one, doing whatever may be necessary to deal with the problem encountered.

The necessary periods of field work, being outdoors and observing the natural world, are satisfying on an aesthetic and spiritual level. The detective work involved in establishing the identity of a taxon challenges the intellect and often generates ideas that reach beyond the limited scope of description to help answer an important biological question. And unlike some pursuits such as active sports, or creative research in disciplines such as physics, in which a person never surpasses the efforts of youth, a systematist can increase in skill and knowledge as long as his or her mental faculties remain, making it a pleasure that can last a lifetime.

Sources

References cited in the text are listed in the *Literature Cited* section at the end of the book. *Sources* are included at the end of the appropriate chapter. They may include books or articles that give more information on topics covered in the text or they may provide bibliographies of references useful in research on groups of organisms, preparation of descriptions, and other topics.

Printed Sources

Biodiversity

Balick, M. J., E. Elisabetsky, and S. A. Laird, eds. 1995. *Medicinal Resources of the Tropical Forest: Biodiversity and Its Importance to Human Health.* New York, Columbia University Press.

Committee on Biological Diversity in Marine Systems. 1995. *Understanding Marine Biological Diversity.* Washington, D.C., National Academy Press.

Eldredge, N., ed. 1992. *Systematics, Ecology, and the Biodiversity Crisis.* New York, Columbia University Press.

Forey, P. L., C. J. Humphries, and R. I. Vane-Wright, eds. 1994. *Systematics and Conservation Evaluation.* Oxford, U.K., Clarendon Press.

Gaston, K. J., ed. 1996. *Biodiversity: A Biology of Numbers and Difference.* Oxford, U.K., Blackwell Science.

Heywood, V. H., ed. 1995. *Global Biodiversity Assessment.* New York, Cambridge University Press.

MacDonald, J. F., ed. 1995. *Genes for the Future: Discovery, Ownership, Access.* Ithaca, N.Y., National Agricultural Biotechnology Council.

Norse, E. A., ed. 1993. *Global Marine Biological Diversity: A Strategy for Building Conservation into Decision-Making.* Washington, D.C., Island Press.

Vogel, J. H. 1994. *Genes for Sale: Privatization as a Conservation Policy.* New York, Oxford University Press.

Wilson, E. O., ed. 1988. *Biodiversity.* Washington, D.C., National Academy Press.

Wilson, E. O. 1992. *The Diversity of Life.* Cambridge, Mass., Harvard University Press.

Systematics

Benson, L. 1962. *Plant Taxonomy, Methods and Principles.* New York, Ronald Press.

Benson, L. 1979. *Plant Classification,* 2d ed. Lexington, Mass., Heath.

Blackwelder, R. I. 1967. *Taxonomy: A Text and Reference Book.* New York, Wiley.

Davis, P. H. and V. H. Heywood. 1965. *Principles of Angiosperm Taxonomy.* Princeton, N.J., D. Van Nostrand.

Goto, H. E. 1982. *Animal Taxonomy.* London, Edward Arnold.
Hawksworth, D. L. 1974. *Mycologist's Handbook. An Introduction to the Principles of Taxonomy and Nomenclature in the Fungi and Lichens.* Kew, U.K., Commonwealth Mycological Institute.
Hillis, D. M., C. Moritz, and B. K. Mable. 1996. *Molecular Systematics,* 2d ed. Sunderland, Mass., Sinauer Associates.
Jeffrey, C. 1982. *An Introduction to Plant Taxonomy,* 2d ed. Cambridge, U.K., Cambridge University Press.
Jones, S. B. and A. E. Luchsinger. 1986. *Plant Systematics,* 2d ed. New York, McGraw-Hill.
Mayr, E. and P. D. Ashlock. 1991. *Principles of Systematic Zoology,* 2d ed. New York, McGraw-Hill.
Minelli, A. 1993. *Biological Systematics: The State of the Art.* London, Chapman & Hall.
Quicke, D. L. J. 1993. *Principles and Techniques of Contemporary Taxonomy.* London, Blackie Academic and Professional.
Radford, A. E. et al. 1986. *Fundamentals of Plant Systematics.* New York, Harper & Row.
Ross, H. H. 1974. *Biological Systematics.* Reading, Mass., Addison-Wesley.
Sivarajan, V. V. 1991. *Introduction to the Principles of Plant Taxonomy,* 2d ed. Cambridge, U.K., Cambridge University Press.
Sneath, P. H. A. and R. R. Sokal. 1973. *Numerical Taxonomy.* San Francisco, W.H. Freeman.
Solbrig, O. T. 1970. *Principles and Methods of Plant Biosystematics.* New York, Macmillan.
Soltis, P. S. et al. 1992. *Molecular Systematics of Plants.* New York, Chapman & Hall.
Stace, C. 1989. *Plant Taxonomy and Biosystematics,* 2d ed. London: Edward Arnold.
Stuessy, T. F. 1990. *Plant Taxonomy: The Systematic Evaluation of Comparative Data.* New York, Columbia University Press.
Talbot, P. H. B. 1971. *Principles of Fungal Taxonomy.* London, Macmillan.
Wiley, E. O. 1981. *Phylogenetics. The Theory and Practice of Phylogenetic Systematics.* New York, Wiley.

Exploration

Blunt, W. 1971. *The Compleat Naturalist: A Life of Linnaeus.* New York, Viking.
Brosse, J. 1983. *Great Voyages of Discovery: Circumnavigators and Scientists 1764–1843.* New York, Facts on File.
Isley, D. 1994. *One Hundred and One Botanists.* Ames, Iowa State University Press.
Keeney, E. B. 1992. *The Botanizers: Amateur Scientists in Nineteenth Century America.* Chapel Hill, University of North Carolina Press.
Linklater, E. 1972. *The Voyage of the Challenger.* Garden City, N.Y., Doubleday.
Moorehead, A. 1969. *Darwin and the Beagle.* New York, Harper & Row.

Reveal, A. L. 1992. *Gentle Conquest: The Botanical Discovery of North America with Illustrations from the Library of Congress.* Washington, D.C., Starwood.

Stap, D. 1990. *A Parrot Without a Name: The Search for the Last Unknown Birds on Earth.* New York, Knopf.

Steinbeck, J. and E. O. Ricketts. 1986. *The Log from the Sea of Cortez: The Narrative Portion of the Book, Sea of Cortez by John Steinbeck and E. R. Ricketts, 1941.* New York, Penguin.

Wilson, E. O. 1994. *Naturalist.* Washington, D.C., Island Press.

Internet Sources

Except as noted all addresses begin with **http://www.** Web sites change frequently; sites are discontinued and new sites and new links to old sites are added. This section lists sites available at the time this book went to press, but a new subject search by the user is recommended.

nbs.gov/nbii
U.S. National Biological Information Infrastructure.

http://darwin.eeb.uconn.edu/systematics.html
This University of Connecticut site provides information on evolutionary biology, systematics, collection management software, and many links to other sites.

biodiversity.uno.edu/
The Biodiversity and Biological Collections Web server has many links to biodiversity and systematic biology resources; it includes directories of biology- and biodiversity-related publications and is good for locating botany resources.

york.biosis.org/index.htm
The Zoological Record home page has a set of links called Internet Resources in Systematics, Taxonomy and Nomenclature.

utexas.edu/ftp/depts/ystbiol/info/links.html
Society of Systematic Biologists home page.

lib.washington.edu/slanat.mus.html
This page includes links to natural history museum Web sites, natural history libraries, bibliographies and reference sources, and other natural history sites.

science.uts.edu.au/sasb/
The Society of Australian Systematic Biologists home page has a Systematics Internet Resources Link and an Internet Directory for Botany.

mobot.org/mobot/molib/
The Missouri Botanical Garden Library includes many links to other sites, including some searchable library catalogs.

rgbkew.org.uk/index.html
Royal Botanic Gardens, Kew. Links to other botanical sites.

Chapter 2

Biological Nomenclature

Biologic categorization is one of the most conspicuous aspects of successful behavior, not only of man, but of all animals, in meeting the requirements for survival in a complex environment.

–DUNN AND DAVIDSON (1968:75)

Linnaean nomenclature came to be universally employed by scientists, though initially it met with little enthusiasm by non-scientists. . . . Even today many people are repelled by scientific names.

–EVANS (1993:12)

Humans as Taxonomists

We are all taxonomists. Pattern seeking and the urge to classify our environment are part of our biology. Like other animals, we need such behavior to tell food from nonfood, predator from nonpredator, and to recognize potential mates, relatives, and offspring. As human animals, we put our conclusions into words and names (Dunn and Davidson 1968; Raven et al. 1971; Pinker 1995). We also use classification in our daily lives to create order and to make things easier to find. We devise our own schemes for household organization–socks in one bureau drawer, shirts in another–to make dressing faster so we can grab an extra 5 minutes of sleep. And we depend on classifications made up by others to make books easier to find in the library, canned peaches in the supermarket, or a new pair of jeans in a department store.

We can observe taxonomic behavior in children's development: the excitement of a toddler as she learns to name the components of her world, or the concentration of a preschooler as he manipulates the pieces of a simple wooden puzzle.

Yet some children like puzzles better than others, and some adults are more interested than others in searching for patterns in their world. This may be at least partially due to innate differences in brain function and learning style. Most students of the mind now believe that intelligence is a not a single property, but rather a number of distinct capacities: linguistic, logical–mathematical, spatial, musical, bodily-kinesthetic, and so on. Children doing puzzles are using and developing their logical–mathematical, spatial, and kinesthetic abilities (Gardner 1983). If they also become interested in biology, young puzzle-lovers might well grow up to become professional taxonomists. Spatial intelligence, in particular, is important to taxonomists, whose science involves the mental sorting and manipulation of characters and taxa, cladograms, and trees.

Unlike exceptional mathematical ability, which usually develops very early, but in which peak ability and contributions to the field come in adolescence and early adulthood, unusual spatial ability can manifest itself early and persist until late in life (Gardner 1983). The importance of spatial intelligence in taxonomy might explain why its mastery has often been likened to the mastery of visual arts such as painting, in which talent is often shown very early and technical mastery and creative expression continue to develop throughout life. Some unconscious perception that an innate combination of abilities is involved may explain the conflicting feelings I have heard expressed by taxonomists and other scientists that taxonomists are born, not made, but that it takes 50 years to become a master of the field. It may also explain why scientists in some other fields have had difficulty accepting systematics as science rather than art.

However, despite our common human heritage as taxonomists, most of us, whether students, scientists, or even practicing systematists, have little patience with the nomenclatural aspect of the field. If we must deal with the codes of nomenclature, we can find ourselves reacting with irritation to their inconsistencies and contradictions, and behaving as if these rules had appeared engraved in stone for no other reason than to plague us.

There is some justification for this reaction. The codes that govern the biological naming of plants and animals are written in a legal style and language. The aim of their authors was to make their terminology so precise that the meaning of each regulation could not be misconstrued. Unfortunately, it seems to have had the opposite effect, producing a legalese often completely impenetrable to the uninitiated. Even when their language has been decoded, reading a code of nomenclature is about as stimulating for most of us as reading a will in which we are not beneficiaries.

However, because the chief purpose of this book is to enable you to research, write, and publish taxonomic descriptions–all activities that involve nomenclature–it must cover the subject in some detail. The rest of this chapter surveys the development of biological nomenclature, showing that much about the way we name things is consistent across cultures and over time. It also summarizes the history of our present codes of botanical and zoological nomenclature, pointing out what basic concepts they share and in what ways they differ. Finally, it discusses the inherent limitations and future of the current system.

Biological Nomenclature

Biologists have divided the living world into groups of organisms, or *taxa,* which they have arranged in a series of levels, or a *taxonomic hierarchy* (table 2.1), with the *species* (chapter 3) as its essential element. Below species are subspecies, populations, and individuals. Above the species level, taxa are grouped into more and more inclusive levels, based on fewer shared features. The levels of this taxonomic hierarchy are given names, or *taxonomic ranks.* Species sharing similar features are grouped into a *genus,* genera sharing traits into a *family,* families into *orders,* orders into *classes,* classes into *Phyla* (animals) or *Divisions* (plants), up to the broadest level of *Kingdom* (table 2.1). All taxa at the same level or rank belong to the same *taxonomic category.* Assigning names to these groupings is the part of taxonomy called *nomenclature.* By agreement among biologists, scientific names are assigned or reassigned according to sets of rules called *codes of nomenclature.*

Biologists first began arranging taxa into hierarchies on the basis of shared features long before there was any understanding of the genetics and evolutionary basis of such similarities. Now, although the present system is not perfect or complete (and never can be because it is a compromise system), most biologists share the goal of attempting to make ensure that the groupings used reflect a common evolutionary history. This is deduced by making use of structures similar in a certain way (by homology), particularly by the use of the shared, derived features that give the best information, as well as by aspects of body chemistry, physiological processes, and genetics at both the organismic and molecular level.

Table 2.1 Folk and Scientific Taxonomic Hierarchies

Folk Classification			Biological Classification		
GROUP	GROUP NAME		GROUP	GROUP NAME	
	Example 1	*Example 2*		*Example 1*	*Example 2*
Life form	yakt[*]	as[†]	Kingdom	Animalia	Animalia
			Phylum	Chordata	Chordata
			Class	Amphibia	Aves
			Order	Anura	Falconiformes
			Family	Hylidae	Accipitridae
Generic	ccp	as	Genus	*Hyla*	*Accipiter*
Specific	ccp kamayket	as jejeg	Species	*Hyla angiana*	*Accipiter melanochlamys*
Varietal		as jejeg km	Subspecies	Unnamed variant[‡]	

Sources: Karam names from Bulmer and Tyler 1968; Bulmer 1970. Scientific names from Bulmer and Tyler 1968; Bulmer 1970; Barnes 1984; Peterson 1980.

[*] Flying birds or bats.

[†] "Soft food," frogs and certain small mammals.

[‡] A variant of this highly variable species with upper surface mostly green.

As mentioned in chapter 1, there are actually five nomenclatural codes, but this book covers only those for plants and animals. Animals are named and described according to the International Code of Zoological Nomenclature (ICZN), of which the latest edition is the third edition, published in 1985. The fourth edition is now in press, with publication expected in mid-1999, to come into effect on 1 January 2000. The ICZN is currently regulated by the International Union of Biological Sciences (IUBS). Plants are named and described following the International Code of Botanical Nomenclature (ICBN), currently under the jurisdiction of the International Botanical Congress. The latest edition is the Tokyo Code, published in 1994. It will be revised again at the next International Botanical Congress, scheduled to be held at the Missouri Botanical Garden, St. Louis, in July 1999.

Far from being written in stone, the present rules are the result of many hard-fought scientific battles over the last 200 years. They represent the best compromise taxonomists have come up with to deal with two diametrically opposing needs: the need to *name,* to provide each organism with a unique and stable name; and the need to *classify,* to provide a system of classification that is explanatory and predictive. They are opposing because if the criterion of stability is to be met, a name once given could never be changed. Yet, as chapter 1 showed, we still have a lot to learn about relationships and even identities of Earth's organisms, so the ability to continue to make changes that reflect increasing knowledge is vital.

Each code is made up of a set of *rules* or *articles* that must be followed when taxa are named or names used or changed. Some of the rules are accompanied by *recommendations,* which point out what is currently considered to be the best procedure to follow in specific instances. There is no enforcement of these rules except the voluntary cooperation of biologists. But these voluntary sanctions are quite effective. No work that goes against them is likely to be published in a reputable journal, and if such work should get published, it will probably be ignored (Jeffrey 1989).

Folk Taxonomy

The system we use for the scientific naming of organisms developed over a very long period of time. Its roots go back to prehistory, maybe even to prehuman history. Some historians believe that the scientific study of living things developed via primitive medicine (Hopwood 1959). In that case, taxonomy may well date back to prehuman primates, for scientists have recently realized that a number of primates dose themselves with medicinal plants to combat various complaints (Gibbons 1992; Clayton and Wolfe 1993).

Ethnobiologists have shown that primitive people often develop sophisticated folk taxonomies. For example, in New Guinea, for the 120 bird species occurring in their territory, the Fore people recognize 110 taxa; 93 are equivalent to scientifically recognized species, 9 are equivalent to species complexes, and 8 represent the dimorphic sexes of the four resident species of bowerbirds and birds of paradise (Diamond 1966).

Organisms very important to a culture may be classified in great detail. For example, frogs are important food items to the Karam, another tribe of the New Guinea Highlands. They distinguish 25 taxa of frogs, 17 of which correspond to biological categories and 16 to separate species (Bulmer and Tyler 1968). Contrast this with the much smaller number of terms for kinds of frogs (e.g., *frog, toad, bullfrog, tree frog*) used in English folk taxonomy. Two examples of Karam folk classification and nomenclature, names for a common frog and a forest bird, are given in table 2.1, along with their biological classification and nomenclature.

Studying the common characteristics of folk taxonomies gives insight into the shared mental and cultural processes involved in nomenclature. Raven, Berlin, and Breedlove (1971) concluded that in all of them, many biologically related groups of organisms (taxa) are recognized. Taxa are arranged hierarchically into a small number of categories: *unique beginner* (original living thing, equivalent to the biological "common ancestor"), *life form* (roughly equivalent to our major biological subdivisions of kingdom [plants] and phylum [animals]), *generic* (more or less equivalent to genus), *specific* (more or less equivalent to species), and *varietal* (subspecies or variety). Members of life form categories are few (5–10) in number and comprehensive, including most of the lesser-ranked taxa, although the taxa of particular economic interest may remain unaffiliated. Generic taxa are the most numerous (usually about 500 in number); specific and varietal taxa are usually much less numerous. Generic taxa with more than two members are usually those of major cultural importance. Intermediate taxa (all those falling between phylum and genus level) are rarely recognized or labeled (e.g., examples in table 2.1).

This makes sense from our own folk taxonomy. If we equate generic taxa with "kinds" and specific taxa with "particular kinds," it is clear that most of us operate on the "kind" level in every day life. For example, unless we have a specialized interest in trees or outdoor life, we speak of an oak, rather than a white oak, a red oak, or a pin oak. Although others have argued that species may actually be more commonly recognized in folk taxonomies than genera (Stuessy 1990), the important points for us to remember are that the folk taxonomic systems developed from prehistoric times have included the idea of a taxonomic hierarchy, similar to that used by biologists (table 2.1) and that naming of groups similar to biologists' genera and species is far from a recent invention.

Binomial Nomenclature

Ancient Greek and Roman naturalists and the medieval herbalists who followed them produced classifications much like those of the New Guinea Highlanders, and for similar reasons. Although they were not constrained like the taxonomists of preliterate societies by the limitations of memorization, they were still limited by that fact that their manuscripts could be distributed only by laborious hand copying. The lists of genera they compiled were similar in size to those of primitive cultures, about 250 to 500, depending on the biological diversity of the region in which the writer lived (Raven et al. 1971).

The expansion of exploration from the fifteenth to the seventeenth century made naturalists realize that a guide produced for one geographic area would not necessarily be useful in another area. Regional studies of animals and plants, printed on the recently invented printing press, began to be produced. In fact, Linnaeus's eighteenth-century taxonomic system was essentially an update of previously existing regional folk taxonomy for the northern European region (Walters 1961; Raven et al. 1971).

Most naturalists of the time wrote in Latin, so the names they used for plants and animals were Latin. But their names were inconsistent and often paragraphs long. Authors of the time expected a name to serve as diagnosis and description, as well as identification (Mayr et al. 1953). Linnaeus is often given the credit for inventing binomial nomenclature, but his real contribution lay in standardizing and simplifying a system already in use.

Common organisms in those regional floras and faunas were often known by a single (*uninomial*) name, corresponding to the genus, as in folk taxonomy. Many such names dated back to Greek and Roman times, such as *Crocus, Viola, Amaryllis,* and *Narcissus.* Pre-Linnaean naturalists, like folk taxonomists before them, also commonly gave genera in which two species were known two-word (*binomial*) names, with the species term an adjective qualifying the genus term (e.g., *Iris Illyrica, Iris sylvestris*). But if more relatives turned up, the additional names might be *trinomial* (e.g., *Iris maritima Norbonensis*) or might expand into a whole descriptive phrase or *polynomial* (e.g., *Iris perpusilla saxatilis Norbonensis a caulis ferme*) (Mayr et al. 1953; Greene 1983).

Dealing with names a paragraph long was inconvenient. It was here that Linnaeus made his breakthrough, coming up with a sort of nickname, the *Nomen triviale,* a unique two-word (*binomial*) combination for every species, printed in the margin of his text (figure 2.1). Linnaeus used his abbreviated binomial name in addition to his longer traditional scientific name, which he also standardized to have a first term part which indicated the genus and a second term part which was a descriptive species name, including the "differentia specifica" or essential characters. His traditional species name was still often polynomial (consisted of several words) and resembled a modern diagnosis in telegraphic style. He first used this system consistently for plants in the 1753 edition of *Species Plantarum* and for animals in the tenth edition of his *Systema Naturae* in 1758 (Mayr et al. 1953; Eriksson 1983; Stace 1989).

Development of Codes of Nomenclature

The Linnaean shortcut of using a unique combination of two words to identify a species caught on rapidly among scientists, who began using his binomial system to replace the longer traditional names. For the next hundred years names proliferated, as new collections kept pouring in from distant lands, and the development of good lenses and microscopes revealed microcosms teeming with new organisms. But the rapid increase in the number of scientific names was not just a result of the increase in newly described species. Botanists and zoologists seized upon previously described species that had been named in a modern language or in nonbinomial terms and renamed them according to the binomial system. Authors of taxonomic revisions also changed Latin species names that had been incorrectly constructed or improperly spelled, or perhaps were not accurately descriptive. Some taxonomists, like Linnaeus himself (1736, 1751), developed their own rules of procedure (Schmidt 1952). But none caught on widely, and well before the middle of the nineteenth century

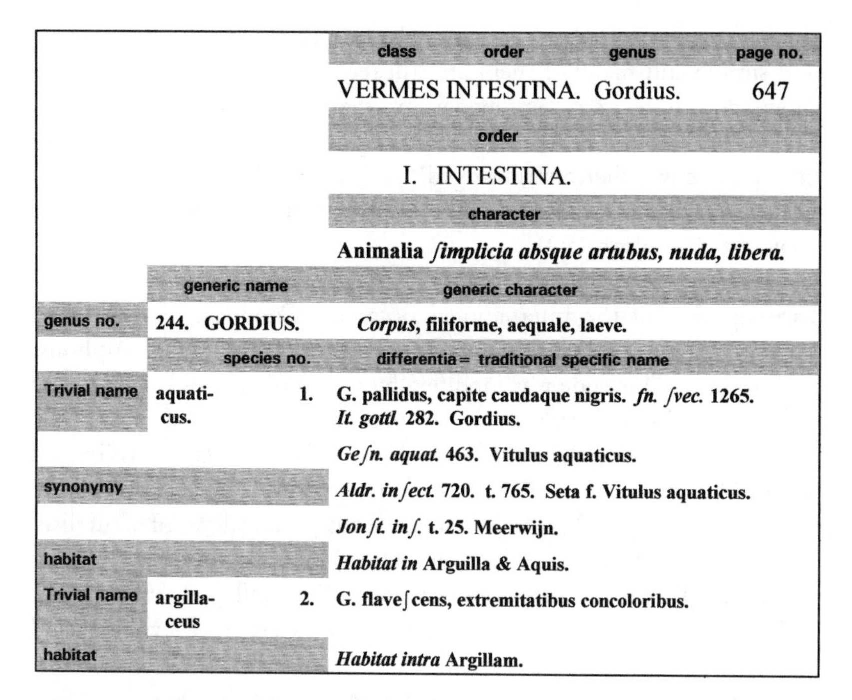

Figure 2.1 This transcription of part of a page from the tenth edition of Linnaeus's *Systema Naturae* (1758) shows the format of his species descriptions, including his new method for labeling any species with just two words: a generic name and a species name. This innovation forms the basis for the binomial system of biological nomenclature we still use today. Linnaeus numbered each genus and species in his book. His taxonomic descriptions included the traditional species name, or "differentia," which described the distinguishing characters of the species. When possible, he also gave a guide to the habitat in which the organism lived and, if the species had been cited previously, a *synonymy,* or list of references to that species (Choate 1912; Heller 1964). What we now use as the species name (also called trivial name or epithet) is his marginal *trivial name.* According to legend, he invented this system as an aid to students rather than as an intentional revolution in nomenclature. *–Source: Modified after figure 5 of Heller (1964).*

it was clear that the nomenclatural chaos from this binge of renaming was hampering biology. Some general rules of nomenclature were badly needed.

But if the first post-Linnaean period saw a proliferation of names, the next stage marked a proliferation of codes. The first attempt to provide a code of nomenclature for general use was the British Association code (Strickland 1842), produced by a committee delegated by the British Association for the Advancement of Science. This Strickland Code was the forerunner of today's international codes, and was meant to cover the naming of both plants and animals. Its recommendations were brief and simple and many of them are still recognized in our current rules. However, one of its recommendations–that the author of a species be the person who published the first description of it (whether or not the genus name was thereafter changed), rather than the person who put it in its current genus–caused a still unresolved split between botanists and zoologists (Nicolson 1991).

Botanists were working on their own set of rules. For the 1867 Paris meeting of the International Botanical Congress, a series of 68 nomenclatural "laws" or rules was developed by Swiss botanist Alphonse de Candolle. This code was modified by a commission of botanists and each article discussed by the congress and passed or rejected. It was then adopted as "the best guide to follow for botanical nomenclature" (Nicolson 1991:208–209).

In 1877 William Dall published a code based on a survey of naturalists' opinions in response to a questionnaire developed for the American Association for the Advancement of Science. This Dall Code attempted to include both botany and zoology, giving separate articles for zoology and botany only where the two systems differed. In 1881, the International Congress of Geology adopted a code for naming fossils (Douvillé 1882), and in 1886 the American Ornithologists' Union created a code for bird nomenclature (Coues et al. 1892). It was obvious that scientists had recognized such codes were needed; however, instituting codes that were truly international took decades longer.

The International Code of Botanical Nomenclature

The beginnings of the ICBN date from that Paris meeting of the International Botanical Congress for which Candolle produced his set of rules (1867). For several decades after that congress, however, controversy reigned between the conservative systematists, who were satisfied with the Candollean laws, and various reformers and radicals (Nicolson 1991).

The first real international botanical code was created by the Vienna Congress of 1905. The code specified that descriptions of new taxa be written in Latin, a rule that caused a split among botanists, chiefly between the Europeans and the Americans, who were followers of influential American botanist N. L. Britton. Many American botanists became so unhappy that they used their own American (or Brittonian) Code of Botanical Nomenclature (Arthur et al. 1907) for another 25 years (Nicolson 1991).

The current International Code of Botanical Nomenclature (ICBN) consists of 62 rules or Articles, with accompanying recommendations and examples. It is based largely on the set of rules agreed upon at the International Botanical Congress held in Cambridge, England, in 1930. On the basis of recommendations by the Nomenclature Sessions and the Permanent Nomenclature Committee, the ICBN is now modified (discussed, emended, ratified) by a plenary session of each International Botanical Congress, which is held every 6 years (Stace, 1989; Nicolson 1991). The most recent congress was held in Tokyo in 1993; the current version of the ICBN was published in 1994.

International Code of Zoological Nomenclature

Zoologists date their international code from the first International Congress of Zoology, held in Paris in 1889. Following discussion and committee reports, the first version of the code was adopted by the Fifth International Congress of Zoology (Berlin, 1901), authorized by the Bern Congress in 1904, and published in 1905 in French, English, and German as the *Règles Internationales de la Nomenclature Zoologique.* This code was amended a number of times by various actions of international congresses and finally replaced in 1961 by the first edition of the International Code of Zoological Nomenclature (ICZN) (Mayr 1969; Ride 1986).

No International Congresses of Zoology were held after 1972. Responsibility for carrying on the code and the duties of the Commission on Zoological Nomenclature were transferred to the International Union of Biological Sciences (IUBS) in 1973. The IUBS is a nongovernment nonprofit organization whose mission includes the coordination of scientific activities that require international cooperation (Younès 1996). The current third edition of the International Code of Zoological Nomenclature (ICZN) was prepared by the International Commission of Zoological Nomenclature, approved in 1983, and published in 1985 (Ride 1985). A draft version of the fourth edition was circulated to zoologists by mail and Internet for comment and discussed at the nomenclature sessions of the IUBS meeting in Budapest, Hungary, in August 1996. Changes were agreed upon by the Commission on Zoological Nomenclature. The Editorial Committee of the Commission then drafted a final version. Publication of the fourth edition is expected in mid-1999.

The Current Codes of Nomenclature

Similarities

Table 2.2 shows the features shared by the current international codes for botanical and zoological nomenclature. The primary purpose of both codes is to provide a unique binomial scientific name for every species of organism. They do so by outlining rules and recommendations for naming, describing, and publishing descriptions of new taxa and by providing a system of continuing review for nomenclatural questions and for the codes themselves.

They are both guided by the principle of *priority,* the idea that the first published name for a particular organism is the correct one to use.

Table 2.2 Similarities Between the Botanical and Zoological Codes

Both the ICBN and the ICZN aim to

- Ensure a unique scientific name for every taxon. These names must be Latin in form, written in the Latin alphabet, and composed in accordance with the rules of Latin grammar.
- Provide rules for publication of new names and determine whether previously published names are acceptable.
- Ensure adequate documentation and dating of names.
- Ensure typification (that each name is linked to a type specimen, specimens, or species).
- By use of types permit names to be assigned to taxa without interfering with scientific freedom to make changes in classifications on the basis of new information and insights.
- Give a chronological starting point to names used (names in publications before the starting date are ignored).
- Provide an administrative system (the International Botanical Congress and its Permanent Nomenclature Committee and the IUBS and its International Commission of Zoological Nomenclature for animals) to ensure and oversee the interpretation and improvement of the codes.

Sources: Jeffrey 1986, 1989; McNeill and Greuter 1986; Ride 1986; Melville 1986.

Priority was not an issue in Linnaeus's day. He had become such an authority that his names were generally accepted by others. However, priority was established as a guideline early in the post-Linnaean period as taxonomists sought an objective method by which to cut through the confusion resulting from the rampant changing and updating of names (Ride 1986; Nicolson 1991).

The two modern codes also share the concept of *typification,* the system that makes objective identification possible by linking the name with an actual specimen. In contrast to priority, which was recognized as an important principle early on, the type concept developed later. In fact, this system, which seems to current taxonomists such a simple means of ensuring stability, could not have come about until after Darwinism had become established and the essentialist concept of species rejected. It was not formally introduced into the codes until the early twentieth century, and is still misunderstood by many people (see chapter 9) (Ride 1986; Nicolson 1991).

Differences

Table 2.3 lists some of the differences between the codes. One important point is that the two codes are *independent;* the same genus name may be used for an animal and for a plant. For example,

Phoebe, a genus of the American tyrant flycatcher family (Tyrannidae) and

Phoebe, an Asian genus of the plant family Lauraceae (laurel family).

Pieris, a genus of butterflies (family Pieridae, or white and sulphur butterflies) and

Pieris, an Asian-American genus of the plant family Ericaceae (heaths).

Other major differences between the two codes involve publication, especially the necessity of a Latin diagnosis in botany, as well as other details concerning valid publication, and criteria for deciding whether a previously published name is acceptable (Jeffrey 1989).

The two codes also have different starting dates. After some argument during the nineteenth century as to whether the tenth or twelfth edition of Linnaeus's *Systema Naturae* was preferable (Verrill 1869; Dall 1877), zoologists settled on the tenth edition (1758) as the starting date for

Table 2.3 Differences Between the Botanical and Zoological Codes of Nomenclature

Independence

The two codes are independent: A genus name can be used by botany, and the identical name used in zoology without violating any rules.

Scope

ICBN Regulates names from Division to subform
ICZN Regulates names from Superfamily to subspecies

Starting Point Dates

ICBN 1753, 1801, 1821, 1848, 1886, 1892, 1900
ICZN 1758

Diagnosis

ICBN Latin diagnosis required for valid publication
ICZN Latin diagnosis not required

Names

TAUTONYMS (BINOMENS WITH IDENTICAL GENUS AND SPECIES TERM)

ICBN Not allowed
ICZN Species name tautonyms allowed, e.g., *Mitra mitra* (L.)

ILLEGITIMACY

ICBN Applies to names that do not meet certain rules and cannot be taken into consideration for the purpose of priority
ICZN Concept not found

SUPERFLUOUS NAMES

ICBN Applies to names belonging to taxa that actually include the type of another name that should be used as the correct name
ICZN Concept not found

Continued

Table 2.3 Continued

Operation of Priority

ICBN Dates and authorship of species affected by change in rank (i.e., date of valid publication of a particular *combination* of genus name and species name [binomen] has priority)

ICZN Dates and authorship of species unchanged by change in rank (i.e., date of publication of a given species name has priority even if genus is changed)

Orthography

DIFFERENT ENDINGS USED FOR HIGHER TAXA

	ICBN	ICZN
Order	*-ales**	Not regulated
Suborder	*-ineae**	Not regulated
Superfamily	Not regulated	(*-oidea*) recommended
Family	*-aceae*†	*-idae*
Subfamily	*-oideae*	*-inae*
Tribe	*-eae*	(*-ini*) recommended

CAPITALIZATION OF SPECIES NAME

ICBN Species names may be capitalized, but practice is not recommended.

ICZN Species names are never capitalized.

-OIDES ENDING

ICBN Feminine

ICZN Masculine

Conservation and Suppression of Names

ICBN Except for certain widely misapplied names and sanctioned family names, only species to family level names may be conserved or rejected. Proposals must be submitted to the Committee on Nomenclature and adopted by an International Botanical Congress.

Continued

Table 2.3 Continued

ICZN Decided by opinions of the International Commission on Zoological Nomenclature and published in *Bulletin of Zoological Nomenclature* and certain other journals.

Fossils

ICBN Names based on a Recent type have priority over name based on fossil type (except for algae).

ICZN No such rule.

Typification

The two codes differ in the way the type method connects names and taxa.

ICBN Type of a name.

ICZN Type of a nominal taxon (one denoted by properly published name).

ICBN Types cannot be living organisms.

ICZN No such restriction.

ICBN As of 1958, type must be named for valid publication of family-level or lower taxa.

ICZN Not a requirement (but a strong recommendation); will be required for species-level taxa in fourth edition of ICZN.

Terminology

In a number of cases the two codes use different terms for similar concepts:

ICBN	ICZN
Later homonym	Junior synonym
Nomenclatural synonym	Objective synonym
Taxonomic synonym	Subjective synonym
Validly published	Available
Correct name	Valid name
Specific epithet	Specific name
Specific name	Binomen, scientific name of a species

Sources: Jeffrey 1986, 1989; McNeill and Greuter 1986; Ride 1986; Melville 1986; International Trust for Zoological Nomenclature 1997.

*Mandatory, but only if based on name of an included family.

†With several exceptions for family names in long use.

zoological nomenclature. That meant that no name given to an animal before 1758 had to be considered in deciding on a name. Botanists use several starting dates for their nomenclatural code. For most plants (seed plants, ferns, one family of mosses, the Sphagnaceae, fungi, and most algae), the starting date used is May 1, 1753 (the date of publication of Linnaeus's *Species Plantarum*). For the rest of the mosses, the starting date is January 1, 1892. Certain algae also have a later starting date. For the Desmidiaceae, it is January 1, 1848, for the Nostocaceae heterocysteae, January 1, 1886, for the Nostocaceae homocysteae, January 1, 1892, and for the Oedogoniaceae, January 1, 1900. All these later dates mark an agreed-upon publication date for an important work on that group (ICBN, Art. 13, pp. 19–20).

One fundamental difference between the two codes is the way in which priority operates with regard to dates and authorship of names. In the zoological code names have *coordinate status,* meaning that if a name is established for a taxon and its taxonomic placement later changes, the date of original publication and the name of the *author* (the person who originally described it) do not change. When a species is put into a new genus, the species name is still good and the species is still considered to have the same author and date of publication as before the change of genus took place. For example, the olive snail was originally given the name *Voluta oliva* by Linnaeus in 1758 (p. 729). *Voluta oliva* Linnaeus 1758 later became *Oliva oliva,* but is still *Oliva oliva* (Linnaeus 1758), with parentheses around the author and date to indicate that it has been moved to another genus since Linnaeus's time. On the other hand, botanists also include the author's name in citing a species, but make changes to reflect changes in placement or rank; for example, *Cucumis chrysocomus* Schumach. (1827) is a species described by Schumacher in 1827. But when Jeffrey placed in a different genus, it became *Raphiodiocystis chrysocoma* (Schumach.) C. Jeffrey (1962).

Botanists prefer their system because they believe it facilitates information retrieval. By their method, taxonomic changes may more readily be traced through the literature. But zoologists claim that their system makes it easier to trace the original description, which is often important in taxonomic decision making (see chapter 4).

As indicated in chapter 1, this book takes a more heavily zoological approach, but some of the ways in which the two codes differ are discussed further, with examples, in the relevant chapters.

Future of the Codes

As mentioned at the beginning of this chapter, our present system can never be perfected because it incorporates two conflicting goals: the need to provide each organism with a unique and stable name and the need to place that organism in a biologically satisfying classification. Until all organisms on earth are identified and described, it is clearly impossible to do both. As we learn more, new names must be added and some old names must be changed to reflect a better understanding of their evolutionary relationships. The current codes of nomenclature are merely the best compromise so far devised to do the job. From the perspective of scientists such as field ecologists or natural product chemists, who may work with both plants and animals in the course of a research project, it is unfortunate that the botanical and zoological codes developed separately over the past hundred years. The changes now required to bring them into complete agreement would be so disruptive to the whole system that it seems unlikely they will ever happen. Just as librarians of large, long-established research libraries have realized the impossibility of recataloging their entire collection (e.g., from Dewey to Library of Congress) and resigned themselves to dealing with two cataloging systems, so biologists who work with both plants and animals (e.g., ecologists) must resign themselves to dealing with two codes of nomenclature, at least in the foreseeable future.

Recently, perhaps because of our sense of crisis over the rapid disappearance of organic diversity, there has been an upsurge of interest in ways to improve our system of nomenclature or even to discard it entirely. Some systematists find the idea of any further compromise offensive and would like to see a completely phylogenetic (cladistic) classification system replace it (Hennig 1981; de Queiroz and Gauthier 1990). But this necessitates developing a new organizing system to take the old system's place in communicating a taxon's rank and membership to others. So far nothing has been suggested that seems to meet taxonomists' needs, and most taxonomists are taking the practical approach and continuing to use the familiar Linnaean system in their work (Smith 1994).

A Systematic Association symposium, "Improving the Stability of Names" was held in 1991. It brought together botanists and zoologists working on both living and fossil organisms to discuss future nomenclatural options (Hawksworth 1991). This meeting was an important step toward reconciling some aspects of the codes or coming up with alternative solutions.

The *Draft BioCode* (1996) was developed by members of the IUBS International Committee on Bionomenclature as a new attempt at a unified nomenclature for plants and animals. It is not intended to be retroactive, but would govern all new names created after the date of its adoption. Its proposals include devising new nomenclatural terminology that would solve the problems that now exist because the two codes use either the same term for two different concepts or different terms for the same concept (e.g., in table 2.3, "validly published" [ICBN] vs. "available" [ICZN]); establishing coordinate status for names (as in the current ICZN); registering all new names; making all new names unique across Kingdoms; and providing authoritative guidance on the formation of new names (Hawksworth 1995; Grueter and Nicolson 1996). According to one of its authors, it could achieve this in a form that would be shorter and simpler than either of the existing codes (Grueter 1996). This proposed BioCode is now available for discussion by biologists (and particularly needs input from zoologists). Whether it will eventually be accepted remains to be seen, but one of its recommendations, its list of revised terms, has already been put to use in the latest *International Code of Nomenclature for Cultivated Plants* (Trehane et al. 1995).

The idea of creating new starting point dates for nomenclature (as in the Draft BioCode) has also been suggested as a way to cut down on changes and problems in the existing botanical and zoological codes. Bacteriologists have already done this. The original code used for bacteria grew out of the botanical code, was separated from it in 1948, and expanded to include viruses. By the late 1950s, microbiologists felt that this code had become too complex and burdensome. The bacteriologists decided to cut down on the instability caused by searches for the earliest possible name by using a new starting date of 1958 (Sneath 1986, 1992). The virologists eventually set up their own new code (Francki et al. 1990).

Lists of names in current use have been also been promoted as a means of increasing stability. These would be internationally accepted lists of genus or species names that could serve as authority files for research, databases, conservation, and commercial applications such as agricultural and horticulture catalogs. Such lists have been resisted by many taxonomists for two reasons. First, like all scientists, they cherish their freedom to incorporate new knowledge into their field, and new knowledge sometimes necessitates changing the position, rank, and names of taxa. Second, for many groups of organisms, so little is known that it would take a major research effort to come up with a list of names to use. Proponents argue that such lists would make searching the biological literature more efficient, and global taxonomic information systems for both research and resource management purposes could be developed more quickly. In fact, lists of standard names for many kinds of organisms have been in use for many years in print form in dictionaries or catalogs, and as online databases continue to grow, their creators are producing their own electronic taxonomic authority files without or without the aid of taxonomists. The first instances of internationally approved lists of names in current use will probably be for cultivated plants and agriculturally important plants and animals (Hawksworth 1991).

Even if botany and zoology can someday be reconciled, and lists of standard names used to please those desiring greater stability, there are other limitations to the system. Almost 30 years ago, Raven et al. (1971) pointed out how much our present system is based on deeply rooted folk principles and limited to communicating information about a restricted number of recognizable units (about 500) to others who already knew something about the field of study. Why this limitation exists is not known, but it may have something to do with the limitations of human memory (Hunn 1994). They advocated the development and use of computerized data banks from which information could be extracted for a variety of purposes, including the construction of classifications, concluding that "The invention of high-speed electronic data processing equipment is seen as analogous to but more important than the invention of movable type in the history of systematic biology" (Raven et al. 1971:1213). Even when I began working on this book in 1987, museums and taxonomists seemed far from the center of such a technological revolution, but they have advanced a great deal in the last 10 years. Many museums now have computerized databases covering at least part of their collections available online via the Internet. World Wide Web searches are familiar procedures to scientists and students, and the best way to communicate with a taxonomist today is usually by e-mail.

Biodiversity conservation is a global problem. Some kind of global taxonomic information system is sure to come. Right now the botanists seem to be ahead; the Tropicos System begun in 1983 by the Missouri Botanical Garden, the Flora of North America and Flora of China Projects, and the Germ Plasm Research Information Network are good examples (Allen 1993), but zoological databases also are under way.

Also exciting for the future are automated methods for routine taxonomic identification: molecular tests that would quickly allow determination of species identity. Such tests are rapidly becoming faster and cheaper, and are already in use in medicine and forensics. But it may take a revolution equivalent to the development of the polymerase chain reaction to develop a molecular species analyzer as economical and indestructible under field conditions as an experienced taxonomist. Meanwhile, in order to communicate our research findings, we will continue to need the present system to describe new taxa, and the even older tradition of journals and books to communicate our findings to others.

Sources

Printed Sources

Codes of Nomenclature

Francki, R. I. B., C. M. Fauquet, D. L. Knudson, and F. Brown. 1991. *Classification and Nomenclature of Viruses.* New York, Springer-Verlag.

Greuter, W. et al. 1994. International Code of Botanical Nomenclature (Tokyo Code) Adopted by the Fifteenth International Botanical Congress, Yokohama, August–September 1993. *Regnum Vegetabile* 131.

Greuter, W., D. L. Hawksworth, J. McNeill, M. A. Mayo, A. Minelli, P. H. A. Sneath, B. J. Tindall, P. Trehane, and P. Tubbs, eds. 1996. Draft BioCode: The prospective international rules for the scientific names of organisms. *Taxon* 45: 349–372; *Bulletin of Zoological Nomenclature* 53: 148–166.

Hawksworth, D. L. 1995. Steps along the road to a harmonized bionomenclature. *Taxon* 44: 447–456.

Hawksworth, D. L. 1997. The new bionomenclature. The BioCode debate. *Biology International* special issue 34: 1–103.

International Committee on Bionomenclature. 1996. *Draft BioCode: The Prospective International Rules for the Scientific Names of Organisms.* Paris, pp. 19–42.

Ride, W. D., C. W. Saborsky, G. Bernardi, and R. V. Melville, eds. 1985. *International Code of Zoological Nomenclature,* 3d ed. London, International Trust for Zoological Nomenclature.

Sneath, P. H. A. 1992. *International Code of Nomenclature of Bacteria.* 1990 revision, ed. by S. P. Lapage et al. Washington, D.C., International Union of Microbiological Societies/American Society for Microbiology.

Trehane, P., C. D. Brickell, B. R. Baum, W. L. A. Hetterscheid, A. C. Leslie, J. McNeill, S. A. Spongberg, and F. Vrugtman. 1995. International Code of Nomenclature for Cultivated Plants, 1995. *Regnum Vegetabile* 133.

Other Reading

Berlin, B. 1992. *Ethnobiological Classification. Principles of Categorization of Plants and Animals in Traditional Societies.* Princeton, N.J., Princeton University Press.

Johns, T. 1990. *With Bitter Herbs They Shall Eat It: Chemical Ecology and the Origins of Human Diet and Medicine.* Tucson, University of Arizona Press.

Knight, D. 1981. *Ordering the World: A History of Classifying Man.* London, Burnett.

Mayr, E. 1982. *The Growth of Biological Thought.* Cambridge, Mass., Belknap/Harvard University Press.

Melville, R. V. 1995. *Toward Stability in the Names of Animals: A History of the International Commission on Zoological Nomenclature 1895–1995.* London, International Trust for Zoological Nomenclature.

Naqshi, A. R. 1993. *An Introduction to Botanical Nomenclature.* Jodhpur, India, Scientific Publishers.

Pinker, S. 1995. *The Language Instinct.* New York, Harper Perennial.

Internet Sources

All addresses begin with **http://www.**

rom.on.ca/biodiversity/biocode/biocode.html
Draft BioCode.

bgbm.fu-berlin.de/iapt/nomenclature/code/tokyo-e/
International Code of Botanical Nomenclature (electronic version of the Tokyo Code available on the International Association of Plant Taxonomy Web site).

iczn.org/code.htm
International Trust for Zoological Nomenclature, 1997, International Code of Zoological Nomenclature, pp. 1–4. This site discusses ICZN, but is not an electronic version of it.

Part Two

RECOGNIZING SPECIES

Chapter 3

Species and Their Discovery

> *There is no question that species have objective reality and that they also have evolutionary reality.* —Stuessy (1990:180)

> *The idea of good species . . . is a generality without foundation—an artifact of the procedures of taxonomy.*
> —Ehrlich and Holm (1962:654)

> *The literature on species concepts is vast, dispersed, and varied. The only thing generally agreed upon is that variation in phenetic parameters is not continuous and that character states are not combined randomly. . . . Most practicing taxonomists take for granted that species can be recognized by inspection, though inspection sometimes has to be extremely close.*
> —Andersson (1990:375)

The first two sections of this chapter discuss species concepts and speciation processes affecting different kinds of organisms. The third section discusses characters biologists use to distinguish species. The final section gives examples of the many ways in which biologists have discovered and described new species in the course of their work.

There is already too much literature on the "species problem" (e.g., Sylvester-Bradley 1956; Mayr 1957; Lewis 1959; Heiser 1965; Slobodchikoff 1976; Andersson 1990; and other references in this chapter's *Sources*). It has been discussed, argued over, and symposiumed to death for years. And it is clear from the quotations above that there is still no final consensus as to what a species is, or even whether species exist. Anyone who wades through that literature must surely be reminded of the fable of the blind men and the elephant, and must come to realize that much of the so-called problem results from the fact that there isn't just *one* elephant and that they aren't all even *elephants*–or solitary sexually reproducing animals. Some are corals or cauliflowers.

A species can mean a number of things biologically, depending on the characteristics of the organisms involved. It is a different phenomenon in solitary sexually reproducing animals (those most like ourselves, and perhaps for that reason, the most closely studied), than it is in sexually reproducing plants, clonal plants, and clonal or colonial invertebrates, let alone bacteria and viruses.

These less-studied kinds of organisms are the interesting ones to many biologists–the ones whose biology must be understood to make evolutionary theory complete. Current evolutionary theory still has many shortcomings in dealing with clonal, colonial, and asexually reproducing organisms. By describing and naming these organisms and studying the way the evolutionary process works for them, biologists can contribute both to systematics and to a more comprehensive evolutionary theory. Even those who have not purposely sought out such organisms will often find that they are the ones needing description, because, as pointed out in chapter 1, they make up a large part of Earth's unstudied biodiversity.

Species Concepts

Species concepts are models of the patterns brought about by the way the evolutionary process works under various conditions. They are attempts to explain how phenetic variation is compartmentalized. Numerous models have been proposed. They can be grouped under five main headings (Table 3.1): phenetic, biological, phylogenetic, ecological, and cohesive.

The *phenetic* species concept is a strictly operational and nonexplanatory concept based only on the observable facts of similarity and discontinuity. The other concepts attempt to address the process of species formation and maintenance. The *biological* species concept is most popular with those who study solitary sexually reproducing organisms such as vertebrates. The evolutionary or *phylogenetic* species concept is favored by many biologists, but has the disadvantage that lineage affiliation cannot be unequivocally proven (without time travel). Another drawback is that this model deals well with speciation by branching (cladogenesis), but not as well with reticulate processes. The *ecological* species concept suggests

Table 3.1 Current Species Concepts

Phenetic Species Concept

Species are the smallest groups that are consistently and persistently distinct and distinguishable by ordinary means (Cronquist 1978; DuRietz 1930; Sokal 1973; Doyen and Slobodchikoff 1974).

Biological or Reproductive Species Concept

Species are groups of interbreeding natural populations that are reproductively isolated from other groups (Mayr and Ashlock 1991).

Phylogenetic Species Concept

Species are a single lineage of an ancestor–descendant population that maintains its identity from other such lineages and has its own evolutionary tendencies and historical fate (Simpson 1961; Wiley 1978, 1981).

Ecological Species Concept

Species are lineages (or sets of closely related lineages) occupying minimally different adaptive zones (Van Valen 1976; Andersson 1990)

Cohesion Species Concept

Species are the most inclusive population of individuals having the potential for phenotypic cohesion mechanisms, i.e., mechanisms that limit the population boundaries by the action of such basic microevolutionary forces as gene flow, natural selection, and genetic drift (Templeton 1989).

---

that species can be explained, although not defined, by the adaptive zones they inhabit. Ecological information can therefore be used in assessing difficult cases (Andersson 1990). The *cohesion* species concept is favored by scientists who work on asexual organisms (Templeton 1989).

No one of these approaches is completely right or wrong. Instead, the different concepts provide models appropriate for different purposes.

Processes Affecting Speciation

Still, most books on systematics and evolution seem to assume that biologists will be dealing with "elephants" or, if not actual elephants, then animals like elephants: solitary, sexually reproducing organisms such as rats, birds, spiders, or fruit flies. The species concepts and speciation processes that apply to them are the ones stressed. Other kinds of organisms and processes are mentioned only briefly, as aberrations from the major pattern. For that reason I discuss here several processes besides sexual reproduction that may be important in describing species in some other groups of organisms, as well those affecting "elephants."

In particular, what is important is whether speciation occurs strictly by branching or cladogenesis (figure 3.1) or whether reticulation is possible. In some groups of organisms other processes may affect or replace speciation by cladogenesis or alter branching patterns: *external transfer of genes,* which should lead to reticulate patterns of evolution; *asexual reproduction,* which should limit variation; and *somatic mutation,* which in certain organisms should increase the possibility of splitting and production of new species.

Sexual Reproduction (The "Elephants")

The biological species concept is defined by reproductive isolation. Species by this definition are groups of "actually or potentially interbreeding natural populations which are reproductively isolated from other groups" (Mayr 1963:19). In sexual or biparental reproduction, haploid gametes form by meiosis. This is followed by fertilization, with recombination and reassortment of chromosomes and genes, producing a new individual containing two copies of genetic information, one from each parent.

Populations of sexually reproducing species form a reproductive community. They are *panmictic,* gametes from different parent individuals combining in offspring (figure 3.2A). Over time this results in a network of genetic relationships (figure 3.1), which provides a substantial amount of genetic variation. Panmictic organisms may be dioecious (having separate sexes) or hermaphroditic as long as there is cross-fertilization between different parent individuals. Within populations of the species, natural selection operates to remove deleterious combinations and promote beneficial ones (adaptation) (Ross 1974; Mayr and Ashlock 1991).

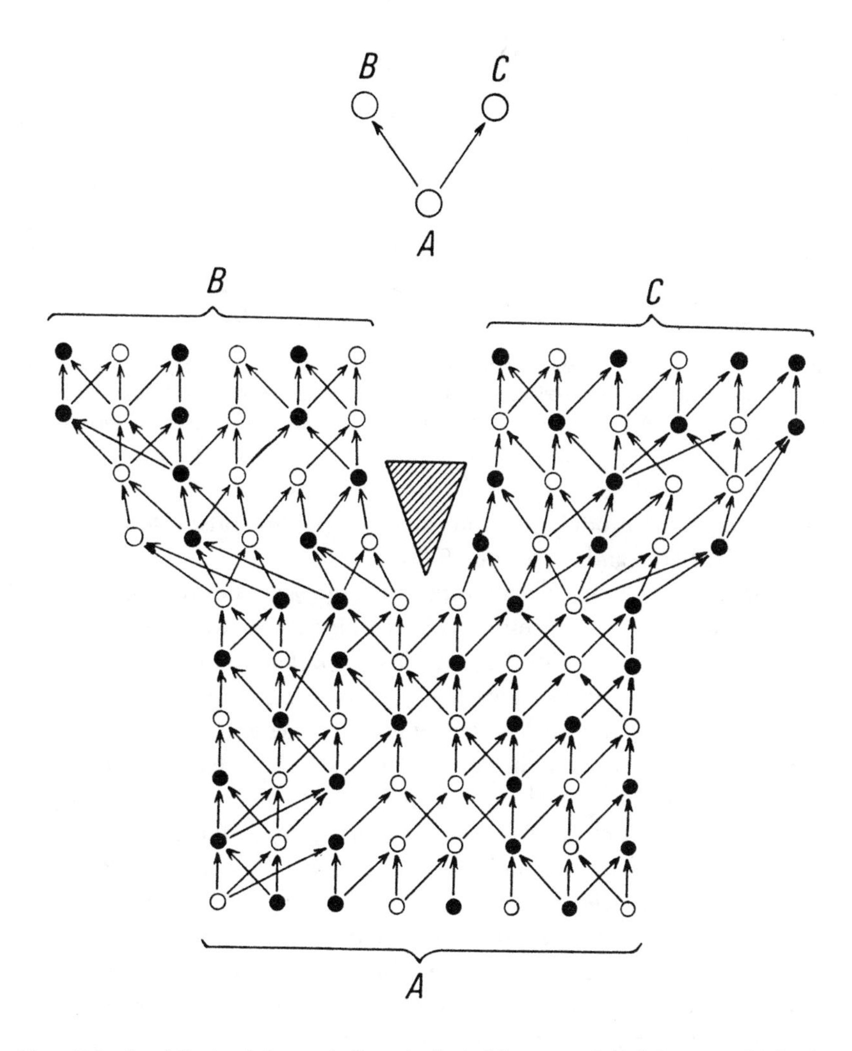

Figure 3.1 Speciation by cladogenesis. Reproductive isolation, caused chiefly by geographic barriers, leads to gaps in the networks of genealogical relationships in populations of sexually reproducing organisms. Two species, B and C, diverge from ancestral species A. *–Source: From figure 4 of Hennig (1966).*

Interbreeding and gene flow take place within populations of the species. Isolating mechanisms protect the gene pool from input from the gene pools of other species. Barriers to gene exchange that lead to reproductive isolation can be either prezygotic (e.g., ecological or habitat separation, breeding behavior, and incompatibility of gametes)

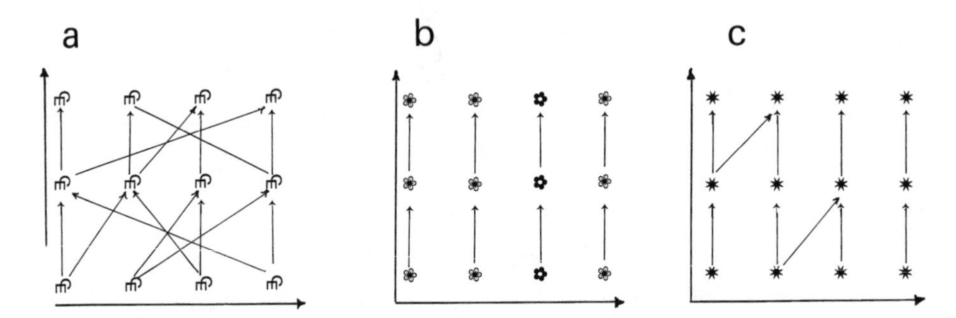

igure 3.2 A, Networks of genetic relationships in populations of a panmictic (sexually reproducing) species. **B,** inear (nonreticulate) pattern of genetic relationships in successive generations of clones of an asexual (apomictic) rganism. **C,** Mixture of linear and reticulate relationships in a facultative apomict. *–Source: Modified from figure '3 of Asker and Jerling (1992).*

or postzygotic (production of inviable or sterile hybrids) (Dobshansky et al. 1977; Byrne and Anderson 1994).

Speciation in sexually reproducing organisms occurs hierarchically by branching. One ancestral species branches to form two or more descendant species (figure 3.1). The first important point of this section is the contrast between the networks of genetic relatedness at the population level and the sharp dichotomies of cladogenesis at the species level. This contrast was recognized and illustrated by Hennig (1966) and others. Phylogenetic analysis depends on the branching process, and the formation of new clades, to detect ancestor–descendant relationships. The second point of this section is that certain other phenomena can obscure or confuse this major pattern: both processes that create networks at the species level and processes that create branching or linear patterns below that level. When speciation proceeds by cladogenesis, with complete and thorough splitting of a lineage, species are clearly demarcated and easy to recognize and describe. But life is not always neat and logical. For the majority of animals and at least a substantial minority of plants, this assumption mostly seems to be met, and phylogenetic analysis works well. But you must be aware of its limitations and the potential impact of these other phenomena on your study organism.

Geographic or allopatric speciation is perhaps the most common mode of speciation, with reproductive isolation gradually built up in populations isolated from each other by geographic barriers. Speciation in isolated founder populations at the edges of a species's range is called parapatric speciation. Other modes require reproductive isolation to be achieved by ecological or behavioral isolation in populations that are sequentially connected or even share a habitat. Arguments persist as to how, and whether, these other modes occur (Mayr and Ashlock 1991; Minelli 1993). Plant biologists (Grant 1981) have recognized a whole range of modes of speciation, at least in theory.

The biological species concept certainly applies to solitary sexually reproducing organisms. It is also useful in colonial groups in which sexual reproduction processes overwhelm asexual ones in importance. But defining species in this way has two drawbacks: The definition cannot be applied to fossil taxa and it is difficult to demonstrate even with living organisms. Few studies have been carried out to determine whether populations are reproductively isolated.

In addition, three other processes (external transfer, incorporation of symbionts, and hybridization) may affect speciation and evolution in a group.

EXTERNAL TRANSFER The whole theory of evolution, as well as the development of the fields of embryology and genetics, is based on the assumption that genes change only by internal processes. Mutation, deletion, duplication, and transposition of genes are what we depend on to deduce a pattern of branching from a common ancestor and to sort living organisms into unambiguous groups. For groups such as vertebrates the system works well because such organisms are genetically closed systems. Introduction of new genetic material into their genomes is likely to be so disruptive that it won't be passed on. In organisms with closed genomes, germ cells become distinct from somatic cells early in development. This means that only material introduced into germ cells can be passed on; genetic material passed into body cells, even if it doesn't kill the invaded cell, cannot be inherited. In contrast, in organisms with open genomes, such as plants and certain groups of invertebrates, almost any dividing cell can give rise to reproductive cells, meaning that any external genetic material that survives introduction into their cells has a chance of being passed on (Buss 1987; Heron 1992).

Prokaryotes with no cell nucleus have the most open genomes, but eukaryotic organisms can also receive transferred genetic material from other organisms in three ways: via incorporation of symbionts, lateral transfer, and hybridization.

INCORPORATION OF SYMBIONTS Incorporation of entire genomes is the rarest kind of external transfer, but it is now accepted as the likeliest explanation for the presence of chloroplasts and mitochondria in eukaryotic cells (Margulis and Schwartz 1988). In at least one plant species a bacterial symbiont is known to be passed on in the seeds. Thus transfer of incorporated symbionts during reproduction could explain why some distinctly unusual secondary metabolites are found in both plants and microorganisms (Heron 1992) or in sessile animals and microorganisms.

HYBRIDIZATION Hybridization occurs when gametes from different species fuse. In sexually reproducing and outbreeding organisms, recognition of other members of the same species as potential sex partners *should* be one objective test of which organisms belong to a species. In actuality, things aren't quite so neat. One of the criticisms of the biological species concept is that, except for animals such as *Drosophila* that can be maintained and bred in the laboratory, it can be pretty hard to determine whether the members of different populations can breed and produce viable offspring. Most field biologists have also observed that when it comes to sexual reproduction, living organisms don't always play the game according to the rules. Some individuals seem to be as sexually adventurous as the people who place personal ads in the *Village Voice.* For example, on the New England coast in springtime, three species of *Littorina* snails, which during summer months inhabit separate regions of the intertidal, are clumped together on rocks near low water. During that time I've seen small *L. obtusata* males attempting to copulate with much larger *L. litorea* females. Probably none of their attempts produce viable offspring, as hybrids have not been described. However, Victorian zookeepers and animal breeders, who had a strong interest in creating "novelties," described many instances of hybrid offspring resulting from crosses of large vertebrates: wolf-dogs, zebra-ponies, zebra-donkeys, cattle-bison, and so on (Ritvo 1997).

In animals, particularly vertebrates, offspring of such crosses, even if viable, are usually infertile. In plants, new species with one or more extra chromosome sets often originate. Such allopolyploid speciation is characteristic of plants. It is particularly common in long-lived perennial herbs, but it occurs in many angiosperm groups. A recent analysis suggested that approximately 70 percent of angiosperms have polyploidy in their evolutionary history (Masterson 1994). For example, *Spartina townsendii,* a saltmarsh grass, was created by hybridization between *Spartina maritima,* the native British species, and *Spartina alterniflora,* an introduced American species. First discovered in southern England in 1870s, it spread rapidly around Britain and has now replaced the native species in most localities (Davis and Heywood 1963). Polyploidy is rare in gymnosperms but extremely common in ferns, and is also common in certain algae and mosses (Grant 1981, 1991)

Polyploidy was thought to be very rare in animals, but it is now being documented in an increasing number of clonal invertebrates such as *Lasaea* clams (O'Foighil and Thiriot-Quiévreux 1991), ostracodes (Havel et al. 1990), freshwater snails (Jarne and Delay 1991), and even vertebrates such as lizards (Cole 1979; Bogart 1980) and fish (Schultz 1969, 1980; Vrijenhoek et al. 1989; Abramenko 1990). Some fish groups (including goldfish, spiny loaches, and desert stream fish of the genus *Poeciliopsis*) contain species consisting of females that are polyploidal and reproduce mainly by gynogenesis. In this form of reproduction the females breed with males of a closely related species, but all male chromosomes are eliminated early in embryogenesis; offspring are genetic copies of the female parent and so are strictly clonal. Other all female lineages reproduce by hybridogenesis. In hybridogenesis, genetically identical females crossed with males of a closely related bisexual species produce hybrids that are hemiclones (clonal female parent, nonclonal male parent).

Polyploidy is also fairly high in anuran amphibians (frogs and toads), being found in 12 genera and 9 families. A phylogenetic analysis of mtDNA variation in geographically separated diploid–tetraploid pairs of tree frogs *Hyla chrysoscelis* and *Hyla versicolor* indicated that the tetraploid species had at least three separate origins (Ptacek et al. 1994).

Cases in which species exchange genes through hybrid backcrossing with their parents are common in plants. Groups of species may form hybridization chains in which species at the geographic extremes can be morphologically quite different, but still share genes via a chain of intermediate species, each hybridizing with its neighbors (Heron 1992). Recent studies have also confirmed the importance of hybridization in the aftermath of invasions of an area by exotic plant species, sometimes leading to rapid evolution of new taxa (Abbott 1992).

LATERAL TRANSFER Hybridization occurs between closely related organisms. *Lateral transfer* is a mechanism by which genes from related or unrelated organisms may be incorporated into a genome. Most commonly, pathogenic or parasitic bacteria or viruses temporarily splice their genome onto that of the host. In the process they may incorporate adjacent host genes. The incorporated host genes may be transferred to the next host, and so on. We are learning to take advantage of lateral transfer in the burgeoning field of biotechnology. For example, genes from a bacterium that produce polyhydroxybutyrate (a biodegradable thermoplastic) have been successfully introduced into a plant, paving the way for growing plastics (Poirier et al. 1992). But lateral transfer has been occurring in the natural world all along, via parasites that insert their genetic material where it does not interfere with the hosts cells' vital functions, although they may have a strong influence on the host.

Lateral transfer can occur between unrelated organisms or even from animals to plants, but it is more likely between groups of related organisms sharing the same parasites. For example, a study by Houck et al. (1991) showed possible horizontal transfer of a *Drosophila willistoni* gene to *Drosophila melanogaster* by a semiparasitic mite *Proctolaelaps regalis.* The mite acquired 18S rDNA and P element *Drosophila* DNA when inhabiting fruit fly cultures and transferred the genetic material from *willistoni* to *melanogaster* in the course of feeding on *melanogaster* eggs.

Theoretically, lateral transfer chains, similar to hybridization chains, are possible: Strains of microorganisms affecting mainly species in one family might transfer genetic material to another microorganism capable of infecting members of another family, and so on. It has been hypothesized that lateral transfer via bacteria might be the reason the animal oxygen-carrying hemoglobins are also found in the roots of leguminous plants (Heron 1992).

Somatic Mutation

An option open to some organisms is the production of new variants or even species via somatic mutation. Like hybridization, somatic mutation occurs commonly in plants (Witham and Slobodchikoff 1981). A somatic mutation that arises in a bud meristem cell can grow along with the plant, turning that individual plant into a *genetic mosaic* or *chimera.* The mutation could then be passed on sexually in the gametes produced in that part of the plant, or asexually in new individuals produced by fragmentation or budding of the mutated portion. The process has long been known to be important in horticulture, where, before the use of chemicals or radiation to induce mutations, such natural *bud sports* were the source of new varieties for cultivation. For example, the pink grapefruit we eat today arose as a somatic mutation of a single branch of a tree (Hartmann and Kester 1975).

Somatic mutations are believed to be ecologically important, aiding plants in defense against herbivores and providing close adaptation to local environments. They could also play a role in speciation by increasing the potential for branching. Somatic mutation has been thought to be of little importance in animals, but for animal groups with open genomes and modular construction, such as sponges, cnidarians, and bryozoans (see Buss 1987 for a complete list), somatic mutation is certainly possible and might well have some of the same ecological and evolutionary consequences it has in plants.

Asexual Reproduction

In asexual or uniparental reproduction, offspring are produced without a meiotic division via parthenogenesis, spore formation, fission, fragmentation, or any method that results in offspring genetically identical to their parent. Theoretically, asexual reproduction should lead to reduced genetic variability, with whole populations being made up of one or a few clones (White 1978), but there is considerable evidence that such populations can be genetically variable (Asker and Jerling 1992). Although the biological species (as defined in table 3.1) cannot exist in completely asexual organisms, biologists who work with such organisms believe that there is some equivalent of species (Templeton 1989; Mishler and Budd 1990) and have been working on developing a concept that applies to them. They define species in such organisms as new lineages originated by a breakdown the integrative or cohesive mechanisms of parent lineages. There is still much to be learned about the circumstances under which asexual reproduction is advantageous and about the phylogenetic patterns found in asexual organisms. Even if completely asexual species are rare in nature, they should still be studied, for they may provide the clue to understanding the sexual process, one of the most basic phenomena in biology (Mishler and Budd 1990). In addition, a large number of invertebrates reproduce, either cyclically or sporadically, by both means. The significance of the many phylogenetic patterns they may exhibit still needs much more research.

Taxonomic Characters

All organisms contain an array of biological information, some of which is useful in a comparative or taxonomic sense, some of which is useless, and some of which may even be misleading. Quicke (1993:263) defines a *taxonomic character* as "any physical structure (macroscopic, microscopic or molecular) or behavioral system that can have more than one form (character state), the variation in which potentially provides phylogenetic information." Mayr (1969:121) defines a taxonomic character as "any attribute of a member of a taxon by which it differs or may differ from a member of a different taxon."

Most taxonomists favor using as many characters as can be found for an organism, believing that the useful will outweigh the misleading (Minelli 1993). In seeking useful characters they attempt to avoid individual aberrations and mutations and features subject to environmental influences. They look for genetic, heritable changes, particularly those that result from the influence of many genes and metabolic interactions.

In particular, they seek *homologies,* or conditions of essential similarity in a group of organisms due to evolution from a common ancestor (Quicke 1993). The most useful are shared, derived characters, or *synapomorphies,* evolutionary novelties that arose in a common ancestor and are shared by its descendants. Taxonomists try to avoid using characters that are nonhomologous (*homoplasous* or *analogous*), superficially resembling each other but having arisen independently and not indicative of relationship by descent from common ancestry, but instead reflecting evolutionary convergence. Taxonomic characters can be morphological, chemical, physiological, behavioral, genetic, reproductive, or ecological. Their use is illustrated in the next section.

Doyen and Slobodchikoff (1974) created an operational diagram of the logical procedures biologists follow in describing species. Figure 3.3 is a modification of their flow chart. The first step, in most cases, is the recognition of phenetic groups on the basis of morphological or physical distinctions. Phenetically different groups of specimens can be sorted out immediately. Phenetically similar groups may still be recognized as species based on distinctions made at succeeding levels of geographic, reproductive, and ecological criteria. Whereas collection-based taxonomists usually attack the problem from the top of the chart down, as in the original diagram, a field or laboratory biologist might enter the diagram at any level. For example, the field biologist notes an oddity: two corals that look similar but seem to have different modes of reproduction. She might then look more carefully to see whether there are subtle morphological or ecological differences between them. Depending on the starting point, the chain of reasoning leading to determination of a species can proceed in either direction in the diagram, as the examples in the next section show.

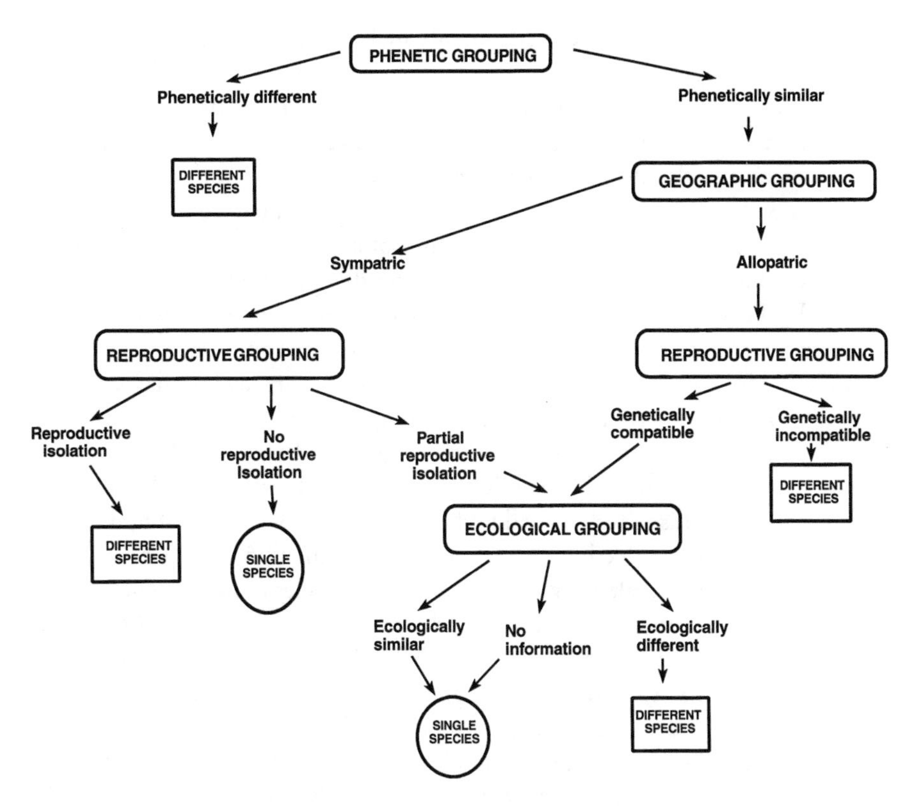

Figure 3.3 Schematic representation of the decision-making process in species description. A museum taxonomist would usually enter the diagram at the top of the chart, with a group of phenetically similar specimens. An ecologist, field biologist, or geneticist might start almost anywhere. Source: Redrawn from Doyen and Slobodchikoff (1974).

Examples of Ways in Which Biologists Have Discovered New Species

Community Studies and Surveys

New species may be found when an unfamiliar area is surveyed for the first time. Sometimes their striking morphology makes their presence known immediately. For example, crawling on sponges that ecologists diving in McMurdo Sound, Antarctica, were studying was a pale yellow fist-sized blob from which two delicate branching tentacles projected. Unable to recognize even the phylum to which the strange creature belonged, the ecologists took specimens back to the laboratory to observe. They preserved some of them as best they could for later identification. Back at home they examined the specimens carefully and, after searching the literature, decided that the animal they had found was an undescribed species of platyctenean ctenophore. Most members of this group are tropical in distribution. Not much biological information was available even for the described species. Learning this gave Gordon Robilliard and Paul Dayton the incentive to describe the species and relate what they had been able to discover about its habits (Robilliard and Dayton 1972).

More often, new species are found by those already familiar with other members of a group. For example, ecologist Kerry Clark was conducting a preliminary inventory of the previously unstudied intertidal community at Sebastian Inlet, Florida (Clark and Goetzfried 1976). Some of the specimens his survey turned up belonged to a new species of nudibranch, which he found feeding on one of the branching hydroids that grew in the crevices of a breakwater. Like many of the species ecologists encounter, it was cryptic, its cerata mimicking the clusters of hydranths of the hydroid and its brown stripes blending with the hydroid's branches. Because he was familiar with nudibranchs and knew the habitats they frequent, he no doubt had his eye on that particular hydroid as good "nudibranch country" and therefore examined it closely, rather than (as most of us would have done), avoiding that particular hydroid forever after his first encounter because of its powerful sting.

But your new find may not even be the kind of organism you were hunting. In 1962, Malcolm Edmunds was a Ph.D. student at the University of the West Indies, studying defensive mechanisms in nudibranchs. While unsuccessfully searching for nudibranchs on the *Caulerpa* algae he had collected in the mangrove swamps of Port Royal, Jamaica, he found instead a new species of bivalved opisthobranch gastropod or clam-snail, a member of an unusual family of mollusks whose shells had been known as fossils, but that had only recently been discovered alive anywhere in the world and certainly were not known from Jamaica (Edmunds 1963).

New species are often collected in community studies, area surveys, and oceanographic sampling in which the primary intent of the project is to describe community diversity or the distribution and abundance of major organisms. In such cases the specimens that cannot be identified to species level are often given numbers or letters to distinguish them as separate taxa when the project results are initially written up for publication. Later they may be described formally as new species, often by collaboration with a taxonomist, and attention called to their status in previous publications. For example, copepod crustaceans are usually a dominant group in pelagic communities, but the larger calanoid species have been studied much more intensively than the smaller cyclopoids. Even in studies in which fine mesh nets were used, the cyclopoids have often been neglected, at least partly for taxonomic reasons; although there are only a small number of genera, they include a large number of species. Ruth Böttger-Schnack, a biological oceanographer from the Institüt für Meereskunde in Kiel, Germany, and Geoffrey Boxshall, a taxonomist from the British Museum, collaborated on a description of two cyclopoid species that began, "The present paper describes another two *Oncaea* species based on females found in plankton taken with fine mesh nets. . . . Both species are common and had formerly been described as *Oncaea* sp. F and *Oncaea* sp. E/F in the published quantitative analysis of cyclopoids" (Böttger-Schnack and Boxshall 1990:861).

But new species can also turn up in familiar environments, including those studied by biologists for years. Shrimp of the genus *Crangon* are abundant in California waters, even contributing to a small local fishery. But, although four *Crangon* species had been known to occur in California for over a hundred years, a fifth, new and distinctive *Crangon* species, *Crangon handi,* was described in 1977 (Kuris and Carlton 1977). The steps leading to the description of this species are typical of the way many new species are described. First, a new species is noticed but is thought to be rare; perhaps only one or two specimens have been found. Later studies turn up more specimens, or a locality is found in which the organisms are more abundant. *Crangon handi* was first captured in 1965 at one locality, but only a few specimens were found there. Only 10 years later did the authors find a coarse-grained sandy habitat in low intertidal tide pools where the species could be easily and abundantly collected. Then specimens could be measured and compared with material of related species, and the description written up and published. It is not unusual in such cases for the entire process to take 10 or 15 years.

Even species of commercial value can be misidentified or undescribed. For years fishermen in Great Britain recognized two forms of the lugworm, *Arenicola marina,* which they called "blow lug" and "black lug." But despite real differences in habitat and morphology, traditional taxonomists persisted in lumping the two forms into a single species. It was 1990 before a paper was published validating the fishermen's experience by giving genetic evidence (based on electrophoresis) indicating that there were indeed two species (Cadman and Nelson-Smith 1990).

Often, when a biologist starts to study what has been described as a single common and widespread species, it turns out that several cryptic species are involved. Leo Buss and Phil Yund were studying the biology of *Hydractinia* hydroids, which live symbiotically on the shells of pagurid hermit crabs. Western Atlantic *Hydractinia* were considered to be identical with European *Hydractinia echinata.* However, colleagues pursuing molecular studies called Buss and Yund's attention to noncomplementary alleles at the MDH locus in *Hydractinia* material, and suggested that there might be a problem. Looking at distribution of *Hydractinia* along the northeastern U.S. coast, Buss and Yund then noticed that *Hydractinia* colonized shells occupied by three different pagurid crabs: *Pagurus acadianus, P. longicarpus,* and *P. pollicaris.* They also observed some morphological differences. *Hydractinia* colonies are polymorphic, containing three types of zooids. Some of the hydroid colonies had an unusual nutritive zooid. On some colonies, the chitinous skeleton projected from and enlarged the gastropod shell the crab occupied. Subsequent breeding experiments between hydroids associated with each of the three crabs, plus Scottish *H. echinata* (which was associated with a fourth crab, *Pagurus bernhardus*) showed that although some hybridization occurred, hydroids associated with different crabs did not interbreed freely. In addition, all the western Atlantic hydroids were distinct from *Hydractinia echinata.* They augmented the breeding experiments with further electrophoresis and with studies of larval and adult characters, all of which upheld the conclusion that different species were involved. Two of the species, *Hydractinia symbiolongicarpus* and *H. symbiopollicarus,* were described as new. The third, *Hydractinia polyclina,* had actually been described by Louis Agassiz in 1862, but because he made his diagnosis on the basis of hydranth size and tentacle number, characters later shown (Fraser 1912) to be quite plastic and therefore taxonomically unreliable, his species was later synonymized back under *Hydractinia echinata* (Buss and Yund 1989).

Sometimes the morphological characters traditionally used are so uninformative that new ones (perhaps at the cellular or ultrastructural level) must be discovered. For example, Muricy et al. (1996) found cytological evidence for cryptic speciation in Mediterranean *Oscarella* sponge species. Members of the demosponge genus *Oscarella* lack any skeleton (skeletal morphology being important in sponge taxonomy) and have few other morphological characters available for use. Reproductive and internal anatomical characters are also uniform in the genus, but cytological characters of mesohylar cells proved good diagnostic characters for use in distinguishing species. Using these characters as well as the small morphological differences that did exist, the authors were able to describe three new species collected from caves and vertical walls along the coast of Provence, France.

New species are sometimes even known to invade the laboratory to confront the biologist. In August 1986, at the Helsingør Marine Laboratory, Denmark, a mysterious dorvilleid polychaete was first noticed in an aquarium containing coarse sand and burrowing bivalves. A few weeks later it was also observed in another aquarium. Two biologists working at the station found the worms easy to culture and became interested in the reproductive biology of these tiny (1- to 3-mm-long) polychaetes, which appeared to link two species groups within the genus. Eventually they published a formal description of the species, together with their biological observations on it (Ockelmann and Åkesson 1990), but at the time of publication the species had still not been collected outside the laboratory.

The undescribed species may not be the one the biologist intended to study, but an associate of some kind. As a graduate student, Leo Buss had been studying competitive interactions between *Antropora tincta* and *Onychocella alula,* two encrusting bryozoans from the Panamanian intertidal, when he realized that the bryozoans had an isopod predator, which fed upon the colonies a zooid at a time. The juvenile forms of the isopod lived on the coralline algae that grew adjacent to the bryozoans on the rocks and had patterns of coloration that seemed to mimic the phoronid and barnacle bore holes common on the algae. He started to study the biology of the isopod, too, and discovered that it was a sequential hermaphrodite with a socially mediated sexual transformation. When his study was over he teamed up with an isopod systematist from the Alan Hancock Foundation to describe the isopod, which turned out to represent a new genus as well as a new species (Buss and Iverson 1981).

In all these cases, a biologist noticed a new or unusual organism directly. In other cases the field biologist works backward from significant or disturbing differences in ecology or behavior and finds corroborating morphological differences later (or not at all).

The presence of a new or previously unrecorded species may sometimes be predicted by ecological detective work. Entomologist Hans Bänziger found that a rare fruit-piercing noctuid moth, *Adris okurai,* he had studied in Nepal occurred in Thailand only on the upper reaches of Doi Suthep, a mountain in Pui National Park. From his earlier work on the life history of Nepalese populations of the species, he had learned that its brightly colored caterpillars fed only on lianas of the family Lardizabalaceae and would die rather than accept substitutes, even from closely related plant families. Members of the Lardizabalaceae had never been found in Thailand, but it seemed to him that they must be there. His suspicion was confirmed when a botanist collecting on Doi Suthep came across a flowering *Parvatia brunoniania,* a member of the family in question (Bänziger 1989).

Behavior Differences

Behavioral differences, especially those related to mate recognition, often provide the first clue that a biologist is dealing with an undescribed taxon. For example, one evening Chris Corben, a Forest Service biologist in Queensland, heard an unfamiliar wavering tree frog call, an unusual sound that reminded him of the sound of a high-voltage electrical arc. Locating and collecting the frog, he also noticed subtle differences in color and body form between his specimen and *Litoria rubella,* the common species in the area. Eventually he and Glenn Ingram, a herpetologist from the Queensland Museum, found these frogs at several other localities, and described the new species as *Litoria electrica,* on the basis of its distinctive morphology and mating call (Ingram and Corben 1990).

Behavioral differences may be very important in distinguishing species in invertebrate groups also. In fiddler crabs of the genus *Uca,* for example, behavioral differences have been instrumental in detecting species (Salmon and Atsaides 1968; Salmon et al. 1979; Salmon and Kettler 1987). These crabs, found in saltmarsh communities around the world, are important research organisms. They are used in ethology, endocrinology, genetics, and neurophysiology, as well as for bioassay organisms to assess pollution effects. Although many species show clear morphological differences, some groups of *Uca* species show distinct differences in behavior (and often in physiology and ecology) while appearing almost identical morphologically. For example, *Uca spinicarpa* was once described as a distinct species, but was synonymized by Crane (1975) as a subspecies of *Uca speciosa* because the two could not be distinguished morphologically on the basis of preserved material. But Salmon et al. (1979) found that *Uca speciosa* and *Uca spinicarpa* could be distinguished on the basis of behavior by male crabs. In the daytime, male fiddler crabs attract females by "waving," rhythmically lifting, extending, and retracting their large claws. The two species differed in duration of waves, as well as in synchronization, and in speed of movements of the smaller claw. At night courtship is prolonged by means of pulses of sound produced by rapping the large claw on the substrate (an activity they can carry on for several hours at a time). *U. speciosa* males typically made sounds consisting of three or four pulses, whereas *U. spinicarpa* males produced sounds of 5 to 10 pulses. Such behavioral displays have been shown to be genetically determined and correlated with mate choice by females (Salmon et al. 1978; Salmon and Hyatt 1979). It makes sense that characters involved in sexual selection would be useful taxonomically. However, the behavioral differences were also backed up by electrophoretic studies and by morphology; the crabs could be easily distinguished on the basis of color (when alive) and, although there was some overlap, on size as measured by carapace width.

Differences related to mating behavior have enabled biologists to describe many sibling species (closely related species that are genetically isolated but differ very little morphologically). For example, ecologists Nancy Knowlton and Brian Keller described a group of four sibling species of alpheid shrimp that live associated with two Caribbean sea anemones (Knowton and Keller 1983, 1985). Juvenile shrimp are found with small anemones. Large anemones or anemone clusters host male and female pairs of shrimp. The four species they studied showed only a few morphological differences: tiny differences in rostrum shape, in setae of the antennular flagellum, and in uropod spines of males. The most noticeable difference between them was in color patterns (which are lost when specimens are preserved). However, reproductive isolation by mating behavior was complete. Mixed-species pairs were never found in the field, and the ecologists could not force shrimp of different species to pair in the laboratory. Shrimp of any one species behaved aggressively toward potential mates of the other species, and would not pair, even though it was easy to induce pairing in conspecifics.

Genetic Evidence (Chromosomal and Molecular)

When it is possible to carry them out, breeding experiments can often confirm the existence of two cryptic species in morphologically similar forms. For example, Langan-Cranford and Pearse (1995) used egg-laying and breeding experiments to confirm the specific identity of *Lacuna marmorata* and *Lacuna unifasciata,* two very similar snails found on algae and surfgrass in California. Both species possessed many of the same shell colors and patterns, and there was overlap in shell shape and radular characters. However, breeding experiments showed that not only were there species-specific differences in the number of eggs produced by females mated with conspecific males, but also that interspecific pairs produced no fertilized eggs at all, indicating complete reproductive isolation maintained by some form of gamete incompatibility.

Studies of karyotype (chromosome number and structure) and molecular genetic studies of differences detected by protein electrophoresis or DNA sequencing are playing an increasing role in the detection of cryptic and sibling species. Plant biologists have found cytological studies of chromosome number to be essential for sorting out species of plants in groups in which polyploidy occurs (Grant 1981; Stuessy 1990). The information these studies yield is complex and often requires careful study and comparison with phenotypic traits to determine which reflect common ancestry and which result from convergence (see Stuessy 1990). Electrophoretic studies are also useful in separating members of problematic plant groups, including algae. For example, Lindstrom and Cole (1990a) used starch gel electrophoresis to separate species of the red algal *Porphyra perforata* complex. Results gave a way to clearly distinguish species in this group (which includes species of potential commercial value) even though morphological characters in the groups varied within, as well as between, species.

In marine invertebrates molecular genetic work is being used to distinguish species and to resolve relationships in groups in which taxonomic confusion has long been apparent. For example, marine bryozoans of the genus *Alcyonidium* are common intertidally and subtidally in cool water regions. They are often conspicuous faunal elements in ecological surveys, but their gelatinous skeletons and morphological variability made their species identification a nightmare to taxonomists. Using protein electrophoresis, John Thorpe and colleagues at the University of Wales, Swansea, were able to detect at least 10 genetically distinct British species, where only 2 or 3 had been recognized before. Careful reexamination of these species in the field and lab showed that some of them were actually detectable also by morphological and ecological characters (e.g., intertidal zonation, or substratum preference) but others were indistinguishable by these means (Thorpe, Ryland, and Beardmore 1978; Thorpe, Beardmore, and Ryland 1978; Thorpe and Ryland 1979). This early work demonstrated the possibility that speciation events for many marine invertebrates may take place in far shorter geographic distances than had been previously supposed.

The scleractinian coral *Montastraea annularis* is abundant on Caribbean reefs. On many reefs it is the dominant coral at a variety of depths and exposures. Marine ecologists had considered it to be a morphologically plastic, environmentally variable species. However, Ernesto Weil and Nancy Knowlton (1994) found that genetic variation (as documented by protein electrophoresis), behavior (as determined by aggressive reactions to other species), and corallite morphology all supported the recognition of three sibling species: *Montastraea annularis,* the species actually represented by the type specimen, and two previously synonymized species, *Montastraea faveolatea* and *M. franksi.* Under the name *Montastraea annularis* these corals had been used in many studies of reef ecology, coral biology, physiology, and biochemistry whose value is now questionable. Perhaps most importantly, results of studies of their growth rates and skeletal isotope ratios had been the basis for conclusions about paleooceanography, environmental degradation, and global climate change. Significant differences in growth rate and oxygen isotope ratios between two of the species may have confused studies of past climate conditions, and unusual coloration in the third species may have sabotaged coral bleaching research (Knowlton et al. 1992). Their study shows how critical it is that species be correctly identified and distinguished in applied work.

Many other recent studies have used molecular genetic techniques to document the occurrence of cryptic species in marine invertebrates. For example, Stobart and Benzie (1994) used allozyme electrophoresis to distinguish two species of the Australian scleractinian coral *Montipora digitata.* Vrijenhoek et al. (1994) detected cryptic species of vesicomyid deep-sea clams from hydrothermal vent and cold-water seep environments. McKinnon et al. (1992) used allozyme electrophoresis to detect sympatric sibling species in Australian copepods.

Molecular genetic methods have been extremely useful to scientists attempting to trace the provenance and spread of exotic species. For example, Ward and Andrew (1995), using allozyme methods to track the history of the introduction of the northern Pacific starfish *Asterias amurensis* into Tasmania, found that Tasmanian populations could not be separated from each other genetically and were more similar to populations from central Japan than to those from northern or southern Japan, indicating a recent origin from a source on the central Japanese coast. Studies of the colonization of European seas by two American species of spionid polychaetes of the genus *Marenzelleria* indicated that the species found in the Baltic Sea had probably come from American populations occurring between Chesapeake Bay and Currituck Sound, whereas populations in the North Sea corresponded genetically to U.S. populations found between Barnstable Harbor and Delaware Bay (Röhner et al. 1996).

Molecular techniques can also provide evidence of gene flow (or lack of gene flow) between populations in separate habitats or localities. An allozyme study of the branching Great Barrier Reef coral *Seriatopora hystrix* by David Ayre and Sandra Duffy (1994) was used to test for variation in the relative contributions of asexual and sexual reproduction in recruitment and to gain a greater understanding of larval dispersal and gene flow. They found that widespread dispersal did occur in this species, but that its direction and magnitude were influenced by currents and weather, with each separate reef having a partially isolated and genetically highly subdivided population. Palumbi and Metz (1991) used morphological and molecular (mitochondrial DNA and single-copy nuclear DNA) to show the presence of at least four independent gene pools in the tropical sea urchin *Echinometra mathaei.* On the other hand, work by Silberman et al. (1994) based on mitochondrial DNA showed no evidence of genetic structure among and between widespread populations of the spiny lobster *Panulirus argus* in the Caribbean, Florida, and Bermuda, indicating high levels of gene flow in this species, which has an extended planktonic larval stage.

Palumbi (1992) recently reviewed marine speciation in general, stressing the roles that transient isolation, gamete ecology, and gamete recognition may play in the process. He pointed out the parallels between the isolating mechanisms of marine animals and those of plants and called for further comparisons between all groups (marine animals, terrestrial animals, and plants) in order to gain greater insight into the relationships between dispersal, speciation, and reproductive ecology. The discussion in this chapter is intended as encouragement to think globally even as you act locally in describing a particular species. As you start the process of species description, you will have a grasp of one pattern, just a piece of the evolutionary "elephant." Your practical goal is publishing an accurate written description according to nomenclatural codes, so your definition of species will necessarily be an operational one. But as you describe your species according to these constraints, keep in mind the varied and interacting biological and ecological processes by which the pattern you observe might been produced; you may find yourself on the path to solving some basic biological problem.

Sources

It would be impossible to list here all the books and articles that deal with populations and species–major topics of evolutionary biology. Those listed here were selected to give different perspectives and opinions.

Printed Sources

Abbott, R. J. 1992. Plant invasions, interspecific hybridization and the evolution of new plant taxa. *Trends in Ecology and Evolution* 7: 401–405.

Andersson, L. 1990. The driving force: Species concepts and ecology. *Taxon* 39: 375–382.

Archibald, D. 1994. Metataxon concepts and assessing possible ancestry using phylogenetic systematics. *Systematic Biology* 34: 27–40.

Ball, I. R. 1983. On groups, existence, and the ordering of nature. *Systematic Zoology* 32: 446–451.

Baum, D. A. and M. J. Donoghue. 1995. Choosing among alternative "phylogenetic" species concepts. *Systematic Botany* 20: 560–573.

Baum, D. A. and M. J. Donoghue. 1995. Genealogical perspectives on the species problem, pp. 289–303 in P. C. Hoch and C. D. Stephenson, eds. *Experimental and Molecular Approaches to Plant Biosystematics.* St. Louis, Missouri Botanical Garden.

Brandon, R. N. 1996. *Concepts and Methods in Evolutionary Biology.* New York, Cambridge University Press.

Brower, A. V. Z., R. Desalle, and A. Vogler. 1996. Gene trees, species trees, and systematics: A cladistic perspective. *Annual Review of Ecology and Systematics* 27: 423–450.

Bush, G. L. 1975. Modes of animal speciation. *Annual Review of Ecology and Systematics* 6: 339–364.

Buss, L. W. 1987. *The Evolution of Individuality.* Princeton, N.J., Princeton University Press.

Carson, H. L. 1985. Unification of speciation theory in plants and animals. *Systematic Botany* 10: 380–390.

Claridge, M. F., H. A. Dawah, and M. R. Wilson. 1997. *Species: The Units of Biodiversity.* London, Chapman & Hall.

Coyne, J. A. 1992. Genetics and speciation. *Nature* 355: 511–515.

Cracraft, J. 1983. Species concepts and speciation analysis, pp. 159–187 in R. F. Johnston, ed. *Current Ornithology 1.* New York, Plenum.

Cronquist, A. 1978. Once again, what is a species? *Beltsville Symposium on Agricultural Research* 2: 3–20.

Cullis, C. A. 1987. The generation of somatic and heritable variation in response to stress. *American Naturalist* 130: 62–73.

Darwin, C. 1859. *On the Origin of Species.* London, Murray.

Davis, J. I. and K. C. Nixon. 1992. Populations, genetic variation, and the delimitation of phylogenetic species. *Systematic Biology* 41: 421–435.

Dobshansky, T. F., J. Ayala, G. L. Stebbins, and J. W. Valentine. 1977. *Evolution.* San Francisco, W.H. Freeman.

Donoghue, M. J. 1985. A critique of the biological species concept and recommendations for a phylogenetic alternative. *Bryologist* 88: 172–181.

Doyen, J. T. and C. N. Slobodchikoff. 1974. An operational approach to species classification. *Systematic Zoology* 23: 239–247.

Du Rietz, G. E. 1930. The fundamental units of biological taxonomy. *Svensk Botanisk Tidskrift* 24: 333–428.

Endler, J. A. 1977. *Geographic Variation, Speciation, and Clines.* Princeton, N.J., Princeton University Press.

Erlich, P. R. 1961. Has the biological species definition outlived its usefulness? *Systematic Zoology* 10: 167–176.

Ehrlich, P. R. and R. W. Holm. 1962. Patterns and populations. *Science* 137: 652–657.

Erwin, D. H. and R. L. Anstey, eds. 1994. *New Approaches to Speciation in the Fossil Record.* New York, Columbia University Press.

Ghiselin, M. T. 1975. A radical solution to the species problem. *Systematic Zoology* 23: 536–544.

Ghiselin, M. T. 1977. On paradigms and the hypermodern species concept. *Systematic Zoology* 26: 437–438.

Gingerich, P. D. 1985. Species in the fossil record: Concepts, trends, and transitions. *Paleobiology* 11: 27–41.

Grant, V. 1981. *Plant Speciation,* 2d ed. New York, Columbia University Press.

Grant, V. 1991. *The Evolutionary Process: A Critical Study of Evolutionary Theory,* 2d ed. New York, Columbia University Press.

Graybeal, A. 1995. Naming species. *Systematic Biology* 44: 237–250.

Heiser, C. N. 1965. Modern species concepts: Vascular plants. *Bryologist* 66: 120–124.

Hutchinson, G. E. 1968. When are species necessary? pp. 177–205 in R. C. Lewontin, ed. *Population Biology and Evolution.* Syracuse, N.Y., Syracuse University Press.

Iwatsuki, K., P. H. Raven, and W. J. Bock, eds. 1986. *Modern Aspects of Species.* Tokyo, University of Tokyo Press.

Knowlton, N. 1993. Sibling species in the sea. *Annual Review of Ecology and Systematics* 24: 189–216.

Lawton, J. H. 1994. What do species do in ecosystems? *Oikos* 71: 367–374.

Levin, D. A. 1979. The nature of plant species. *Science* 204: 381–384.

Levinton, J. 1988. *Genetics, Paleontology, and Macroevolution.* Cambridge, U.K., Cambridge University Press.

Lewis, W. H. 1959. The nature of plant species. *Journal of the Arizona Academy of Science* 1: 3–7.

Lovtrup, S. 1979. The evolutionary species: Fact or fiction? *Systematic Zoology* 28: 386–392.

Luckow, M. 1995. Species concepts: Assumptions, methods and applications. *Systematic Botany* 20: 589–605.

Mallet, J. 1995. A species definition for the modern synthesis. *Trends in Ecology and Evolution* 10: 294–299.

Mayr, E., ed. 1957. *The Species Problem.* Washington, D.C., American Association for the Advancement of Science.

Mayr, E. 1963. *Animal Species and Evolution.* Cambridge, Mass., Harvard University Press.

Mayr, E. 1970. *Populations, Species, and Evolution.* Cambridge, Mass., Harvard University Press.

McDade, L. A. 1992. Hybrids and phylogenetic systematics II. The impact of hybrids on cladistic analysis. *Evolution* 46: 1329–1346.

Mishler, B. D. and A. F. Budd. 1990. Species and evolution in clonal organisms: Introduction. *Systematic Botany* 15: 79–85.

Mishler, B. D. and M. J. Donoghue. 1982. Species concepts: A case for pluralism. *Systematic Zoology* 31: 491–503.

Nixon, K. C. and Q. D. Wheeler. 1990. An amplification of the phylogenetic species concept. *Cladistics* 6: 211–223.

Otte, D. and J. A. Endler, eds. 1989. *Speciation and Its Consequences.* Sunderland, Mass., Sinauer Associates.

Platnick, N. I. 1982. Defining characters and evolutionary groups. *Systematic Zoology* 34: 282–284.

Reynolds, D. R. and J. W. Taylor, eds. 1992. *The Fungal Holomorph: Mitotic, Meiotic and Pleomorphic Speciation in Fungal Systematics.* Wallingford, U.K., CAB International.

Rieppel, O. 1994. Species and history, pp. 31–50 in R. W. Scotland, D. I. Sierber, and D. M. Williams, eds. *Models in Phylogeny Reconstruction.* Oxford, U.K., Clarendon.

Riesberg, L. H. 1997. Hybrid origins of plant species. *Annual Review of Ecology and Systematics 1997* 28: 359–389.

Schmid, B. 1990. Some ecological and evolutionary consequences of modular organization and clonal growth in plants. *Evolutionary Trends in Plants* 4: 25–34.

Scowcroft, W. R. 1985. Somaclonal variation: The myth of clonal uniformity, pp.

217–245 in B. Hohn and E. S. Dennis, eds. *Genetic Flux in Plants.* Vienna, Springer-Verlag.

Scudder, G. G. E. 1974. Species concepts and speciation. *Canadian Journal of Zoology* 52: 1121–1134.

Slatkin, M. 1984. Somatic mutations as an evolutionary force, pp. 19–30 in P. J. Greenwood, P. H. Harvey, and M. Slatkin, eds. *Evolution: Essays in Honour of John Maynard Smith.* Cambridge, U.K., Cambridge University Press.

Slobodchikoff, C. N., ed. 1976. *Concepts of Species.* Stroudsburg, Pa., Dowden, Hutchinson & Ross.

Sokal, R. R. 1973. The species problem reconsidered. *Systematic Zoology* 22: 360–374.

Sokal, R. R. 1985. The continuing search for order. *American Naturalist* 126: 729–749.

Sokal, R. R. and T. J. Crovello. 1970. The biological species concept: A critical evaluation. *American Naturalist* 104: 107–123.

Sylvester-Bradley, P. D. 1956. *The Species Concept in Palaeontology.* Systematics Association publication no. 2: 1–145.

Templeton, A. R. 1989. The meaning of species and speciation: A genetic perspective, pp. 3–27 in D. Otte and J. A. Endler, eds. *Speciation and Its Consequences.* Sunderland, Mass., Sinauer Associates.

Tuomi, J. and T. Vuorisalo. 1989. Hierarchical selection in modular organisms. *Trends in Ecology and Evolution* 4: 209–213.

Tuomi, J. and T. Vuorisalo. 1989. What are the units of selection in modular organisms? *Oikos* 54: 382–385.

Van Valen, L. 1976. Ecological species, multispecies and oaks. *Taxon* 25: 233–239.

Vrana, P. and W. Wheeler. 1992. Individual organisms as terminal entities: Laying the species problem to rest. *Cladistics* 6: 67–72.

Vrba, E. S., ed. 1985. *Species and Speciation.* Transvaal Museum monograph no. 4.

Wiley, E. O. 1978. The evolutionary species concept reconsidered. *Systematic Zoology* 27:-17–26.

Williams, G. C. 1992. *Natural Selection: Domains, Levels, and Challenges.* New York, Oxford University Press.

Witham, T. G. and C. N. Slobodchikoff. 1981. Evolution by individuals, plant–herbivore interactions, and mosaics of genetic variability, the adaptive significance of somatic mutation in plants. *Oecologia* 49: 287–292.

Internet Sources

In addition to the Internet resources on evolutionary biology given in Chapter 1, a search using an Internet search engine may turn up additional material of interest. A search using carried out on May 23, 1998, using **http://www.mamma.com,** a site that submits searches to several search engines at once, found 90 sites for the search phrase *species concepts.* These included course lectures (some with illustrations), definitions of many species concepts, bibliographies, essays, and papers by biologists and philosophers of science.

Discussion lists (listservs) such as TAXACOM (the Biological Systematics Discussion Group) can be a great source of discussion of species concepts and speciation processes. Previous discussion threads can be traced in the listserv's archives. Addresses of the listservs can be found on several systematic and evolutionary biology sites.

To subscribe to TAXACOM (you have to be a subscriber to post a message), send an e-mail message to usobi.org with no subject and "subscribe taxacom Your first name Your last name" as the text of the message. Replace "Your first name Your last name" with your real name. You may also subscribe via web interface: **http://usobi.org/archives/taxacom.html.** Follow the online instructions to subscribe, post a message or search the TAXACOM archives.

Chapter 4

Establishing Identity: The Literature Search

Field guides are instruments of the pleasure of pure knowledge.
–von Baeyer (1984:24)

There is no better way to learn taxonomic procedure than to try to identify material with the help of a good monograph.
–Mayr et al. (1953:72)

Chapter 3 illustrated some of the ways in which biologists have discovered new species in the course of their research. All of the new species given as examples in chapter 3 were recognized by biologists because they differed from their relatives in some significant characteristic: morphological, behavior, or reproductive. Similar clues may have led you to believe that you have also found an undescribed species. This chapter and chapter 5 explain how to make sure.

Mistakes and Bad Examples

One reason some taxonomists strongly discourage species description by nonspecialists is that they are aware of the awful mistakes that have been made (by both amateurs and professionals) who rushed into print without stopping to make sure that what they were describing was both new and real. Some of those authors didn't think straight, some couldn't see straight, and some were just seeing what they believed should be there. Their errors are cautionary tales to remind us that taxonomic research is a serious undertaking in which mistakes can have a negative impact on an author's career or even an area of research. Just about anyone who has tried to identify organisms has been misled from time to time. Mistakes are especially easy to make when using incomplete material, such as fossils or smashed and fragmented specimens from dredge or trawl hauls, or through not recognizing that some item doesn't belong in a sample at all. I still remember the amazing bright blue nematode I *almost* found while sorting through a benthic sample. It was a fantastic discovery until I realized it was exactly the same color as my sweater–it *was* a microscopic thread of yarn from my sweater.

One of the best (or worst) examples of how people see what they expect to see is the story of the Burgess Shale fossils as explained by Steve Gould in *Wonderful Life* (1989). These incredible fossil life-forms were discovered in 1909 by Charles Doolittle Walcott. We now know they represent an early burst of diversity. Many of them are unrelated to living groups, but because Walcott expected them to fit into then-current views of life's evolution, he interpreted them all as primitive members of extant phyla. Their real significance was not discovered until the 1970s; research on them and other very early faunas continues to be an exciting area of paleontology.

Some nineteenth-century authors were notorious for excessive splitting of taxa, creating names for the slightest variation. For example, thousands of names have been applied to the few European species of freshwater mussels. One author described 208 *Unio* and 250 *Anodonta* species from France alone (Simpson 1901), and one *Anodonta* species was given more than 500 names (Emerson and Jacobson 1976). This is not to disparage the real richness of the group, which includes some of our most diverse and endangered freshwater organisms, but to point out that both their diversity and their susceptibility to overnaming stem from their ecology. Their life history patterns, with sedentary adults and dispersal via parasitic glochidia larvae that hitch rides on fish, results in differentiation into species endemic to different rivers and drainage systems. Once the young mussel settles to the bottom, it attaches; once mature it remains in one spot, unable to reattach if disturbed. Shell growth is strongly influenced by environmental conditions. Shells of the same species that have grown in different environments can look very different; shells of mussels from hard bottoms are shorter and rounder than those from muddy or sandy bottoms, and mussels found in small streams have shells less swollen than those in rivers. Early taxonomists had no knowledge of this environmental plasticity and one result was the proliferation of unnecessary names (Emerson and Jacobson 1976).

Some great mistakes have resulted from poor observation. Echinoderm parts have engendered some particularly outrageous examples. The pedicellariae (defensive pincer organs) of the Mediterranean starfish *Asterias tenuispina* were once described as a new ectoparasitic stalked sponge, *Microcordyla asteriae.* The description was complete with a comparative illustration of a real stalked sponge (vaguely similar in outline, but much, much larger, for one thing). The hard parts of pedicellariae were also mistaken for "an extremely minute Vertebrate lower jaw" by Wallich (1862:304), who even conjured up a body to go with it: "The extreme length is 1/100 inch; so that, assuming the body to have been five times as long as the jaw, we have here evidence of the existence of a vertebrate animal measuring only 1/20 inch in length–a size considerably below that of many of the organisms usually regarded as microscopic" (figure 4.1). Theodor Mortensen, the echinoderm worker who corrected these two mistakes, also described fossil pedicellariae remains being mistaken for Foraminifera and living pedicellariae for some new echinoderm parasite of unknown affinity (Mortensen 1932).

Establishing Identity

When your biological knowledge of your unknown organism seems to indicate that it does not match any described species, you must consider all of the following possibilities:

It may be an individual genetic variant or mutant.

It may be a variant produced by growth under unfavorable conditions (e.g., a plant subjected to heavy grazing, an animal stunted by an inadequate food supply).

Especially with plants, it may be a hybrid between two known species.

It may have been described so recently that it is not in the book or article you used for identification.

It may be a known species, but one that has never been described before from the region where you found it. Maybe it is an introduced species, or maybe it is just rare or inconspicuous and was overlooked by earlier workers.

It may indeed represent a completely new species, which must be named and described.

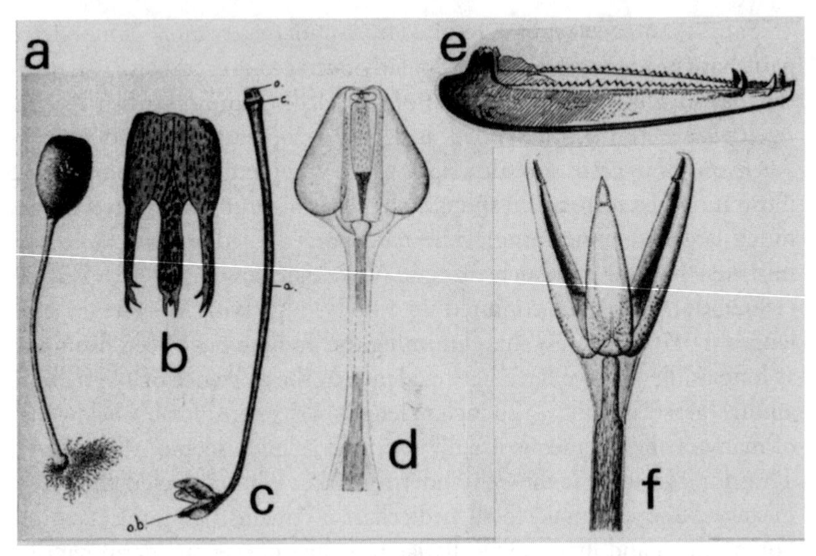

Figure 4.1 Bad examples. Pedicellariae of echinoderms have been mistakenly described as, among other things, a stalked sponge and a "tiny vertebrate jaw." **A,** A genuine stalked sponge, *Stylocordyla borealis.* **B,** "Basal organ" of pedicellaria "sponge." **C,** Entire pedicellaria "sponge" "*Microcordyla asteriae.*" **D,** A stalked pedicellaria of *Paracentrotus lividus* for comparison. **E,** "Tiny vertebrate jaw." **F,** A pedicellaria of *Eucidaris clavata* for comparison (illustrations from Mortensen 1932). Similar mistakes could result if you are not careful in observing and studying your unknown organism or in carrying out adequate library and museum research before describing it as a new species.

The species it represents may have been described before, but some later worker on the group mistakenly decided that it was not a good species and *reduced it to synonymy.* In other words, that specialist decided that the species actually fell within the range of variation of a species that had been named earlier, and in a taxonomic paper put its name into the list of alternative names for the earlier species (see chapter 7). So, after that point, its original distinctiveness was no longer recognized and its original name no longer used. In this case, it must be redescribed and its original name reinstated.

It may have been known previously but called by the wrong name. For example, early naturalists gave many New World species names belonging to species distributed in Europe and Great Britain; later, more detailed study found that they were distinct. In such cases, redescription and a new name are indicated.

An ecologist is more likely to recognize the first three possibilities than a taxonomist working from very limited material (such as a single specimen sent for identification), and you may already been able to discard them, but all possibilities except the last three must be considered and eliminated before making a decision to describe the unknown as new.

To give satisfactory proof that you are dealing with a new entity, you must go through a two-step process of elimination. First, you must track down descriptions of all closely related species in the literature. To accomplish this, you must learn where to find the taxonomic literature for the group of organisms with which you are dealing and learn how to read it once you have found it. Then, if at all possible, you must locate and examine actual specimens. This chapter covers the literature research part of your background search. Chapter 5 deals with research on specimens in collections.

Where to Find the Taxonomic Literature

Your college or university library is the place to start your search. The level at which you start depends on the level of knowledge at which you begin. Your search pattern also will vary with your purpose, but a few general principles of searching (Woodford 1986) are worth stating here.

Be sure to allow enough time. Even the largest university library may not have everything you need, and to obtain all the references by regular interlibrary loan may take several weeks or months. There are excellent fast document delivery services that will mail or fax you articles almost instantly, if you have the funds available, but they do not cover the older literature you will probably need for a thorough taxonomic search.

Verify your references. Jot down bibliographic information right away; make sure you include the book title, article title, year, serial volume, and page numbers (or place and publisher for a book) for the reference you got the citation from as well; this bibliographic information may be required for obtaining interlibrary loans.

Sharpen your question. Limit it by organism, geographic area, and so on.

Know your resources. If you haven't had a tour of your library's science reference section, ask a librarian today.

Discriminate. As you look at titles of possibly relevant books or articles in catalogs and indexes, try to judge which of them will be most relevant before making your selection.

Put the information you've acquired into a useful form, whether cards, notebooks or computer files; if you don't do it immediately, you're going to have to go back and do it later, and it can be much harder to retrace your steps.

General Guides

At the beginning your search may be exploratory. If you know nothing about the literature on a particular group of organisms, you may have to begin with a general guide to the biological literature. Sources listed at the end of this chapter give several recently published general guides to zoological, biological, and botanical literature. Guides such as these are a starting point. However, they usually give only a few references for each group. It would take a book much longer than this one to list the essential taxonomic literature for every major group of eukaryotic organisms.

Books and Textbooks

Another way to find a starting point is to look in the library OPAC or catalog for books on the particular taxonomic group in which you are interested. For example, if you were trying to check the identity of a nemertean worm and knew very little about nemerteans, you might go first to some invertebrate textbooks and then to the sources they cited. Or you might try a subject search through the library catalog, where you might find three items listed:

Advances in Nemertean Biology. 1993. *Hydrobiologia* 266: 18–8158.
Gibson, Ray. 1972. *Nemerteans.* London, Hutchinson University Press.
Gibson, Ray. 1994. *Nemerteans: Key and Notes for Identification of the Species,* 2d. ed. Synopses of the British Fauna, new series, no. 24. London, Linnaean Society of London.

You decide to begin by reading Gibson (1972) to learn more about nemertean biology and terminology. The book's bibliography should lead you to some of the important taxonomic works on the group. The first item, the compilation of meeting papers, will have many references to taxonomic research papers and faunal studies. Even if you are not working in Great Britain, you should look at the third item also because it will have definitions of terminology and citations of much more primary taxonomic literature. It may even help narrow down your identification search.

Many college and university libraries are part of regional networks so that you can search electronic union catalogs of other large libraries, and your library staff may be able to search for you on WORLDCAT, part of the large OCLC system to which most academic libraries belong. This database has more than 36 million records (including 50 listed under the subject *nemerteans*).

Field Guides

Another way to begin is with field guides or handbooks. The first step in your original identification process was probably a field guide or handbook to the creatures of the region in which you were working, such as *The Audubon Guide to Seashore Creatures* or *A Field Guide to the Moths of Eastern North America.* This preliminary search has given you some possible names for your unknown organism, or at least has helped you narrow down the possibilities. Most such guidebooks list the references the authors used in compiling their books. Say, for example, that you find your possibly unknown nemertean during a benthic study you were doing at Chincoteague, Virginia. You attempt to identify it using Kenneth Gosner's *A Field Guide to the Atlantic Seashore* (1978). This guide covers invertebrates and algae from the Bay of Fundy to Cape Hatteras. Your animal seems to most closely resemble *Lineus bicolor* or *Lineus socialis* (Plate 37), so you turn to the descriptions of those species on page 104. But the descriptions are pretty brief, and you are still unsure; it could be either one of them, or neither. Sometimes geographic ranges will help eliminate one possibility, but in this case, both species are listed as being found as far south as Chesapeake Bay, so that does not help. Next you check the references Gosner used and find this citation:

> McCaul, W. E. 1963. Rhynchocoela: Nemerteans from marine and estuarine waters of Virginia. *J. Elisha Mitchell Sci. Soc.* 79(2): 111–124.

The next step is to obtain a copy of this article. In the article you find a more detailed description of *Lineus bicolor.* Morphological characters and measurements given seem to agree well with those of your specimens.

Moreover, its occurrence is given as being subtidal and associated with hydroids, algae, and tunicates, and its habitat as marine and estuarine. This fits the description of the site where you have collected the species, and you decide to go with that identification, but there is one more step: checking the literature from the time of those citations up to the present to see whether there are additional records of the species or any changes in its taxonomic placement and nomenclature.

Abstracts and Indexes

References gleaned from your own guidebooks, textbooks, and your institution's library will give you a point of entry into the literature for the organism or group in which you are interested. Once you have the name of a potential family, genus, or species you want to know more about, you can conduct a search of the biological literature for that group. For that search you must turn to the abstracting and indexing serials. *Biological Abstracts* and *Zoological Record* are the major English language indexes for zoology and botany.

Zoological Record covers the period from 1864 to the present. Its strength is its comprehensive coverage of the biological and taxonomic literature on animals. It indexes other topics, but selectively. Currently it covers more than 6,000 scientific journals, in addition to magazines, newsletters, books, and conference proceedings. About 72,000 references are added each year. From 1864 to 1977 it was produced by the Zoological Society of London. From 1978 to the present it has been a joint project of the Zoological Society of London and BIOSIS, the world's largest indexing and abstracting service. Each yearly volume of *Zoological Record* is now divided into 27 sections (but note that both the number and organization of sections has varied over time). Twenty five sections deal with animal groups, one covers articles and books on general zoology, and one lists all the new generic and subgeneric names indexed in that volume. The size of each section depends on the size of the group; for example, Section 5 contains a whole phylum, "Echinodermata," but Section 13, " Insecta" is divided into subsections on various orders of insects, such as 13 F, "Hemiptera."

Biological Abstracts, also published by BIOSIS, is a comprehensive biological abstracting service covering more than 6,500 serials, with both bibliographic information and abstracts of articles. Two volumes, each consisting of 12 biweekly issues, are published each year. Its biological coverage is broader than that of *Zoological Record,* including botany, ecology, zoology, biochemistry, clinical medicine, pharmacology, and instrumentation. However, a comparative study (Chisman 1989) concluded that its taxonomic coverage for zoology is not as thorough as that of *Zoological Record.* That study found that at least 25 percent of the zoological literature citations found in *Zoological Record* were not picked up by *Biological Abstracts.*

This second level of searching can be accomplished much more quickly if you have access to a library equipped for online or CD-ROM searches. Before such electronic databases existed, it could take weeks to build up a bibliography of potentially useful references. If you don't have institutional access to such services, it may still take weeks, but if you do (as most university students and faculty do), you can rapidly obtain any relevant material from the last 25–30 years by having a librarian conduct an online search for you. In many libraries you can use CD-ROM versions of the databases to conduct your own search. Both *Zoological Record* and *Biological Abstracts* have online (*Zoological Record Online* and *BIOSIS Previews*) and CD-ROM versions. Figure 4.2 shows a sample record from *Zoological Record Online.*

Most important for taxonomic searches are the taxonomic data section ("Taxonomic Categories"), which gives the complete taxonomic hierarchy for the cited organism, and the "Taxa Notes," which gives geographic

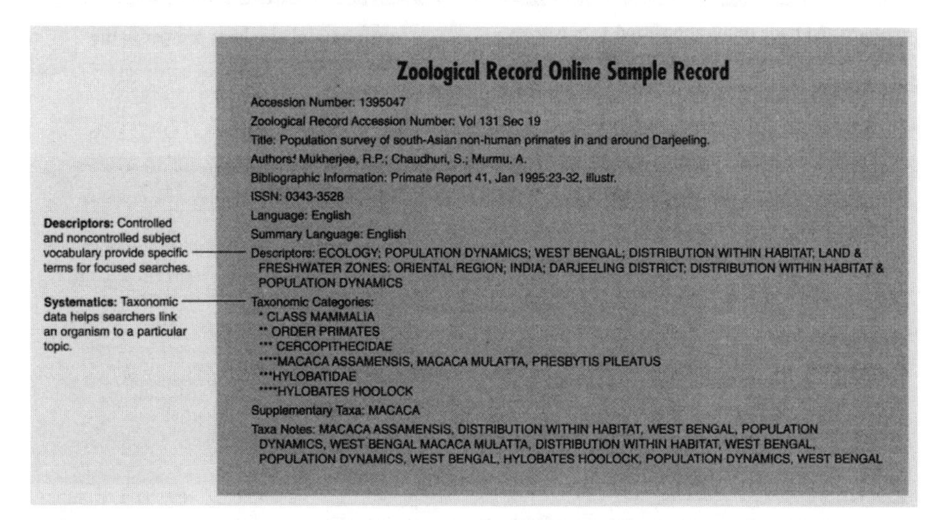

Zoological Record Online Sample Record

Accession Number: 1395047
Zoological Record Accession Number: Vol 131 Sec 19
Title: Population survey of south-Asian non-human primates in and around Darjeeling.
Authors: Mukherjee, R.P.; Chaudhuri, S.; Murmu, A.
Bibliographic Information: Primate Report 41, Jan 1995:23-32, illustr.
ISSN: 0343-3528
Language: English
Summary Language: English
Descriptors: ECOLOGY; POPULATION DYNAMICS; WEST BENGAL; DISTRIBUTION WITHIN HABITAT; LAND & FRESHWATER ZONES: ORIENTAL REGION; INDIA; DARJEELING DISTRICT; DISTRIBUTION WITHIN HABITAT & POPULATION DYNAMICS
Taxonomic Categories:
* CLASS MAMMALIA
** ORDER PRIMATES
*** CERCOPITHECIDAE
****MACACA ASSAMENSIS, MACACA MULATTA, PRESBYTIS PILEATUS
***HYLOBATIDAE
****HYLOBATES HOOLOCK
Supplementary Taxa: MACACA
Taxa Notes: MACACA ASSAMENSIS, DISTRIBUTION WITHIN HABITAT, WEST BENGAL, POPULATION DYNAMICS, WEST BENGAL MACACA MULATTA, DISTRIBUTION WITHIN HABITAT, WEST BENGAL, POPULATION DYNAMICS, WEST BENGAL, HYLOBATES HOOLOCK, POPULATION DYNAMICS, WEST BENGAL

Figure 4.2 *Zoological Record Online* sample record of an article on population biology of some Indian primate species. Searching for any of the species cited (e.g., *Macaca assamensis*) would have retrieved this citation. Note that the record includes the class, order, and family-level classification of the species studied.

information. *BIOSIS Previews* is the online version of *Biological Abstracts.* It began with publications from 1969 and now contains more than 10,000,000 records. It is also available on CD-ROM from 1980 on. Figure 4.3 shows a sample record from *Biological Abstracts on Compact Disc.* For the purpose of taxonomic literature searches, note especially the species names of organisms (appearing under "Keyword") and the "Biosystematic Codes" (which give the higher taxonomic level placement of the organism being cited).

Working forward from the references you used for your initial identification by searching the electronically available records will enable you to make certain that the species hasn't been described since that identification work was published. It will also make you aware of any changes in taxonomic placement of the species. To return to the nemertean example, an electronic search of the BIOSIS Previews database for *Lineus* brought up five items, including this taxonomic work:

Riser, N. W. 1993. Observations on the morphology of some North American nemertines with consequent taxonomic changes and a reassessment of the architectonics of the phylum. *Hydrobiologia* 266: 141–157.

Sample Record from Biological Abstracts on Compact Disc

TI: Sea turtles in North Carolina waters.
AU: Epperly-S-P; Braun-J; Veishlow-A
CS: NOAA, Natl. Mar. Fish. Serv., Southeast Fish. Sci. Cent., Beaufort Lab., Beaufort, NC 28516, USA
SO: Conservation Biology 9(2): 384-394
PY: 1995
IS: 0888-8892
LA: English
LS: English Spanish
AB: Until the turn of the century the inshore waters of North Carolina harbored populations of sea turtles large enough to support a commercial fishery. Based on a 4- to-5-year record of sighting reports by the public, interviews of recreational fishermen, and records kept by commercial fishermen the waters continue to

⋮

excluder devices in shrimp trawls to inshore areas during the entire year. Full implementation of these requirements was achieved by December 1994.
DE: RESEARCH ARTICLE; CARETTA CARETTA; CHELONIA MYDAS; MALACLEMYS TERRAPIN; LEPIDOCHELYS KEMPII; COMMERCIAL FISHERY; SEASONALITY; THREATENED SPECIES; ENDANGERED SPECIES; USA
MJCC: CC00512 (General-Biology-Conservation-Resource-Management);
CC07504 (Ecology-Environmental-Biology-Bioclimatology-and-Biometeorology);
CC07512 (Ecology-Environmental-Biology-Oceanography);
CC07516 (Ecology-Environmental-Biology-Wildlife-Management-Aquatic);
CC62516 (Chordata-General-and-Systematic-Zoology-Reptilia)
BC: BC85402 Chelonia
ST: Animals; Chordates; Vertebrates; Nonhuman vertebrates; Reptiles
JA: Biological Abstracts Vol. 99, Iss. 12, Ref. 170299.
UD: 9502

Descriptors: Keywords added by BIOSIS enhance the title.

Concept Codes: Used to index broad life science categories. The 571 Concept Codes ensure easy retrieval of subject areas.

Taxonomic Information: Biosystematic Codes are provided to indicate taxonomic information at the higher classification levels. Special indications of new taxa are also provided. Super Taxa allow easy searching of broad taxonomic groups, such as nonhuman primates, vertebrates, and plants.

Figure 4.3 *Biological Abstracts on Compact Disc* sample record. *DE,* "Descriptors," keywords or names of species cited in the article (e.g., *Caretta caretta*). *BC,* Biosystematics codes, which give some information on classification (e.g., Chelonia, the order of reptiles to which these sea turtles belong).

The paper includes a detailed description of *Lineus bicolor,* with additional references to its occurrence. It also describes a new genus, *Tenuilineus,* with *bicolor* as its type species. So you have identified your species and verified its scientific name and current taxonomic placement: *Tenuilineus bicolor* (Verrill, 1892). In this case your identification job is done. If, after reading this material, you still had doubts about the species identity, you would have had to search backward to the original description of each similar species, and possibly look at specimens.

Once your search takes you back before the limits of the electronic tools, you are limited to manual searches of the print versions of the indexes (figure 4.4). You may not have to search every year of every index; your search of the past 25 years may have given all the citations you need to earlier descriptions, including the original description, but you must examine those older references, particularly the reference containing the original description. Because of their age and that fact that they appeared in obscure journals, many of these works probably won't be available in most college libraries. To find them, you may have to travel to a major university or museum library (many museum libraries, such as the library of the American Museum of Natural History in New York, are open to the public for research purposes) or obtain the references via interlibrary loans through your own institution. Note that this doesn't necessarily mean a ordering a copy of an entire monograph. The synonymy section of a species description gives the page numbers of earlier descriptions of the species and often the figure number of any illustration (but don't forget that plates in older journals are often at the back of the volume rather than incorporated with the text). Therefore, you may need to request only a page or two of a long monograph or article to get the information you need.

For zoologists, searching back through the taxonomic history of a species usually means going through all the *Zoological Record* sections for that group. Before the *Zoological Record,* the *Royal Society of London Catalog of Scientific Papers* (1800–1900) and various bibliographies compiled for the literature of particular groups of organisms are be your best sources (see this chapter's *Sources*).

In botany also, your first step in identification may have begun with a field guide or a key to plant families. More technical studies, floras (studies of the plants of a particular region), monographs, and revisions followed. To carry out a thorough botanical taxonomic search, *Biological Abstracts* is the most useful general index, but botanists also use several

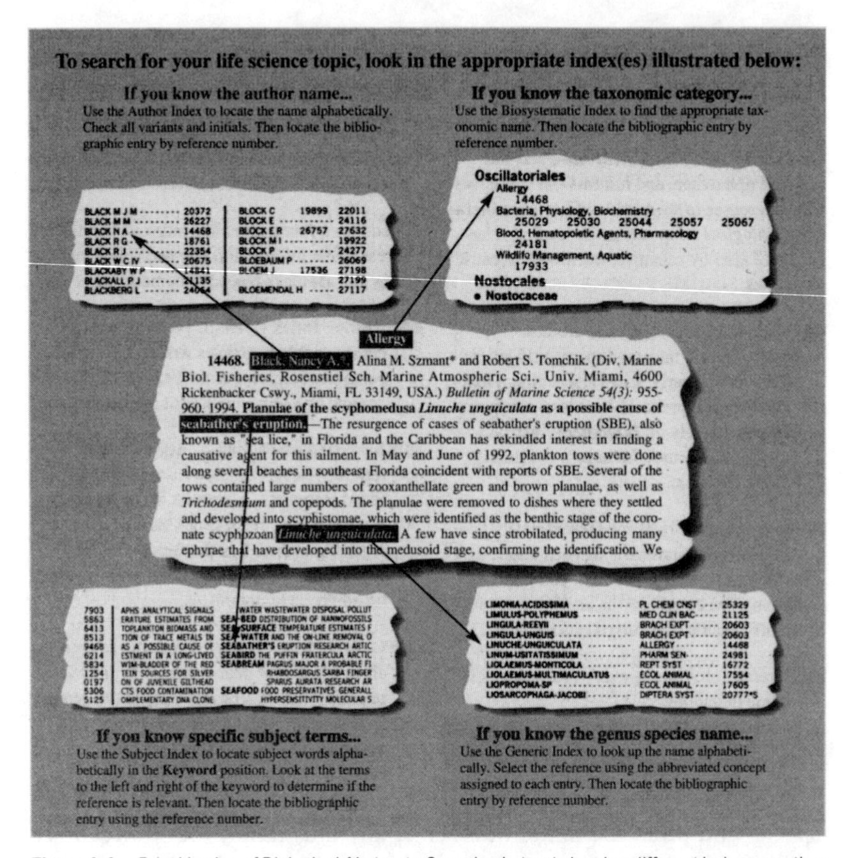

Figure 4.4 *Print Version of Biological Abstracts.* Sample abstract showing different indexes: author name, taxonomic category, subject, and generic (coded by genus and species).

additional sources that have no zoological equivalent. The *Kew Record of Taxonomic Literature* covers the literature on vascular plants from 1971 to the present. The *Bibliography of Systematic Mycology* covers the literature on fungi and lichens from 1943 to the present. Other specialized indexes are *Current Advances in Plant Sciences* (1972–) and *Excerpta Botanica* (1959–). *Botanical Abstracts* (1918–1926) was the botanical predecessor to *Biological Abstracts.* As in zoology, if your search takes you back before the two most inclusive indexes, there are a number of guides. Start with Stafleu and Cowan (1976).

In botany there are also a number of useful dictionaries of plant names; some, such as the *Index Kewensis* (genus and species names of flowering plants from 1753 on), provide citations to the original descriptions. See this chapter's *Sources* for other titles.

Paleontologists also use the botanical or zoological indexes. Both *Zoological Record* and *Biological Abstracts* index most of the journals in which papers on the systematics of fossil organisms might appear. *Zoological Record* has a "Palaeontological Index" that lists entries under geological time period and geographic area. Its coverage of paleontology has varied over the years, and in its early years it completeness depended a lot on the indexer for that particular group at that time. Luckily, a number of its indexers were paleontologists (Bridson 1968).

How to Read the Taxonomic Literature

The first step in your background research was to compile as many citations as possible that include descriptions of species that may match your unknown. When you made your initial identification, you may have relied more on illustrations in the field guide or handbook than on the accompanying text. But, to make a thorough search, you must make a list of taxonomic (and important biological) references to all close relatives or species with which yours could possibly be confused. You must obtain copies of those references, read the descriptions carefully, and interpret them. You must also scrutinize the illustrations and decipher any clues they can provide.

Species Descriptions

To finish your literature search, you must learn to read taxonomic descriptions, understand the parts of a description, and know why they are included. This topic is covered in depth in chapters 7–15, but is summarized here. The style in which taxonomic descriptions are written varies somewhat from group to group. It depends also on the generally accepted style of the time period in which they were written. Earlier descriptions are much briefer and often do not include information that would be required of any description published today. However, there is a common basic structure: a *heading* that consists of *scientific name, name, author,* and *date,* followed by a *synonymy* (a list of previous references to that species), and then the *main body* of the description, which may include *etymology, diagnosis, taxonomic discussion, ecology,* and *distribution* sections. After the somewhat looser style of textbooks and research articles, taxonomic descriptions may at first seem mystifying, rather like a racetrack program in which each entering horse's form is described in tiny print and strange abbreviations. In fact, the standard taxonomic description bears a strong resemblance to the information given on each horse entered in a particular race, giving similar vital statistics (animal's name, parentage, date of "birth," description, and past performance) and packing a considerable amount of information about the organism into a very small space.

Figure 4.5 shows the parts of a species description for a new and a known animal species.

Name. As the examples show, the scientific name of the species is given first. Above the name there may be additional headings, names of higher categories that describe its place in the taxonomic hierarchy.

Author. After the binomial genus and species name in a taxonomic work is a person's name. This is the name of the person who first published a description of the species, called the *author* of the name. For animals, if the name is still used as originally published, the author's name directly follows the organism's name with no punctuation; if a later taxonomist has placed it in another genus, then the author's name follows in parentheses. The author of a plant name is also usually given after the name. If the original name is still in use, it follows without parentheses, as in zoology. One difference is that botanists abbreviate many authors' names, whereas zoologists abbreviate only the names of a very few well-known authors such as Linnaeus (L.). But, as noted in chapter 2, changes in placement and rank are handled differently in botany. If the species has been transferred to another genus (has a new generic name), the author making the transfer is cited as the author of that combination, as in *Linaria spuria* (L.) Mill. 1768, although the original author of the species name is still acknowledged. The author of the original combination is placed in parentheses and the new name and date (name of the transferring authority and the date of publication of the work in which the transfer was made) are added.

Date. If you are looking at the original description of the species, there will be no author or date, but following the name will be *new species* (sometimes abbreviated *n. sp.*) or *species nova* (*sp. nov.*). In the case of described species you will see a date after the name and author. This is the date of publication of the original description.

a

Headings

Name

Figure

FAMILY UMBONULIDAE CANU, 1904
GENUS *HIPPOPLEURIFERA*
CANU AND BASSLER, 1924

Hippopleurifera belizae, new species
Figures 29, 30

Diagnosis

DIAGNOSIS: Colonies are encrusting. Zooids are rhomboidal, with a row of marginal pores and with additional pores occurring proximal and lateral to the orifice. The frontal calcification forms transverse ridges in the area of the accessory pores and is thickened into tubercles proximally. The orifice is hoof-shaped and is surrounded by a flattened margin from which six to eight spines arise. Triangular avicularia are lateral, usually paired and directed proximolaterally, but their number and position is variable. Ovicells are prominent, globular, and perforated by small pores arranged in a radiating pattern.

Type Material

HOLOTYPE: USNM 376788.

Etymology

ETYMOLOGY: Named for the country of Belize in which the species was found.

Description

DESCRIPTION: The colony is encrusting, forming a small patch on coral rubble. Zooids are rhomboidal, with a row of marginal pores. The proximal part of the frontal surface is thickened by calcified tubercles. Lateral to the orifice there are additional pores, the innermost row elongated so that several transverse ridges are formed just below the orifice. The orifice is hoof-shaped, rounded anteriorly, with two large condyles and a broad, shallow posterior portion. It is surrounded by a flattened margin bearing six to eight spines, laterally and distally. Avicularia are suboral, with cross-bars and a triangular mandible. They are paired, placed at midlength, and directed proximolaterally on most zooids; on some zooids a single avicularium occurs at midlength and an additional single, distolaterally directed avicularium is lateral to the orifice on the opposite side of the zooid; on a very few zooids, paired lateral avicularia only are present. The ovicells are globular, thickly calcified, with perforations arranged in more or less radial rows.

Discussion

DISCUSSION: This species appears to be a Recent representative of the genus *Hippopleurifera*, a genus with a number of species in the Tertiary of the S.E. United States (Cheetham, 1963). *Hippopleurifera belizae* most closely resembles the Eocene-Oligocene species *Hippopleurifera crassicollis* (Canu and Bassler, 1920). It differs from *H. crassicollis* in that avicularia and areas of thickest calcification are shifted proximally, and areas with the greatest numbers of pores and grooves are distal and suboral rather than proximal.

Occurrence

OCCURRENCE: Spur and Groove Zone.

Distribution

DISTRIBUTION: Carrie Bow Cay, Belize.

b

Headings

Name

Figure

FAMILY CALLOPORIDAE NORMAN, 1903
GENUS *CRASSIMARGINATELLA* CANU, 1900

Crassimarginatella tuberosa
Canu and Bassler, 1928
Figures 12, 13

Synonymy

Aplousina tuberosa Canu and Bassler, 1928b, p. 21.
Crassimarginatella tuberosa, Hastings, 1945, p. 85. Cheetham and Sandberg, 1964, p. 1017. Cook, 1968, p. 151.

Description

DESCRIPTION: Colonies are encrusting, forming a lacy white to yellowish meshwork on dead coral surfaces. Autozooids generally ovoid and somewhat irregular in size and shape, reflecting underlying irregularities of the substratum. Zooids are separated from each other by distinct furrows. Most of frontal surface membranous, edged by a narrow band of underlying cryptocyst, bordered by smooth textured gymnocyst. Carrie Bow specimens lack the two distal tubercles described in this species by Cheetham and Sandberg (1964) and by Cook (1968). Avicularia are in the form of B-zooids containing functional polypides. The latter are as large or larger than autozooids and are usually more elongated, with two pivotal prongs and a distal shelf for the support of the enlarged toenail-shaped operculum. Ovicells are very small, roofed merely by rectangular pillows of calcification perched on the distal rim of fertile zooids.

Discussion

DISCUSSION: Cheetham and Sandberg (1964) have pointed out the respects in which this species is intermediate between *Aplousina* and *Crassimarginatella*. The B-zooid avicularia are considered to link it with *Crassimarginatella*, whereas the ovicells are more similar to those of *Aplousina*. The avicularia differ somewhat from those in West African specimens described by Cook (1968), and are much fewer in number per colony (some colonies having none).

Occurrence

OCCURRENCE: Spur and Groove Zone, Outer Ridge (15 and 20 m). One of the most abundant species in terms of number of colonies collected. However, colonies were often in poor condition, partially scraped away by grazers, or covered by foulers, and colony life expectancies may be relatively short compared to those of the other abundant species.

Distribution

DISTRIBUTION: Gulf of Mexico. Caribbean. West Africa.

Figure 4.5 Parts of a species description for **(a)** a new and **(b)** a known species (Winston 1984).

Illustration. Often the next line will give a figure or plate number showing where in the publication the species is illustrated.

Synonymy. This is equivalent to the part of the racetrack program that tells how the horse did in previous races: What tracks he raced at (what publications the species appeared in) and how he finished (his taxonomic placement at that time). Styles for synonymies vary from publication to publication, but the essentials are the name under which the species was listed in each publication, along with brief bibliographic information; the name of the author of the publication; the date published; and the page number or numbers and figure or plate numbers.

Etymology. This section occurs only when a species (or higher category) is first described. It gives the derivation of the name.

Diagnosis. This section gives a brief comparative description of the organism, pointing out how it differs from close relatives (why it is a horse of a different color).

Description. This portion contains the straightforward description of the species (2-year-old, chestnut, 14 hands, etc.). Again, style varies from group to group and publication to publication (see chapter 12), but modern journals often favor the abbreviated telegraphic style, in which nonessential grammar is left out. Here the taxonomically significant characters are covered. How well this is done depends on the observational and rhetorical talents of the author, but it should at least provide the information you need to determine whether your unknown could belong in that species.

Discussion. If there have been problems with or changes in the placement of the species being described, there will often be a section for explanation or discussion of that fact.

Ecology. Modern works are likely to have a section giving ecological information, habitat preferences, community association, and other aspects of species biology, such as reproductive season, seasonal changes in growth form, or pelage.

Distribution. In a description of a new species this may just be one locality. For a common or wide-ranging species it may be a long list of localities. If the work is regional in scope (e.g., a flora or a faunal survey) the author may list only its occurrences in the part of the world being studied.

Taxonomic Literature Searching on the Internet

The rapidly increasing amount of information available on the Internet has changed the way we do research. There are many ways Internet searches can help with taxonomic problems, and they are covered in the appropriate chapters. However, the Internet is still not very productive for taxonomic literature searches. There are no free substitutes that can replace the commercial online indexes such as the *Zoological Record* Online or *BIOSIS Previews.* The few free databases you can access for searching the primary literature of journal articles do not cover taxonomy well. Web searches are extremely incomplete; a recent study showed they cover less than half of the pages available on the Web (Brake 1997) and they are not yet very useful for finding journal citations on a subject, or even by a particular author. But it can still be worth doing a search for information on the group you are interested in. More and more bibliographies on a particular group of organisms (e.g., corals, palms, polychaetes) are being put on the Web, either by scientific societies or by individual scientists or students. If you strike one of these, it will give you a good entry into the literature for that group.

If you have Web access and time to spare you can also check quite a few large library catalogs, one at a time, for citations to books on a subject. Doing a Web search on *library catalogs* gave addresses for 183 sites (as of November 28, 1997). You can use this technique to access the Web sites of libraries around the world, and you may be able to make a preliminary list of books or monographs on a subject. For example, the Norwegian University system's catalog (http://www.Bibsys.nor) had one listing under *nemerteans*: the 2d edition of the British Synopsis volume. The Library of Congress site (http://www.LCweb.loc.gov) is one of the most useful sites, but it had only three references on *nemerteans,* compared to the 50 on WORLDCAT. Similar searches can be done for botanical or paleontological topics, of course; a subject search for *trilobites* on the Australian Geological Survey Organisation's online library (http://www.agso.gov.au) brought up 30 matches. However, although there are several electronic full-text projects under way, and you may already be able to read and download some of the world's literary classics, it will be a while before this effort is extended to the classics of taxonomy. You may be able to compile a preliminary list of references online, but you won't yet be able to read them that way.

Sources

Printed Sources

General Guides to the Literature

Bell, G. H. and D. B. Rhodes. 1994. *A Guide to the Zoological Literature: The Animal Kingdom.* Englewood, Colo., Libraries Unlimited.

Blackwelder, R. E. 1972. *Guide to the Taxonomic Literature of Vertebrates.* Ames, Iowa State University Press.

Davis, E. B. and D. Schmidt. 1995. *Using the Biological Literature: A Practical Guide,* 2d ed. New York, Marcel Dekker.

Gilbert, P. and C. J. Hamilton. 1990. *Entomology: A Guide to Information Sources,* 2d ed. London, Mansell.

Sims, R. W., ed. 1980. *Animal Identification: A Reference Guide.* New York, Wiley. 3 vols. (Includes citations to bibliographies of taxonomic literature for various groups of animals, arranged by phylum, class, and family, up to the end of the 1970s.)

Wood, D., J. Hardy, and A. Harvey. 1989. *Information Sources in the Earth Sciences.* London, Bowker-Saur. (Has paleontology chapter.)

Wood, D. N. 1973. *Use of Earth Sciences Literature.* London, Buttersworth. (Has paleontology chapter.)

Wyatt, H. V. 1997. *Information Sources in the Life Sciences.* 4th ed. New Providence, N.J., Bowker-Saur.

Dictionaries of Scientific Terms

Allaby, M., ed. 1992. *The Concise Oxford Dictionary of Botany.* New York, Oxford University Press.

Bates, R. L. and J. A. Jackson. 1987. *Glossary of Geology.* Alexandria, Va., American Geological Institute.

Hawksworth, E. L., B. C. Sutton, and G. C. Ainsworth. 1983. *Ainsworth and Bisby's Dictionary of the Fungi,* 7th ed. New York, French and European Publications.

Henderson, I. F. 1989. *Henderson's Dictionary of Biological Terms,* ed. by E. Lawrence, 10th ed. New York, Wiley.

Margulis, L., H. I. McKhann, and L. Olendzenski, eds. 1993. *Illustrated Glossary of Protoctista. Vocabulary of the Algae, Apicomplexa, Cliates, Foraminifera, Microspora, Water Molds, Slime Molds, and Other Protoctists.* Boston, Jones & Bartlett.

McGraw-Hill Dictionary of Biology. 1984. New York, McGraw-Hill.

Walker, P. M. B. 1990. *Cambridge Dictionary of Biology.* New York, Cambridge University Press.

Periodical Indexes, Abstracts, and Bibliographies: Current

Bibliography of Fossil Vertebrates. 1928–. Los Angeles, Society of Vertebrate Paleontologists.

Bibliography and Index of Geology. 1933–. Alexandria, Va., American Geological Institute. (Before 1969 excludes North America. Includes subject headings for "Paleontology" and "[Geologic Age]-Paleontology.")

Bibliography and Index of Micropaleontology. 1972–. New York, Micropaleontology Press.

Bibliography of North American Paleobotany. 1969–. In Paleobotanical Section, Botanical Society of North America, Kokomo, Ind.

Bibliography of Systematic Mycology. 1943–. Wallingford, U.K., CAB International Mycological Institute. (Print, online, diskette, and CD-ROM versions; world list of books and papers on taxonomy of fungi, with author and classified indexes.)

Biological Abstracts. 1926–. Philadelphia, BIOSIS.

Biological Abstracts/RRM (Reports, Reviews, Meetings). Philadelphia, BIOSIS. vols. 18–.

CAB Abstracts. 1972–. Farnham Royal, Slough, U.K., CAB International. Database of 26 abstracting journals containing abstracts of agricultural and biological articles from more than 8,500 journal, books, theses, conference proceedings, etc.; available online and in a CD-ROM version from 1984 onward (CABCD). Includes

*Helminthological Abstracts.*1932–. Wallingford, U.K., CAB International.

Protozoological Abstracts. 1977–. Wallingford, U.K., CAB International.)

Current Advances in Plant Science (CAPS). vols. 1–, 1972–. Oxford, U.K., Elsevier. (Print, online, diskette, and CD-ROM versions. Monthly. Abstracts of botanical literature.)

Excerpta Botanica. vols. 1–, 1959–. New York, VCH. (Two sections: *Taxonomica and chorologia* [covers articles on systematic botany, herbaria, botanical gardens, etc.] and *Sociologica* [covers monographs on plant ecology and geography].)

Geological Abstracts. 1989–. Norwich, U.K., Elsevier/Geo Abstracts. (Separate chapter for paleontology.)

Georef on CD-ROM. 1785–. Boston, SilverPlatter. (CD-ROM equivalent of *Bibliography and Index of Geology* and *Bibliography of North American Geology.*)

Kew Record of Taxonomic Literature. 1971–. London, HMSO. (World literature for vascular plants.)

Zoological Record. 1865–. Philadelphia, BIOSIS; London, Zoological Society of London.

Periodical Indexes, Abstracts, and Bibliographies: Retrospective and Historical

Bay, J. C. Bibliographies of botany. A contribution toward a Bibliotheca Bibliographica. *Progressus Rei Botanicae* 3: 331–456. (Bibliographies arranged by topic.)

Bibliography and Index to Palaeobotany and Palynology. 1950–1975. 2 vols. Stockholm, Swedish Museum of Natural History.

Bibliography of North American Geology. 1785–1970. Alexandria, Va., American Geological Institute. (For paleontology.)

Botanical Abstracts. 1918–1926. vols. 1–15. Baltimore, Williams & Wilkins. (Continued by *Biological Abstracts.*)

Catalogue of Botanical Books in the Collections of Rachel McMasters Miller Hunt. 1958–1961. Pittsburgh, Hunt Botanical Library. (Early botany books 477–1800.)

Gilver, P. and C. J. Hamilton. 1990. *Entomology: A Guide to the Literature.* London, Mansell.

Hay, O. P. 1929–1930. *Second Bibliography and Catalogue of the Fossil Vertebrata of North America.* 2 vols. Washington, D.C., Carnegie Institution.

International Catalogue of Scientific Literature. 1901–1914. London, Royal Society of London. (Divided into sections; L. is general biology, M. is botany.)

Pfister, D. H., J. R. Boise, and M. A. Eifler. 1990. *A Bibliography of Taxonomic Mycological Literature, 1753–1821.* Berlin, J. Cramer.

Pritzle, G. A. 1972. *Thesaurus Literature Botanicae,* 2d ed. Champaign, Ill., Koeltz (reprint of 1877 ed.).

Royal Society of London. 1867–1902. *Catalogue of Scientific Papers, 1800–1900.* London, Royal Society of London. (Entries by author.)

Stafleu, F. A. and R. S. Cowan. 1976–1988. *Taxonomic Literature,* 2d ed. Utrecht, Bhon, Scheltema, and Holkema. 7 vols. (Covers plant taxonomic literature, by author, from 1753 to 1939; includes information on authors, publication dates, and location of types. Plus Supplement V, 1998.)

White, C. A. 1878. *Bibliography of North American Invertebrate Paleontology.* Washington, D.C., U.S. Government Printing Office.

Indexes and Dictionaries of Animal Names

Amphibian Species of the World. 1985. Allen Press, for the Association of Systematic Collections.

Duellman, W. E. 1993. *Amphibian Species of the World: Additions and Corrections.* Lawrence, University of Kansas Museum of Natural History Special Publication no. 21.

ANI-CD: Arthropod Name Index on CD-ROM. 1995. Wallingford, U.K., CAB International. (For arthropods of economic importance.)

Indexes and Dictionaries of Plant Names and Authors

For searching the botanical taxonomic literature, sources that give authors' names and standard abbreviations are essential. They include Stafleu and Cowan (1976) and the following.

Brummitt, R. K. and C. E. Powell. 1992. *Authors of Plant Names.* Kew, U.K., Royal Botanic Gardens. (Includes all plant groups.)

The botanical literature has much available to guide searchers to original descriptions and publication of plant names, including the following:

*Index Filicum.*1906–. (Gives binomial names of all ferns from Linnaeus on, and of all pteridophytes from 1961 on.)

Index Kewensis. (Print and CD-ROM versions. All flowering plant names at rank of family and below from 1971 on; all binomial and generic names from Linnaeus [1753] on. Useful guide to tracing first publication of a name.)

Index Muscorum. 1959–. Updates published in journal *Taxon.* (Names of mosses with basionyms, synonyms, and places of publication.)

Index of Fungi. 1940–. Kew, U.K., CAB International Mycological Institute.

For fossil plants see various volumes of the *Index of Generic Names of Fossil Plants:*

Andrews, H. N. 1970. *Index of Generic Names of Fossil Plants, 1820–1965.* Washington, D.C., U.S. Geological Survey.

Blazer, A. M. 1975 *Index of Generic Names of Fossil Plants, 1966–1973.* Washington, D.C., U.S. Geological Survey.

Watt, A. D. 1982. *Index of Generic Names of Fossil Plants, 1974–1978.* Washington, D.C., U.S. Geological Survey.

Bonner, C. E. B. (1962–78). *Index Hepaticarum,* Parts 1–9. Cramer. (For liverworts, gives names at all ranks, with places of publication, typification, and geographic distributions.)

Brummitt, R. K. 1992. *Vascular Plant Families and Genera.* Kew, U.K., Royal Botanic Gardens. (Alphabetically lists accepted generic names and selected synonyms, with families. Also lists accepted families alphabetically, with included genera. Lists several classification schemes.)

Crosby, M. R. et al. 1992. *Index of Mosses: A Catalog of the Names and Citations for New Taxa, Combinations, and Names for Mosses Published During the Years 1963 Through 1989 with Citations of Previously Published Basionyms and Replaced Names Together with Lists of the Names of Authors of the Names and Lists of Names of Publications Used in the Citations.* St. Louis, Missouri Botanical Garden.

Farr, E. R. et al. 1979. *Index nominum genericorum.* Utrecht. (Generic names of plants and fungi with place of publication and type.)

Hyam, R. 1995. *Plants and Their Names: A Concise Dictionary.* Oxford, U.K., Oxford University Press.

Mabberly, D. J. 1989. *The Plant Book.* Cambridge, U.K., Cambridge University Press. (Dictionary of flowering plants, including ferns and conifers.)

Stephenson, L. 1992. *What's in a Name? Derivation of Generic Names of Australian Native Plants.* Adelaide, S.G.A.P. South Australian Region.

Wielgorskaya, T. 1995. *Dictionary of Generic Names of Seed Plants.* New York, Columbia University Press. (Most recent compilation for seed plant genera.

Gives detailed geographic distribution, updated lists of species within genera, and corrected names and dates of publication.)

Willis, J. C. 1973. *A Dictionary of the Flowering Plants and Ferns.* Cambridge, U.K., Cambridge University Press. (For flowering plant genera, gives names, authorities, numbers of species and geographic range, plus descriptions and classification for family names.)

Internet Sources

The Web is growing is growing so fast that what is written here is likely to be outdated by the time this book is published. However, changes are in the direction of growth, and of more information and new links being added to current sites. If a particular address no longer works, try a direct search instead. If one search engine comes up empty or without useful information, try the others. They each sample different subsets of sites and information. All addresses begin with **http://www.**

Systematics Societies and Organizations

syst.bot.org
The American Society of Plant Taxonomists site includes a searchable list of members with their research interests.

ascoll.org/
The Association of Systematics Collections site includes databases, TRED, and links of interest.

sicb.org/
Society for Integrative and Comparative Biology.

spnhc.org/
Society for the Preservation of Natural History Collections.

science.uts.edu.au/sasb/links.html
The Society of Australian Systematic Biologists site includes links called Systematic and Internet Resources.

utexas.edu/ftp/depts/systbiol/
The Society of Systematic Biologists site includes links to other systematics-related sites.

nbii.gov/tred
The TRED site includes a searchable list of U.S. systematists, with area of expertise.

nhm.ac.uk/hosted_sites/uksf/
The U.K. Systematic Forum includes a searchable directory of U.K. systematists and good links (e.g., to Royal Botanic Garden at Kew).

nhm.ac.uk/hosted sites/hennig.html

The Willi Hennig Society site lists upcoming meetings and software; you can also check abstracts in the journal *Cladistics.*

Other Sites

york.biosis.org/index.htm

The Zoological Record Home Page includes a Glossary of Nomenclatural Terms in Zoology, the ZR Taxonomic Hierarchy, and links to other Internet resources in systematics, taxonomy, and nomenclature.

ucmp.berkeley.edu/

The University of California, Berkeley Museum of Paleontology Subway Phylogenetic Resources stop includes links to other sites.

biodiversity.uno.edu

The Biodiversity and Biological Collections WWW Server includes directories of biologists and museum/biodiversity-related publications.

darwin.eeb.uconn.edu/systematics/

Systematics Servers includes links to other sites and phylogenetic and biodiversity software.

Chapter 5

Establishing Identity: Using Museum Collections

The physical heart of a museum is its collection, in fact having a collection is what makes a museum a museum, and most activity in most museums is involved with the acquisition, care, understanding, and use of their collections. –EDWARDS (1985:1)

The pedigree of natural history collections is often revealed by their homes; in cupboards and cases of unfashionable design within elderly buildings. . . . Yet natural history collections are undoubtedly part of modern biology, and are arguably one of the most cost-effective research tools. –ARNOLD (1991:25)

Museum collections of plants are . . . the ultimate reference for the identification of unknown plants, they are the foundations for botanical nomenclature and taxonomy, they are the bases of all floras and taxonomic monographs, they are historical archives for diverse botanical studies, and they are a source of materials for education and public information. –OGILVIE (1985:13)

Once you have completed a thorough search of the literature, you will have come to one of two conclusions. Either you will be more convinced that you have found an undescribed species (because you cannot find any description or illustration that agrees in all details with your specimens) or you will have found one or more descriptions or illustrations that represent a known species to which the organism you are studying potentially belongs. If you are working on a well-known group of organisms in a well-studied region of the world and you have been able to make a positive identification, this is probably as far as you need to go in your search for information.

However, if there is any doubt about the identification, even if you have decided that you probably do *not* have a new species, it is now time to look at actual specimens of the potential members of the species in question or those that appear to be similar. There is no substitute for the specimen. If you don't want to be embarrassed for the rest of your career by one of your publications, now is the time to make the extra effort to see reliably authenticated specimens: reference, voucher, even type material if necessary. The literature gives you one kind of information, the specimen itself another. Just as a good homicide detective reads both the victim's diary and the autopsy report (and perhaps even attends the autopsy), you must see and evaluate both kinds of evidence in order to solve your puzzle. And when it comes to deciding whether what you have found is the same as or different from some previously described species, make sure not only that you see material of the putative species or its closest relatives, but that you can recognize its distinguishing characteristics for yourself and explain them to your own satisfaction (even if an expert is there to point them out to you during your museum visit).

Collections, Museums, and Herbaria

Local Collections

Your first stop might be the reference collection of the field station, agricultural research station, state geological survey, or marine laboratory where you have carried out your work. These institutions often have reference collections of local organisms. Unfortunately, they are not always well maintained; their maintenance, if any, may have been the part-time job of a succession of graduate students or of interested visiting investigators who have added specimens for their group of interest, and their holdings may be excellent for one group of organisms and limited or nonexistent for others. Field stations and labs often have libraries and reprint collections related to the flora and fauna of the area, and the research carried out by those who have worked there over the years.

Botanical Gardens

If you are working on a plant species, a visit to the nearest botanical garden to see living examples of related species may be helpful. The larger botanical gardens have research scientists working on their own taxonomic projects–studying plant reproduction, phenetic variation, or hybridization. Discussion with them may also help. Botanical gardens may also have a research collection of dried plants, or *herbarium.*

Museums and Herbaria

Larger collections of preserved specimens are found in university, state, national, and private nonprofit museums and herbaria. Most of these institutions were founded over the last century and have been committed over the course of their existence to two missions: providing for the preservation and conservation of their collections and promoting systematic research on them. Figure 5.1 shows modern compactor shelving, holding over a century's worth of invertebrate collections at the Museum of Comparative Zoology, Harvard University. Most of these museums also contain libraries, catalogs, and electronic databases that document and make their collections accessible. Figure 5.2 shows the way most museums now catalog their collections by entering speciman data into a computerized catalog system.

Although most of the larger and older museums have collections of specimens representing the major groups of living or fossil organisms, they do not have curators actively conducting research on each group. Considering the total number of phyla (more than 60), it would be impossible for even the largest museums to do so, although members of the public seem to believe that each museum is supplied with that many experts. But at a minimum they usually have a collection manager or curator assigned to oversee the use and maintenance of each collection, even if no in-house scientist is currently studying it. Although electronic cataloging has made information retrieval rapid, cataloging the processed and identified specimen is only one step in a long and labor-intensive process. Figure 5.3 shows the result of the hours of work necessary to make a fosil mollusk collection accessible to users.

Locating Material

If you are working with well-known species it may be enough to check them against specimens in one of the local reference collections just mentioned. If you are dealing with a potentially undescribed species you must locate the best collection or collections available for that group of

Figure 5.1 One of the main functions of a museum is the stewardship (preservation and conservation) of the specimens deposited there. Here, modern compactor shelving holds both old and new specimens in a newly renovated collection storage facility in the Department of Invertebrate Zoology, Museum of Cultural and Natural History, Harvard University. Shelving units can be rolled together to conserve space. *–Photo: Dr. Damhnait McHugh, Harvard University.*

organisms. Note that collections for some groups include different kinds of materials, including frozen tissues (figure 5.4). There is no simple way to locate the most complete collection for the group you are studying; the location of a collection depends very much on historical accident (who worked where and developed what collection at which institution). However, your search of the taxonomic literature should have made you aware of the major reference collections and type repositories for that group.

Modern descriptions of new species tell in which institution types are deposited, and often list specimens from various institutions examined during the study. Taxonomic revisions and monographs (those written in the last 25–50 years) also usually provide the accession number, catalog number (if one was assigned), locality information, and the name of the

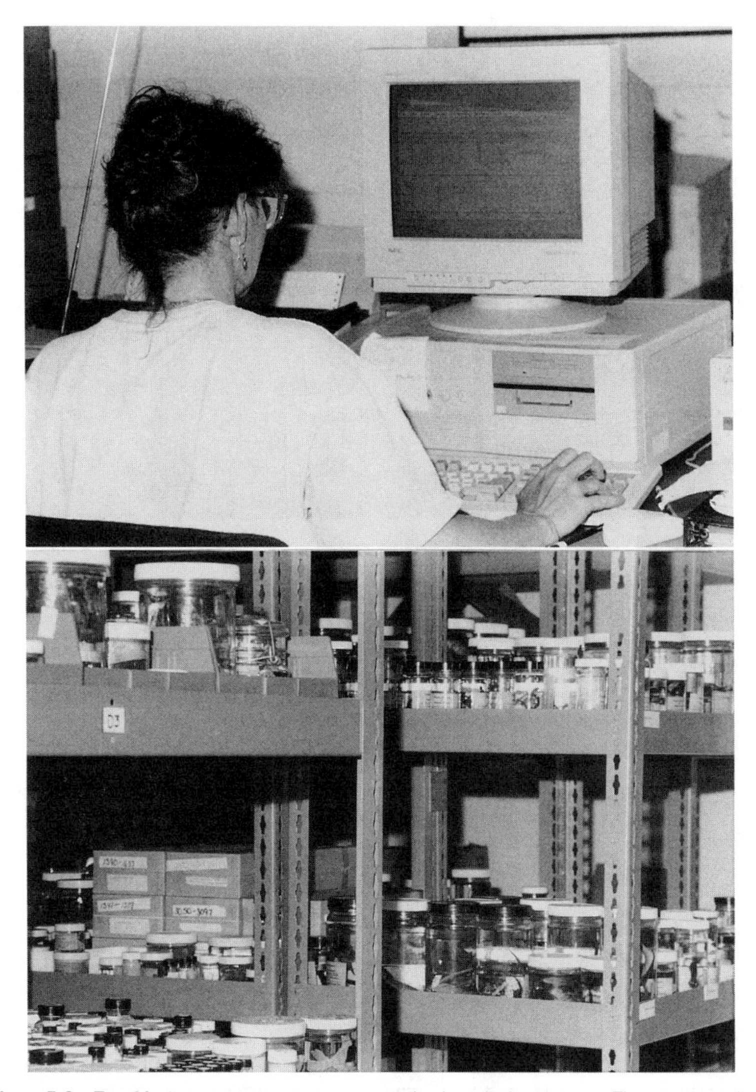

Figure 5.2 Top, Most museums now use computerized cataloging systems. These systems have enabled museums to standardize collection information and to make that information quickly accessible to users. **Bottom,** The specimens themselves cannot be so easily standardized. Even within a collection of a single group of organisms, such as fish or amphibians, specimen container size may range from small vials to gallon jars. Fluid-preserved specimens, those kept in alcohol or formalin solutions, are often segregated in a separate room or building to reduce fire danger. Fluid collections require extra care. Fluid levels must be maintained and preservation monitored. *–Photos: J. E. Winston (J. Lawler entering data in computerized catalog; VMNH fluid collection storage area).*

Figure 5.3 Getting fossils from the quarry into the collection is a labor-intensive process that may take hundreds of hours of sorting, cleaning, and identification. Here resulting collections are stored in nonacidic chipboard trays in modern steel storage units that insulate specimens from changes in temperature and humidity. Note that paleontologists often organize their specimens by time period, stratigraphic unit, and locality. *–Photos: J. E. Winston (Dr. Lauck Ward with part of VMNH Tertiary fossil collection).*

holding institution of specimens examined for each of the species studied (see examples in chapters 9 and 15).

Older papers do not provide this information. Sometimes this means that specimens no longer exist. But often these specimens have been

Figure 5.4 **A,** Molecular systematists may keep frozen tissue collections in ultracold freezer units, in addition to maintaining traditional collections. A single mammal individual may be preserved as several different preparations, such as frozen tissues, **(b)** a prepared skeleton (here, small rodent skeletons in vials), and **(c)** a study skin (e.g., these squirrel skins). *–Photos: J. E. Winston (Dr. Nancy Moncrief, VMNH mammalogist, and portions of the VMNH mammal collection).*

incorporated into a museum collection and can be located through directories of museum collections, more recent papers on the species, written queries to current workers in the field, or even electronic queries to listservs such as TAXACOM. New developments are now making it easier to check some identifications without actually having to visit the museum that holds the specimens. Every day more museums add their Gopher or Web site to the Internet. These sites often include information on collections held and people to contact about them, as well as information on the museum's exhibits and programs. As better and better Web searching tools are developed, the Internet may be the place to begin locating collections of interest.

Unfortunately, museums are still insufficiently funded to make all their collections available online, and they must make choices: first type collections, then comprehensive and well-cataloged collections, then other collections as they can be entered into electronic databases. But online searches, along with the information gained from the literature, will help you narrow your search for relevant material. You may also need to write, fax, or e-mail any taxonomists actively publishing on the taxonomy of the group (obtain their names and addresses from your literature search). Once you have established which taxa you are looking for, you can also post a request to a general systematics listserv such as TAXACOM or a discussion group for particular group of organisms. The systematics and museum sites listed in this chapter's *Sources* provide links to some of the relevant interest groups.

Borrowing Material

You cannot expect that each museum you contact will have a specialist on the Recent or fossil group you are studying, but you can expect on-site access to the institution's collection of the group, if they have one, and you may be able to borrow the specimens you need, either by a written request or during your visit. Museum loan policies vary. Most museums are willing to lend nontype material to any professional biologist who has a need for it. Museums also usually make loans to graduate students, but you may need to do this through your major advisor. Figure 5.5 shows a sample Virginia Museum of Natural History loan form and loan guidelines. The form is similar to that used by many museums, permitting the loan of most materials for 6 months, with extensions approved by the curator. Specimens on loan must be stored and shipped under conditions that meet or exceed those of the home institution. Any form of dissection or sampling of the original specimen always requires advance written approval.

Some natural history museums, such as the California Academy of Sciences and the Smithsonian, also have botany departments with an herbarium on site; fossil plant collections are often located in the paleontology section of natural history museums. Most plant collections are housed in herbaria, which are often part of a botanical garden (e.g., the Missouri Botanical Garden). Increasingly museums and botanical gardens are maintaining frozen tissue and germplasm collections for molecular systematics and preservation of rare species.

Many times, examining specimens from the museum's reference and research collections will answer your questions and enable you to make a positive identification of your unknown specimen. However, if you still think you may have an undescribed species, you must look at the type

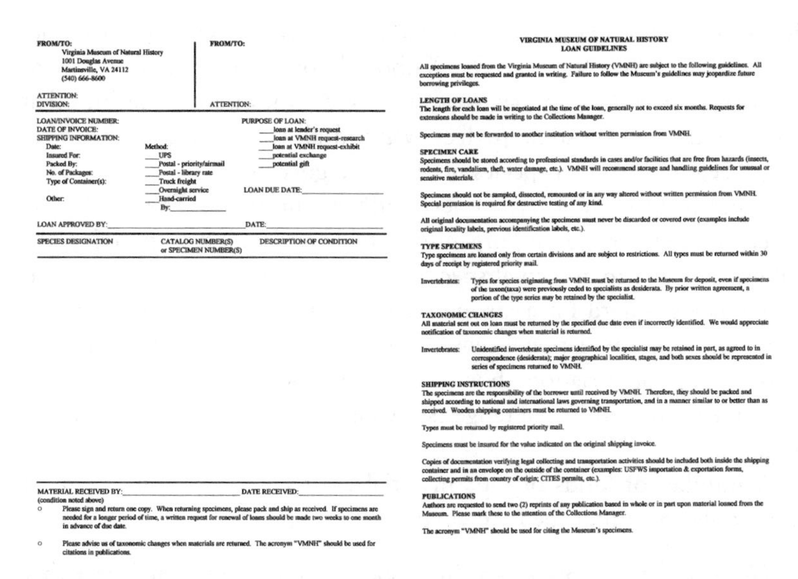

FROM/TO:
Virginia Museum of Natural History
1001 Douglas Avenue
Martinsville, VA 24112
(540) 666-8600

ATTENTION:
DIVISION:

FROM/TO:

ATTENTION:

LOAN/INVOICE NUMBER:
DATE OF INVOICE:
SHIPPING INFORMATION:
Date:
Insured For:
Packed By:
No. of Packages:
Type of Container(s):
Other:

Method:
____UPS
____Postal - priority/airmail
____Postal - library rate
____Truck freight
____Overnight service
____Hand-carried
By:____________

PURPOSE OF LOAN:
____loan at lender's request
____loan at VMNH request-research
____loan at VMNH request-exhibit
____potential exchange
____potential gift

LOAN DUE DATE:____________

LOAN APPROVED BY:____________ DATE:____________

SPECIES DESIGNATION	CATALOG NUMBER(S) or SPECIMEN NUMBER(S)	DESCRIPTION OF CONDITION

MATERIAL RECEIVED BY:____________ DATE RECEIVED:____________
(condition noted above)

- Please sign and return one copy. When returning specimens, please pack and ship as received. If specimens are needed for a longer period of time, a written request for renewal of loans should be made two weeks to one month in advance of due date.
- Please advise us of taxonomic changes when materials are returned. The acronym "VMNH" should be used for citations in publications.

VIRGINIA MUSEUM OF NATURAL HISTORY
LOAN GUIDELINES

All specimens loaned from the Virginia Museum of Natural History (VMNH) are subject to the following guidelines. All exceptions must be requested and granted in writing. Failure to follow the Museum's guidelines may jeopardize future borrowing privileges.

LENGTH OF LOANS
The length for each loan will be negotiated at the time of the loan, generally not to exceed six months. Requests for extensions should be made in writing to the Collections Manager.

Specimens may not be forwarded to another institution without written permission from VMNH.

SPECIMEN CARE
Specimens should be stored according to professional standards in cases and/or facilities that are free from hazards (insects, rodents, fire, vandalism, theft, water damage, etc.). VMNH will recommend storage and handling guidelines for unusual or sensitive materials.

Specimens should not be sampled, dissected, remounted or in any way altered without written permission from VMNH. Special permission is required for destructive testing of any kind.

All original documentation accompanying the specimens must never be discarded or covered over (examples include original locality labels, previous identification labels, etc.).

TYPE SPECIMENS
Type specimens are loaned only from certain divisions and are subject to restrictions. All types must be returned within 30 days of receipt by registered priority mail.

Invertebrates: Types for species originating from VMNH must be returned to the Museum for deposit, even if specimens of the taxon(taxa) were previously ceded to specialists as desiderata. By prior written agreement, a portion of the type series may be retained by the specialist.

TAXONOMIC CHANGES
All material sent out on loan must be returned by the specified due date even if incorrectly identified. We would appreciate notification of taxonomic changes when material is returned.

Invertebrates: Unidentified invertebrate specimens identified by the specialist may be retained in part, as agreed to in correspondence (desiderata); major geographical localities, stages, and both sexes should be represented in series of specimens returned to VMNH.

SHIPPING INSTRUCTIONS
The specimens are the responsibility of the borrower until received by VMNH. Therefore, they should be packed and shipped according to national and international laws governing transportation, and in a manner similar to or better than as received. Wooden shipping containers must be returned to VMNH.

Types must be returned by registered priority mail.

Specimens must be insured for the value indicated on the original shipping invoice.

Copies of documentation verifying legal collecting and transportation activities should be included both inside the shipping container and in an envelope on the outside of the container (examples: USFWS importation & exportation forms, collecting permits from country of origin; CITES permits, etc.).

PUBLICATIONS
Authors are requested to send two (2) reprints of any publication based in whole or in part upon material loaned from the Museum. Please mark these to the attention of the Collections Manager.

The acronym "VMNH" should be used for citing the Museum's specimens.

Figure 5.5 Sample museum loan form (VMNH). Front side of form lists specimens loaned, method of shipping, and other information. When you receive the specimens and the form, you should return one copy to notify the lending museum that you received the shipment. The second copy is returned when you return the specimens. The back of the form gives the museum's guidelines for loan of specimens, including length of loans and instructions for care, shipping, and citation in publications.

material for the species that appear most similar or with which it may have been confused in the past.

Type Material

Biologists are often confused about what taxonomists mean by the word *type*. If your background is in biology, ecology, or paleontology, you were probably taught a species concept that was biological, in which species were recognized to be (or to have been) a population of reproducing individuals, subject to a certain range of genetic variation and to descent with modification. In textbooks and classrooms this is usually contrasted with the pre-Darwinian, medieval, and ancient Greek systems of categorization, in which terms were defined on the basis of their similarity to some *ideal* or *type*. The concepts of essence and properties, which were invented for mathematics and philosophy, were also applied to living things, leading to what is called the typological concept of species (Cain 1958, 1962; Mayr 1969). Without knowledge of the workings of genetics or selection, species were perceived as immutable entities that could be defined by some perfect *type* and all the specimens that were *typical* belonged to that species or other category.

Types in taxonomy have a very different meaning. They make the system we use for naming organisms stable and objective by tying the published name to an actual specimen or suite of specimens, serving as a reference standard for a taxon. They are not expected to be "typical" of a species in a variational sense, not average (at the mean or median of the range of variation shown by the species), although they must fall somewhere within the range (see chapter 9) in order to serve as the standard.

A number of kinds of type material are recognized. You will most commonly encounter the following:

Holotype. The holotype is a single specimen used by an author, either the only specimen he found or one of several found, but the only one designated as a type. When people talk about the type, they are referring to this single specimen.

Syntypes. These are two or more specimens selected from the available material to serve as types.

Lectotype. One specimen of the syntype set (or one specimen known to have been used by the author if no type was identified) may be selected as the lectotype by the author or a later worker. It should be the specimen illustrated in the original description, if not all were depicted.

Neotype. This is a specimen selected (and designated in a paper in the literature) to serve as a type when the original material has been lost or destroyed (chapter 21).

Paratypes. Paratypes are specimens that the person making the original description examined while carrying out the work. They may be from the same or a different locality, but they clearly (at least in the mind of the describer) are members of the new species. Although they are not types in the nomenclatural sense, they are often distributed to other museums as vouchers for the new species, and like holotypes and syntypes they are often cataloged and stored separately from the rest of the collection (figure 5.6).

Topotype. These are specimens collected from the same locality as the type material (although not necessarily at the same time). Again, they have no official standing, but if a museum has such material it is usually well worth seeing. I can vouch for the fact that with marine invertebrates, topotypical material of the species you are seeking is much more likely to belong to that species than material labeled with the same name but from thousands of miles away.

Visiting Collections: What to Expect and How to Behave

If you borrow material by mail, you will get to see some of each museum's material for the taxa of interest, but probably not all that they have. It is far better to visit the relevant collections in person, if at all possible. But when you get there be prepared for a shock: Even the largest and best-curated collections contain uncataloged and mistakenly identified specimens. Not only that, but the specimens you want to see may not all be stored under their current scientific names. This is especially true for specimens in the largest and oldest collections. Just as it becomes prohibitively expensive in terms of time and expense to recatalog a large library, the same is true for a collection of specimens. And most large collections, of course, are aggregations of smaller collections made over time. These may vary in terms of quality of identification and accessibility.

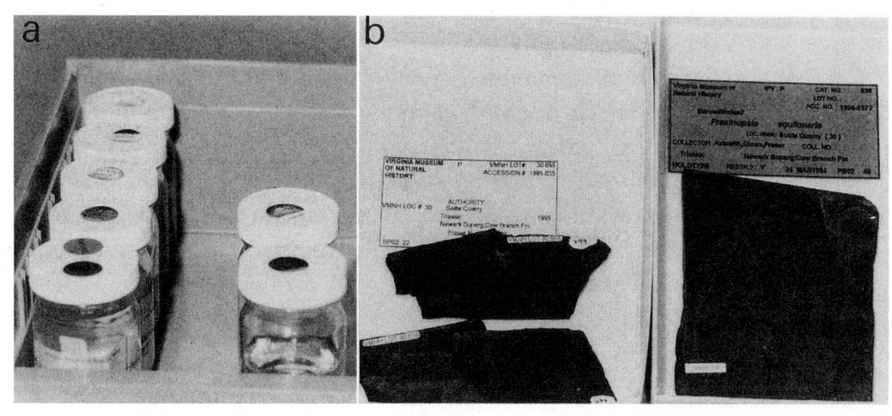

Figure 5.6 Different methods of identifying type material in collections. **A,** Metallic circles mark lids of types preserved in alcohol and kept in a fireproof cabinet. **B,** Red label (right-hand specimen) makes a fossil holotype specimen stand out from regular white-labeled specimens. *–Photo: J. E. Winston.*

They have been obtained from various sources, including expeditions and field work carried out by museum personnel over the years that the museum has existed, similar field collections made by others and donated by them or their heirs, accessions of one to many specimens deposited as vouchers when ecologists or other field workers publish a paper, personal collections donated when a scientist retires or dies, and other "orphaned" collections dumped on the museums when a university discards its teaching collection or another museum closes down.

Often the specimens are kept under the name by which they were known at the time they were collected. So if you are hunting for all the specimens a museum has of a common species, specimens might be filed under half a dozen names. For example, specimens of the subtropical fouling bryozoan *Watersipora subovoidea* might be found under *Cellepora subovoidea* (the name under which it was first described, in 1852), under *Watersipora cucullata,* the name by which it was known in publications of the 1930s and 1940s, under *Dakaria subovoidea,* as it was known in the late 1950s, and under *Watersipora subovoidea,* as it has been called in more recent papers.

In most institutions there is additional material that has been accessioned but never cataloged. You may be allowed to go through such material yourself, and if you are willing to help curate and catalog such a collection, you may get, if not a medal, at least a Research Associateship.

Cooperation with Systematists

If you do find, in addition to a good collection of the group of interest, a professional systematist with expertise on that group, you may find all your problems solved, or you may not. The courtesy (or discourtesy) of your approach to the expert may make all the difference.

In general, systematists like to talk about the organisms they study, and they especially like to talk to people who share that interest. But do call or write ahead; don't just show up on the doorstep of the museum and expect to find the person you want to see available that hour, or even that day. Like university professors, curators attend committee meetings and teach students. They also work on projects with education or exhibit staff. Most systematists also do field work, and in summer, especially, the offices and laboratories of a museum can be deserted. Even though the exhibit halls of museums are open during times such as Christmas break, the research and administrative staff get normal holidays off and may be spending them with family or friends instead of in the lab.

Systematists are research scientists, too, and their career success depends on their productivity and their own research results. The amount of time they can devote to helping other solve taxonomic problems is limited. Although most acknowledge some obligation to identify specimens for others, such service activities are expected to take up only a small percentage of their time (about 3 to 5 percent; Winston 1988).

They are more likely to go beyond the call of duty when there is something of interest in the project from their point of view. It might be the opportunity to keep specimens, gain information about a group in which they are particularly interested, or publish the work as a collaborative effort.

If you do ask a specialist to identify your organism, you are more likely to get a positive response if you remember these guidelines:

Contact the systematist before sending the material (and don't send it until you receive permission to do so, or it may sit unopened indefinitely).

Don't send an excessive amount. Start with one specimen of each species, identified to the best of your ability (even if you can't attach a name, you should have your material sorted to species A, species B, etc.). The systematist will not do the preliminary sorting for you.

Allow some time. Identifying one species may take 10 minutes or 10 hours or 10 days, depending on what has to be done in the library or lab to make the identification. But if a month or two has gone by with no response, do follow up; all of us work under deadlines these days, and systematists are no exception.

Sources

Printed Sources

Guides to Museum and Herbarium Collections

Arnett, R. H. and G. M. Nishida. 1993. *The Insect and Spider Collections of the World.* Boca Raton, Fla., St. Lucie Press.

Ben-Dov, Y. 1993. *A Systematic Catalogue of the Soft Scale Insects of the World.* Boca Raton, Fla., St. Lucie Press.

Holmgren, P. K., N. H. Holgren, and L. C. Banett, eds. 1990. *Index Herbariorum, 8th ed., pt. 1. The Herbaria of the World.* New York, New York Botanical Garden.

Zoological Catalogue of Australia. CSIRO Publishing, Collingwood, Australia. (This series of monographs on nomenclature, taxonomy, and biology of Australian fauna includes location of type material. Twenty groups have been covered so far, including amphibians and reptiles, sponges, echinoderms, nonmarine mollusks, and several orders of arthropods.)

Natural History Collections and Collection Management

JOURNALS AND NEWSLETTERS

ASC Newsletter
The Biology Curator
Collections Forum
Curator
Geological Curator
Insect Collection News
Journal of Conservation and Museum Studies
Journal of the History of Collections
Museum Management and Curatorship
SPHNC Newsletter

BOOKS AND ARTICLES

Arnold, N. 1991. Biological messages in a bottle. *New Scientist* 131(1783): 25–27.

Cato, P. and C. Jones, eds. 1991. *Natural History Museums: Directions for Growth.* Lubbock, Texas Tech University Press.

Duckworth, W. D., H. H. Genoways, and C. L. Rose. 1993. *Preserving Natural Science Collections: Chronicle of Our Environmental Heritage.* Washington, D.C., National Institute for the Conservation of Culture Property.

Herholdt, E. M. 1990. *Natural History Collections, Their Management and Value.* Pretoria, Transvaal Museum.

Impey, O. and A. MacGregor, eds. 1985. *The Origins of Museums. The Cabinet of Curiosities in Sixteenth- and Seventeenth-Century Europe.* Oxford, U.K., Oxford University Press.

Miller, E. H., ed. 1985. Museum collections: Their roles and future in biological research. *British Columbia Provincial Museum Occasional Paper Series* 25: 1–222.

Mound, L. A. 1992. Why collect? Responsibilities and possibilities in a museum of natural history. *Insect Collection News* 7: 8–14.

Rose, C. L. and A. R. de Torres, eds. 1992. *Storage of Natural History Collections: Ideas and Practical Solutions.* Pittsburgh, Society for the Preservation of Natural History Collections.

Rose, C., C. Hawks, and H. Genoways, eds. 1995. *Storage of Natural History Collections: A Preventive Conservation Approach.* Washington, D.C., Society for the Preservation of Natural History Collections.

Schulz, E. 1990. Notes on the history of collecting and of museums in the light of selected literature of the sixteenth to the eighteenth century. *Journal of the History of Collections* 2: 205–218.

Simmons, J. E. 1993. Natural history collections management in North America. *Journal of Biological Curation* 1(3/4): 1–17.

Stansfield, G., J. Mathias, and G. Reid, eds. 1994. *Manual of Natural History Curatorship.* London, HMSO.

Thompson, J. M. A, ed. 1992. *Manual of Curatorship: A Guide to Museum Practice.* Oxford, U.K., Buttersworth.

Methods for Collection, Preservation, and Curation

Many books and articles on various groups of plants and animals also cover aspects of collecting, preparing, and preserving specimens. The following list is just a start; also see references in chapter 4 and chapter 9 sources.

ANIMALS

Barbosa, P. 1974. *Manual of Basic Techniques in Insect Histology.* Amherst, Mass., Autumn Publishers.

Barr, D. W. 1973. *Methods for the Collection, Preservation, and Study of Water Mites (Acari Parasitenogona).* Toronto, Royal Ontario Museum, Life Science Miscellaneous Publications.

Beirne, B. P. 1955. *Collecting, Preparing and Preserving Insects.* Ottawa, Science Service, Entomology Division, Canada Department of Agriculture.

Blackman, R. L. 1974. *Aphids.* London, Ginn.

Brown, P. A. 1997. A review of techniques used in the preparation, curation and conservation of microscope slides at the Natural History Museum, London. *The Biology Curator* 10, Special Supplement: 1–33. (Recommended mounting media for different organisms, recipes, and bibliography.)

Camacho, A. I. and J. Bedoya. 1994. Evaluation of the effects of different preservative and fixative fluids on aquatic invertebrates from interstitial waters. *Collection Forum* 10: 20–31.

Dolling, R. W. 1991. *The Hemiptera.* Oxford, U.K., Oxford University Press.

Foissner, W. 1991. Basic light and scanning electron microscopic methods for taxonomic studies of ciliated Protozoa. *European Journal of Protistology* 27: 313–330.

Gilbert, P. and C. J. Hamilton. 1990. *Entomology: A Guide to Information Sources,* 2d ed. London, Mansell.

Hamond, R. 1969. Methods of studying the copepods. *Journal of the Quekett Microscopical Club* 31: 137–149.

Higgins, R. P. and H. Thiel, eds. 1988. *Introduction to the Study of Meiofauna.* Washington, D.C., Smithsonian Institution Press. (Chapters on different groups, including collection and preparation techniques.)

Huys, R. and G. A. Boxshall. 1991. *Copepod Evolution.* London, Ray Society.

Knudsen, J. W. 1966. *Biological Techniques: Collecting, Preserving, and Illustrating Plants and Animals.* New York, Harper & Row.

Knudsen, J. W. 1972. *Collecting and Preserving Plants and Animals.* New York, Harper & Row.

Noyes, J. S. 1982. Collecting and preserving chalcid waspids (Hymenoptera Chalcidoidea). *Journal of Natural History* 16: 315–334.

Oldroyd, H. 1970. *Collecting, Preserving and Studying Insects.* London, Hutchinson.

Perkins, E. J. 1956. Preparation of copepod mounts for taxonomic work and for permanent collections. *Nature* 178: 1075–1076.

Petersen, A. 1976. *Entomological Techniques: How to Work with Insects,* reprint of 10th ed. Los Angeles, Entomological Reprint Specialists.

Smithers, C. 1981. *The Handbook of Insect Collecting: Collection, Preparation, Preservation and Storage.* Newton Abbott, U.K., David & Charles.

Southey, J. F. 1970. *Laboratory Methods for Work with Plant and Soil Nematodes.* Ministry of Agriculture, Fisheries and Food, ADAS Reference Book 402. London, HMSO.

Steedman, H. F. 1976. *Zooplankton Fixation and Preservation.* Paris, UNESCO Press.

Steyskal, G. C., W. I. Murphy, and E. M. Hoover. 1986. *Insects and Mites, Techniques for Collection and Preservation.* U.S. Department of Agriculture, Agricultural Research Service, miscellaneous publication no. 1443.

Upton, M. S. 1991. *Methods for Collecting, Preserving and Studying Insects and Allied Forms.* Collingwood, Australia, CSIRO Publishing.

Vogt, K. D. 1991. Reconstituting dehydrated museum specimens. *Curator* 34:125–131.

Wagstaffe, R. and J. H. Fidler. 1970. *The Preservation of Natural History Specimens. 1. Invertebrates.* London, Witherby.

PLANTS

Bridson, D. and L. Forman. 1992. *The Herbarium Handbook.* Kew, U.K., Royal Botanic Gardens.

Eggli, U. and B. E. Leuenberger. 1996. A quick and easy method of drying plant specimens, including succulents, for the herbarium. *Taxon* 45: 259–261.

Garner, J. B. and C. V. Horie. 1984. The conservation and restoration of slides of mosses. *Studies in Conservation* 29: 93–99.

Knudsen, J. W. 1966. *Biological Techniques: Collecting, Preserving, and Illustrating Plants and Animals.* New York, Harper & Row.

Knudsen, J. W. 1972. *Collecting and Preserving Plants and Animals.* New York, Harper & Row.

Russell, S. J. 1989. Glass slide preservation of SEM mounted diatoms. *Diatom Research* 4: 401–402.

FOSSILS

Brunton, C. H. C., T. P. Besterman, and J. A. Cooper. 1985. *Guidelines for the Curation of Geological Materials.* Miscellaneous paper no. 17. London, Geological Society.

Chapman, J. L. 1985. Preservation and durability of stored palynological specimens. *Pollen et Spores* 27: 113–120.

Collinson, M. E. 1987. Special problems in the conservation of paleobotanical material. *Geological Curator* 4: 439–445.

Green, O. R. 1995. Pitfalls, problems and procedures in micropalaeontological preparation and conservation. *Geological Curator* 6: 157–166.

Grimaldi, D. 1993. The care and study of fossiliferous amber. *Curator* 36: 31–49.

Leiggi, P. and P. May. 1994. *Vertebrate Paleontological Techniques,* Vol. 1. Cambridge, U.K., Cambridge University Press.

Internet Sources

INFORMATION ON SPECIMENS IN COLLECTIONS

Like taxonomic literature on the Web, information on specimens in collections is far from complete, but sources are constantly being added. Museums have made great strides in making collection databases (especially type specimen databases) accessible electronically. Start with one of the main Web sites, such as ASC or ASPT, which list many museum, herbarium, and botanical garden Web sites. All addresses begin with **http://www.**

MUSEUM-RELATED WEB SITES

ascoll.org/
The Association of Systematics Collections page lists museum Web pages.

syst.bot.org
American Society of Plant Taxonomists.

COLLECTION MANAGEMENT AND CONSERVATION WEB SITES

spnhc.org/
The Society for the Preservation of Natural History Collections page includes information on meetings, publications, searchable archive of SPHNC listserv, and links to other conservation-related sites such as CoOL (Conservation on Line, **palimpsest.stanford.edu/**).

PUBLISHERS OF TAXONOMIC CATALOGS AND OTHER RESOURCES

publish.csiro.au/
Zoological Catalogue of Australia.

EQUIPMENT AND SUPPLIES

global-expos.com/tools.html/tools.html
A catalog of tools for fossil preparation.

Part Three

WRITING SPECIES DESCRIPTIONS

Chapter 6

Species Descriptions in Taxonomy

Objects without names cannot well be talked about or written about; without descriptions they cannot be identified and such knowledge as may have accumulated regarding them is sealed.

–GAHAN (1923:73)

Communication–information exchange–among zoologists is the core of zoological nomenclature; everything else pales in the light of the importance of communication. –BOCK (1994:8)

Reasons for Writing Species Descriptions

This book describes the process of researching and writing the original description in detail because its purpose is a practical one. Once you have followed the steps in chapters 4 and 5 and have satisfied yourself that the organism you are studying does indeed represent an undescribed species, your aim is publication. Only if it is named and described acceptably in a scientific publication will the species name be available for you and others to use. Descriptions of new species are still an important part of publication in the field of taxonomy. A survey of the taxonomic literature produced over the last 28 years (see figure 1.4) showed that the number of taxonomic papers containing at least one new species was increasing up until 1988, and remains high, as figure 6.1 shows, with at least one new species described in 35.7 percent of all taxonomic papers over that time period (Winston and Metzger 1998).

The following chapters should enable you to write good descriptions for a range of types of taxonomic work. Many biologists and paleontologists find it necessary to produce papers that are at least partially taxonomic

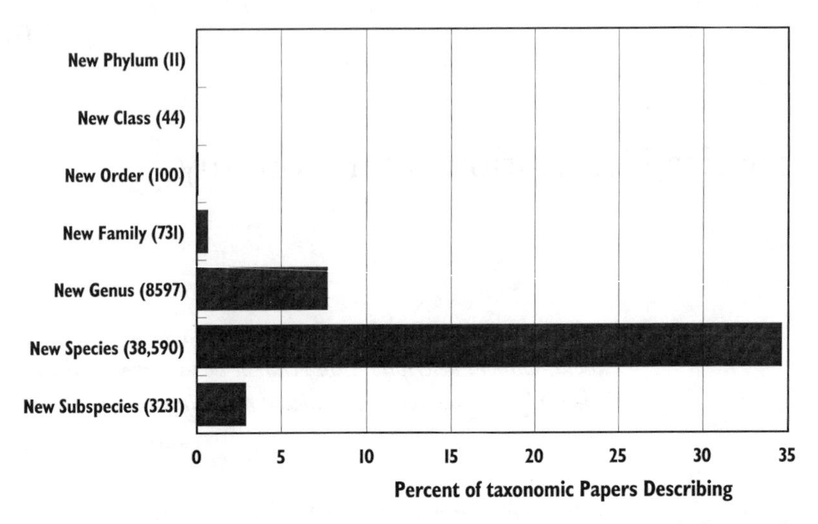

Figure 6.1 Percentage of taxonomic papers containing descriptions of new taxa of different ranks over the last 28 years. New species are described most frequently, followed by new genera and new subspecies. New higher-level taxa are much less common. *–Source: Data from Winston and Metzger (1998).*

over the course of their careers. Professional biologists who work in applied fields (wildlife management, conservation, forestry, state environmental or agricultural programs, environmental consulting companies), especially, need to become skilled at writing descriptions, for they may be called on to produce faunal or floral checklists, reports, surveys, and manuals for their organizations.

Different Kinds of Taxonomic Publications

There are many different kinds of taxonomic publications, ranging from short notes describing a single species to multivolume monographs and treatises. They are all important resources to the taxonomist, and your literature searches will probably include most of them. This section describes some of the major categories. Please note that these categories are not hard and fast. Many papers mix approaches or grade into one another in intensity of effort. Other papers have titles that imply that they belong in one category when they actually fall into another in terms of content.

Species Descriptions

A description of a single new species can often be published by itself, as a short paper or note, as in these examples:

McGinty, M. M. 1969. *Batillipes gilmartini,* a new marine tardigrade from a California beach. *Pacific Science* 23: 394–396.
Rothwell, G. W. 1982. *Cordaianthus duquesnensnis* sp. nov., anatomically preserved ovulate cones from the Upper Pennsylvanian of Ohio. *American Journal of Botany* 69: 239–243.
Erséus, C. and O. Giere. 1995. *Olavius nicolae,* a new gutless marine tubificid species (Oligochaeta) from Belize. *Proceedings of the Biological Society of Washington* 108: 491–495.
Mlíkovsky, J. 1997. A new tropic bird (Aves: Phaethontidae) from the late Miocene of Austria. *Annalen des Naturalhistorischen Museums in Wien* A98A: 151–154.

Species descriptions are often published in "mixed" papers, in which the new species is described in conjunction with the results of research on its physiology, morphology, reproduction, ecology, and so on. Such papers are quite common because so many new species are found in the course of other work, and they are accepted by a wider range of journals than strictly taxonomic papers, as in the following examples:

Gee, J. M. and J. W. Fleeger. 1990. *Haloschizopera apprisea,* a new species of Harpacticoid Copepod from Alaska, and some observations on sexual dimorphism in the Family Diosacchidae. *Transactions of the American Microscopical Society* 109: 282–299.
Goodbody, I. and L. Cole. 1987. A new species of *Perophora* (Ascidiacea) from the Western Atlantic, including observations on muscle action in related species. *Bulletin of Marine Science* 40: 246–254.
Roe, P. and D. E. Wickham. 1984. *Poseidonemertes collaris,* n. sp. (Nemertea: Amphiporidae) from California, with notes on its biology. *Proceedings of the Biological Society of Washington* 97: 60–70.

Redescriptions

A redescription involves examination of existing material and information in order to make a more complete description of a species or other group. It is often carried out when a new species has been found and its generic placement must be evaluated. Higher-level taxa (e.g., genera, families) may also be redescribed. Sample titles include the following:

Hove, H. A. ten and R. S. Smith. 1975. A re-description of *Ditrupa gracillima* Grube, 1878 (Polychaeta, Serpulidae) from the Indo-Pacific, with a discussion of the genus. *Records of the Australian Museum* 42: 101–118.

Grimaldi de Zio, S., M. D'Addabbo Galo, M. R. M. De Lucia, and A. Troccoli. 1990. New description of *Neostygarctus acanthophorus. Cahiers de Biologie Marine* 31: 409–416.

Robinson, H. 1993. A review of the genus *Critoniopsis* in Central and South America (Vernonieae: Asteraceae). *Proceedings of the Biological Society of Washington* 106: 606–627.

Descriptions of Higher Taxa

As figure 6.1 shows, far fewer taxonomic papers include description of at least one new genus (only about 8 percent). New taxa at the family level and above are even scarcer. But although descriptions of higher taxa are rare in the literature, they still occur. If you work on invertebrates, in particular, you could find yourself working on something completely baffling, an organism that might turn out to represent a new family, order, or phylum, as the authors of these sample publications discovered:

Holsinger, J. R. 1989. Allocrangonyctidae and Pseudocrangonyctidae, two new families of holarctic subterranean amphipod crustaceans (Gammaridea), with comments on their phylogenetic and zoogeographic relationships. *Proceedings of the Biological Society of Washington* 102: 947–959.

Bowman, T. E., S. P. Garner, R. R. Hessler, T. M. Illiffe, and H. L. Sanders. 1985. Mictacea, a new order of Crustacea Peracarida. *Journal of Crustacean Biology* 5: 74–78.

Kristensen, R. M. 1983. Loricifera, a new phylum with Aschelminthes characters from the meiobenthos. Zeitschrift für Zoologische Systmatik und Evolutionsforschung 21: 163–180.

Subspecies Descriptions

New subspecies are described in about 3 percent of taxonomic papers published (figure 6.1). Very few people who work on well-known groups such as birds and large mammals will ever be called upon to describe a new species, but anyone active in field work on these groups may encounter new subspecies. The other people who are likely to deal with subspecific variation are graduate students and professionals in genetics and conservation biology, who often end up studying variation within a species. They are almost the only biologists now doing "gamma" taxonomy, once considered to be the highest form of systematics. Some sample titles of papers dealing with subspecies follow:

Morafka, D. J., L. G. Aguirre, and R. W. Murphy. 1994. Allozyme differentiation among gopher tortoises (*Gopherus*): Conservation genetics and phylogenetic and taxonomic implications. *Canadian Journal of Zoology* 72: 1665–1671.

Cronin, M. A., S. Hill, E. W. Born, and J. C. Patton. 1994. Mitochondrial DNA variation in Atlantic and Pacific walruses. *Canadian Journal of Zoology* 72: 1035–1043.

Quinn, T. W. 1992. The genetic legacy of mother goose: phylogeographic patterns of lesser snow goose *Chen caerulescens caerulescens* maternal lineage. *Molecular Ecology* 1: 105–117.

Works of greater scope include synopses, reviews, and catalogs. These summarize current knowledge of a group. They may also include descriptions of new species.

Synopses

Synopsis means a general view. Synopses summarize current knowledge of morphology, ecology, terminology, and classification of a group of organisms. Compiled largely for the purpose of species identification, they give descriptions and illustrations of species known (or species known from a particular geographic area). They may contain practical information on collection and preservation, as well as keys to species. Some sample titles:

Millar, R. H. 1970. *British Ascidians. Tunicata: Ascidiacea. Keys and Notes for the Identification of the Species.* Synopses of the British Fauna, no. 1. London, Academic Press.

Yamaguti, S. 1971. *Synopsis of Digenetic Trematodes of Vertebrates.* 2 vols. Tokyo, Keigaku.

Hayward, P. J. 1995. *Antarctic Cheilostomatous Bryozoa.* Oxford, U.K., Oxford University Press.

Reviews

A review is a publication in which an author critically examines previous work and material on a group. It brings together current information on the group, but does not include the detailed examination of relationships that is involved in a revision or monograph. It is often carried out in conjunction with description of new species.

Sinclair, B. J. and B. R. Stuckenberg. 1995. Review of the Thaumaleidae of South Africa. *Annals of the Natal Museum* 36: 215–253.

Ota, H., M. W. Lau, T. Weidenhöfer, Y. Yasukawa, and A. Bogadek. 1995. Taxonomic review of the geckos allied to *Gekko chinensis* Gray 1842 (Gekkonidae Reptilia) from China and Vietnam. *Tropical Zoology* 8: 181–196.

Opresko, D. M. 1997. Review of the genus *Schizopathes* (Cnidaria: Antipatharia: Schizopathidae) with a description of a new species. *Proceedings of the Biological Society of Washington* 110: 157–166.

Catalogs

In general terms, a catalog is a complete list of items arranged in an organized way, each item accompanied by descriptive details (as in a college catalog or a mail-order catalog).

A taxonomic catalog (often spelled *catalogue*) usually describes the specimens or species of a group of organisms that are found in the collection of particular museum or herbarium. It may just list the species present; it may include annotations on the taxonomy of those species, or the condition of the specimens; it may even include descriptions for each species covered. *Catalog* may also be used for a taxonomic publication discussing all the species reported from a region, cruise, or expedition. Titles of some taxonomic catalogs include the following:

Busk, G. 1852. *Catalogue of Marine Polyzoa in the Collection of the British Museum, Pt. I.* London, British Museum.

Ibarra, M. and D. J. Stewart. 1987. Catalogue of type specimens of Recent fishes in Field Museum of Natural History. *Fieldiana, Zoology,* n.s. n.35, publication 1377:1–112.

Silva, P. C., E. G. Meñez, and R. L. Moe. 1987. Catalogue of the benthic marine algae of the Indian Ocean. *University of California Publications in Botany* 79: 1–259.

Revisions

A revision involves restudy of a group to correct or improve its diagnosis, description, or phylogeny. As Mayr (1969) pointed out, this is one of the most important types of scientific publication for groups in which many new species are still being described. Taxonomic revisions include many papers in which new arrangements, shifts in the rank or position of some of the included taxa, are proposed. In a generic or family revision, complete descriptions are usually given for every species, whether or not they have been described before. You can't always tell whether a paper is a revision without reading through it, although many do have *Revision* in the title. Some typical titles follow:

Ackland, D. M. 1995. Revision of Afrotropical *Emmesomyia* Malloch, 1917 (Diptera: Anthomyiidae), with descriptions of seven new species. *Annals of the Natal Museum* 36: 21–86.

Peirera, L. A. and R. L. Hoffman. 1993. The American species of *Escaryus*, a genus of holarctic centipeds (Geophilomorpha: Schendylidae). *Jeffersoniana* 3: 1–72.

Bock, P. and P. L. Cook. 1996. Concatenellidae, a new family of Catenicelloidea (Cheilostomatida) for *Concatenella airensis* (Maplestone) from the Tertiary of Victoria, pp. 47–53 in D. P. Gordon, A. M. Smith, and J. A. Grant-Mackie, *Bryozoans in Space and Time.* Wellington, New Zealand, National Institute of Water and Atmospheric Research.

In botany there is another category below the level of a revision, called a *conspectus,* defined by Stace (1989) as an outline of a revision. It contains a list of taxa, usually with synonymies, and sometimes brief diagnoses and distributional information. Linnaeus's *Species Plantarum* (1753) is an example of a conspectus.

Monographs

Monographs treat the systematics of a group in the most complete detail possible, usually including, along with full descriptions, whatever is known of the biology, ecology, and distribution of a group. In addition to summarizing all published knowledge on the group, they document the results of the author's own research efforts. A monograph on a genus, with introductory chapters, plus a detailed descriptions of all species included, might well represent a lifetime's worth of research by an author. Often they are published as books or in the monograph series of a museum. They may be book length, or even take up several volumes. Often, but not always, they are worldwide in geographic coverage. Some examples of monographs follow:

Abbott, R. T. 1974. *American Seashells: The Marine Mollusca of the Atlantic and Pacific Coasts of North America,* 2d ed. New York, Van Nostrand Reinhold.

Hunt, D. J. 1993. *Aphelenchida, Longidoridae and Trichodorida: Their Systematics and Bionomics.* Wallingford, U.K., CAB International.

Manucci, W. 1986. *Tardigrada* (Fauna d'Italia, v. 24). Bologna, Edizionia Calderi.

Rehn, J. A. G. and H. J. Grant. 1961. *A Monograph of the Orthoptera of North America,* vol. 1. Monographs of the Academy of Natural Sciences of Philadelphia, no. 12.

Classifications and Phylogenies

These publications synthesize the evolutionary relationships of a group. They may or may not also describe new taxa. Sample titles include the following:

Livezey, B. C. 1997. A phylogenetic classification of waterfowl (Aves: Anseriformes), including selected fossil species. *Annals of the Carnegie Museum* 66: 457–496.

Oygur, S. and G. W. Wolfe. 1991. Classification, distribution and phylogeny of North American (North of Mexico) species of *Gyrinus* Müller (Coleoptera: Gyrinidae). *Bulletin of the American Museum of Natural History* 207:1–97.

Szumik, C. A. 1996. The higher classification of the Order Embioptera: A cladistic analysis. *Cladistics* 12: 41–64.

Floras, Faunas, Field Guides, and Checklists

Full descriptions are also given in many floral and faunal surveys. The descriptions in these differ from those in revisions chiefly by stressing the characteristics of local populations and by limiting discussion of distributions to a particular geographic region. As Stace (1989) points out, however, floras (which are often limited to one group of plants within that region) and monographs on a particular plant group can be hard to distinguish in some cases; the difference is in emphasis. The flora is written mainly for the purpose of identifying all the plants of the *region,* whereas the monograph is written to synthesize all the taxonomic information on the *group.* This distinction applies equally well to works on animal groups. Some sample titles include the following:

Clark, A. H. 1921. Report on the ophiurians collected by the Barbados–Antigua Expedition from the University of Iowa in 1918. *University of Iowa Studies in Natural History* 9: 29–63.

Mathis, W. N. 1995. Shore Flies of the Galápagos Islands (Diptera: Ephydridae). *Annals of the Entomological Society of America* 88: 627–640.

Littler, D. S. and M. M. Littler. 1997. An illustrated marine flora of the Pelican Cays, Belize. *Bulletin of the Biological Society of Washington* 9: 1–149.

Field guides and identification handbooks are similar to floras and faunal surveys in that they cover a limited geographic area, but because they are designed for field or amateur use, they emphasize illustrations and keys and often use simplified descriptions stressing coloration, behavior, vocalizations, or other aspects of biology that can be observed in life.

For example, here is part of the species description of *Ulvaria oxysperma* from Littler et al. (1989), *Marine Plants of the Caribbean:*

> Superficially resembles *Ulva,* but is much paler in color, thinner and more delicate. It is limp and tears when handled out of water, much like wet tissue paper. (p. 20)

Or this example, from the description of the purple finch in Robbins et al.(1983), *Birds of North America:*

> Female is told from the female House Finch by the broad white line back of the eye and the larger beak. The sharp musical chip of the Purple, often given in flight, is distinctive. Song is a long, loud, rich warbling, 4–8/min. (p. 316)

Checklists may merely list the species found, or they may contain discussion, synonymy, distribution, or brief descriptive information:

Coleopora tubulosa (Canu and Bassler), 1924. FIGURED: Canu and Bassler (1928b), pl. ed, figs. 1–6. HABITAT: coral rubble. DEPTH: 40 m or less. AREAS: Florida (11). (Winston 1986:16)

Checklists usually have unambiguous titles:

Coile, N. C. and S. B. Jones, Jr. 1988. Checklist of the vascular flora of St. Catherine's Island, Georgia. *American Museum Novitates* 2920: 1–14.
Rhoads, A. F. and W. M. Klein. 1995. The vascular flora of Pennsylvania: Annotated checklist and atlas. *Memoirs of the American Philosophical Society* 207: 1–636.
Banta, W. C. and J. C. Redden. 1990. A checklist of the Bryozoa of the Galapagos Islands. *Proceedings of the Biological Society of Washington* 103: 789–802.

ATLASES An atlas illustrates all the species in a particular taxonomic group. A distributional atlas provides maps of the distribution of taxa within a certain geographic region:

Harrison, C. 1982. *An Atlas of the Birds of Western Palaearctic.* Princeton, N.J., Princeton University Press.
Burgess, N. R. H. and G. O. Cowan. 1993. *A Colour Atlas of Medical Entomology.* New York, Chapman & Hall.

Miscellaneous

There are many other research publications that are all or partly taxonomic in nature: notes on taxonomy, distribution, or biology, presented either alone or in conjunction with other biological research on a taxon; new records, new additions to floras or faunas, and so on. They all may add valuable information for the description of a species. The following are just a few examples:

Silva, P. C. 1972. Remarks on algal nomenclature. V. *Taxon* 21: 199–205.
Gravina, M. F. and A. Somaschini. 1991. Observations on the genus *Branchiomaldane* (Polychaeta: Arenicolidae) with a new record for the

Italian fauna: *Branchiomaldane vincenti* Langerhans, 1881. *Oebalia* n.s. 17:159–166.
Buhlmann, K. A. and J. C. Mitchell. 1997. Ecological notes on the amphibians and reptiles of the Naval Surface Warfare Center, Dahlgren Laboratory, King George County, Virginia. *Banisteria* 9: 45–50.
Wehrtmann, E. S. and A. Carvacho. 1997. New records and distribution ranges of shrimps (Crustacea: Decapoda: Penaeoidea and Caridea) in Chilean waters. *Proceedings of the Biological Society of Washington* 110: 49–57.

Form of the Descriptive Paper

A descriptive paper is not difficult to write. It is important to remember from the start, though, that taxonomic papers still follow the standard general format of the scientific research paper. In other words, a taxonomic paper, even a simple description of one species, should have an abstract, introduction, materials and methods section, results, discussion, conclusions, and bibliography. As a member of the Scientific Publication Committee at the American Museum of Natural History, I occasionally saw papers written by experienced taxonomists in which some of these sections were left out; they always went right back to the author for repair.

Abstract

Most editors prefer a "content" abstract, one that summarizes the findings of the paper rather than just describing them. In this information age, you will share their preference once you realize that your abstract, whether printed or electronic, is all most readers will ever see of your paper. If you don't put a good summary of your findings into your abstract, you have only yourself to blame if your work never gets used or cited.

Introduction

For the description of a single species this can be very brief (a sentence or two). It should give the reason for writing the paper–the discovery or hypothesis that impelled the research–and the historical background necessary to place the work in perspective.

Methods or Methods and Materials

This section covers techniques used to capture, preserve, and measure the study organism. If there is no section for material (specimens) examined in the main description, it should be included in this section.

The Description

In a species description this is the equivalent of the results or results and discussion section, or main body of the paper. Full descriptions include headings, synonomy (if necessary), etymology (if necessary), material examined or type material, diagnosis, description, taxonomic discussion, ecology, and distribution.

Conclusion

For the simplest papers a summary may be unnecessary; the abstract provides all the summary information needed. But for "mixed" papers, those that discuss biological or other results, in addition to the species description, some kind of conclusion should be drawn or summary included.

Acknowledgments

Although it is optional, an acknowledgments section gives you a chance to thank others for their assistance: colleagues and students, technicians, taxonomists consulted, helpful librarians, your mother, and so on. People like to see their help acknowledged in print. It is also the proper place to acknowledge funding from grants, fellowships, or institutions.

References or Bibliography

Make sure to follow a specific journal's format in this (and all other) sections. Pick out a journal before starting to write, to cut down on rewriting and to avoid having to restructure your whole list of references (see chapter 16).

The next 10 chapters cover the components of the species description and the process of publication of species and other taxonomic descriptions. This chapter lists the similarities between taxonomic papers and other scientific papers, but in reading the rest of the book and in writing your own descriptions, you should also keep in mind the major way in which taxonomic papers differ from most of the rest of the scientific literature. Most research papers are quickly outdated by new work; in 5 or 10 years they are obsolete, never to be cited again. But because of the historical nature of taxonomic research, taxonomic publications will be used by generations of taxonomists. A colleague once said to me, "If Darwin had never written the *Origin of Species,* he'd still be known today–for his four monographs on barnacles." And that is no exaggeration. A search for citations to those four monographs in *Science Citation Index* showed that they had been cited 49 times in the last 5 years alone–not bad for papers published between 1851 and 1854. Will your work stand the test of time as well as Darwin's?

Chapter 7

Headings and Synonymies

Clarity and brevity are among the essential attributes of a good synonymy; clarity should not, however, be sacrificed for the sake of brevity. –SCHENK AND McMASTERS (1936:17)

Uncritical citation of synonyms may lead to a repetition of errors. The monographer should accept nothing on trust that he can confirm personally. –DAVIS AND HEYWOOD (1963:294)

This chapter covers the introductory sections of the species description: headings and synonymies. As pointed out in chapters 4 and 6, styles of description have varied over time and continue to vary according to differences in practice by specialists working on various groups of organisms. However, they all fit into the general form of a research paper, with the taxonomic description or descriptions comprising the results or results and discussion section. Some journal formats label them this way. In other journals description sections may be called "systematics" or "systematic account" (or "accounts" if more than one taxon is described), or the description headings may just follow the similarly untitled introduction section (as is done in the *Proceedings of the Biological Society of Washington*).

Description Headings

Name, Author, Date

A species description is introduced by a heading giving the currently accepted scientific name of the species (i.e., the one that you, the author, believe to be correct), followed by the names of the person or persons who originally published the original description (for animals) or combination (for plants), plus the year of publication. This second part, the *authority citation,* is not part of the scientific name, but just what it says it is: a brief bibliographic reference to the original publication of the name. For example,

Eunice bipapillata Grube, 1866

tells us that the polychaete *Eunice bipapillata* was described by Grube in a paper published in 1866, and that he described it as a member of the genus *Eunice,* the genus in which it is still classified today.

In the case of an author with a common last name, initials may be added:

Ophiosyzygus disacanthus H. L. Clark, 1911
Orobanche minor J. E. Smith

For animal names, if the author of the name is extremely well known (such as Linnaeus), the name may be abbreviated:

Capillaster multiradiatus (L.)

but the name is written out completely for most other authors.

For plant names, as noted in chapter 4, author's names are commonly abbreviated:

Australagus mollissimus Torr.
Iva xanthifolia Nutt.

If the abbreviations are unfamiliar, you may have to consult a dictionary of authors of plant names to find the author's full name and bibliographic references.

When a name has been published jointly by two authors, the names of both are cited, joined by *and* (ICZN), *et* (ICBN), or an ampersand:

Verbena bracteata Lag. & Rodr.
Milnesium tetralamellatum Pilato & Binda 1991
Gigantodax multifilis Wygodzinsky and Coscarón

Three authors are less common. In zoological taxonomy the names of all three may be used:

Macrobiotus snaresensis Horning, Schuster & Grigarick 1978

The ICBN (46C.2) recommends that the citation be restricted to the name of the first author, followed by *et al.* or *& al.* For more than three authors (very rare), the first author's name with *et al.* or *& al.* is always used.

For animals, if the organism has been placed in a different genus since the time of the original description, the original author's name is given in parentheses, as in the bryozoan *Carbasea carbasea* (Ellis and Solander), 1786. This common boreal species was described as *Flustra carbasea, Flustra* being the genus used at the time for all flexible frondose cheilostomes. A later worker created the genus *Carbasea* for species with zooids only on one side of the frond, and *Flustra carbasea* was moved to that new genus, becoming *Carbasea carbasea* (Ryland and Hayward 1977).

Or, in the case of the polychaete worm,

Eunice elegans (Verrill, 1900)

A. E. Verrill's name is placed in parentheses because he described the species as *Leodice elegans* (Fauchald 1992).

Most commonly the parentheses surround only the name, which is followed by a comma and the date. But some journal formats place the entire citation in parentheses, as in the preceding example. In other formats, only the author's name and not the date is given in the heading. However, the date of the original description can be found in the synonymy section, which follows the heading, or in the bibliography or reference list.

Botanists use a different system to deal with changes in nomenclature, as chapter 2 pointed out. The authority citation for a plant name that has remained unchanged since it was originally introduced looks like a zoological citation, with the original author's name following the species name:

Clematis virginiana Linnaeus

But in the case of a plant species that has undergone one or more nomenclatural changes, the author of the *accepted combination* is recognized in the citation. For example, *Sargassum muticum* is an introduced seaweed species that has spread to the shores of northern Europe and the Pacific coast of North America from its native Japan. Its authority citation

Sargassum muticum (Yendo) Fensholt

tells us that a botanist named Yendo first described this species, but is not the author of the currently accepted combination. In fact, although Yendo was the first to describe the seaweed now known as *Sargassum muticum,* he did not recognize it as a new species. Instead, he thought it was merely a variety of a species he had described earlier, and so called it *Sargassum kjellmanianum* forma *muticus* (Yendo 1907). Fensholt, a later worker on *Sargassum,* decided that it was a distinct species and named it *Sargassum muticum* (Fensholt 1955), elevating the form to specific status (Critchley 1983).

Some other types of citations are also used. Both botanists and zoologists may use the word *in* to show that a description by one author has been published in a work by another author:

Viburnum ternatum Rehder *in* Sargent
Parsmittina parsevali, Audouin *in* Savigny

Botanists also use *ex* to show when a name that was proposed but not validly published by one person has been published by another person and ascribed to the first author. For example,

Euphorbia supina Raf. ex Boiss.

explains that the scientific name for the native weed called prostrate spurge was first proposed by the prolific but unorthodox naturalist explorer Constantin Rafinesque, but not validly published by him. Boissier later validly published and described the species, giving credit for the name to Rafinesque.

For a New Species

If you are describing a completely new species, the authority citation is not usually used. The species name is simply followed by a comma, and then *new species, species nova, sp. nov.,* or *n. sp.* (depending on the format used in the journal to which your description is being submitted). Subsequent workers who cite your new species will add your name and the date of your publication as the authority citation. In a few journals (e.g., in entomology, *Pan-Pacific Entomology* and the *Journal of the Kansas Entomological Society*), the name of the describer is mandatory in the original description. In others, its inclusion in an original description is optional, as in these headings of descriptions from *The Coleopterists Bulletin:*

Heterocerus insolens Miller, **new species**
Augyles blanda Miller, **new species** (Miller 1994)

Classification

Description headings may also include the classification of the species being described, placing it within the taxonomic hierarchy by listing the higher categories of which it is a member. How complete a classification is used is depends on several factors, including the protocols used for that group of organism, the requirements of the journal, how much of the classification has already been given in title or abstract, and the state of knowledge in that group or of that species (whether its taxonomic placement can be determined at the time of publication). For example, the title of a description of a new species of lizard by Walter C. Brown and Ely L. Alcala (1995:392), "A new species of *Brachymeles* (Reptilia: Scincidae) from Catanduanes Island, Philippines," gives its classification to family and genus level. The species description following the introduction and materials and methods sections begins simply with

Brachymeles minimus, new species

Another author might choose to give a full classification. For example in a paper for the same journal by Stephen D. Cairns (1987:141), "*Conopora adeta,* new species (Hydrozoa: Stylasteridae) from Australia, the first known unattached stylasterid," the species description begins with these classifying headings:

Class Hydrozoa Owen, 1843
Subclass Athecatae Hincks, 1868
Order Filifera Kuhn, 1913
Superfamily Hydractinioidea
Bouillion, 1978
Family Stylasteridae Gray, 1847
Conopora Moseley, 1879

followed by a diagnosis of the genus, and then by the heading for the species description

Conopora adeta, new species

In a paleontological example, from a paper by Lawrence J. Flynn, "Late Cretaceous mammal horizons from the San Juan Basin, New Mexico" (1986), the systematics section begins with this entry:

SYSTEMATICS

ORDER MULTITUBERCULATA COPE, 1884
FAMILY, GENUS, AND SPECIES INDETERMINATE

followed by a description of two fossil premolars found and an explanation that "no close identification is possible at present due to the poor representation of this taxon and the lack of knowledge of multituberculate anterior premolars in general" (Flynn 1986:9). Another taxon found in the same study was better represented by fossil material and could be more completely identified. Its entry reads,

SUBORDER PTILODONTOIDEA
(GREGORY AND SIMPSON, 1926)

FAMILY NEOPLAGIAULACIDAE
AMEGHINO, 1890

Mesodma Jepsen, 1940

Mesodma formosa (Marsh, 1889)

Figures

Figure or plate numbers illustrating the taxon described often directly follow the species name and authority citation, although they may also be included in the text of the description. If you want people to use and accept your work, you had better include illustrations. If a specimen is worth a thousand pictures, a picture is still worth a thousand words. Unfortunately, even today many published descriptions are worthless because the species in question cannot be recognized from the written description and the author did not include an illustration, even though an illustration would have completely prevented any problems in its recognition.

Synonyms

Following the headings in a species description comes the synonymy. The dictionary defines a *synonym* as "a word having the same or nearly the same meaning as another word" (*Random House Webster's Dictionary,* New York, Ballantine, 1993, p. 670); synonyms are different words with the same meaning. This is the definition we learned in school. But in systematics the term *synonym* has a more specialized meaning. It is defined as a different name used for the same entity (taxon); thus synonyms are the different names that have been used for a taxon over time. There are several types of synonym.

An *objective synonym* is one based on the same type material as another (earlier) named form. Objective synonyms are also called *homotypic synonyms, obligate synonyms,* and *absolute synonyms.* Because they are based on type material, they cannot be disputed (Jeffrey 1989).

For example, the name *Alogostreptus nattereri* (Diplopoda: Spirostreptidae) was proposed by Attems in 1950 from a single specimen in the Vienna Museum, a specimen that was already labeled "*Spirostreptus cultratus* Humber & Saussure," a species described in 1870, from that same specimen. Either Attems overlooked the original label or ignored it because the word *type* was not written on it; whatever his reasoning, the type of *nattereri* is unquestionably the type of *cultratus,* so these two names are automatically objective synonyms (R. L. Hoffman, personal communication 1997).

Subjective synonyms are two or more different names of taxa (e.g., genera, species) that a specialist considers to belong to the same taxon. This kind of synonym is also called *heterotypic* or *taxonomic.* Because they are based on opinion, such decisions may be disputed or changed. For example, Yendo also described other *Sargassum* species, including typical *Sargassum kjellmanianum* and *Sargassum miyabei.* Later workers had trouble distinguishing the two. Yoshida (1978), who looked carefully at their reproductive patterns, concluded that they were conspecific and gave the name *Sargassum miyabei* (described earlier) priority, with *Sargassum kjellmanianum* as a junior synonym (Critchley 1983). A *junior synonym* is the later published of two (or more) names based on identical specimens considered to be conspecific. Usually when taxonomists speak of a synonym with no qualification they mean a junior synonym, one of the names given a species at a later date than the name by which it is properly known.

Synonymies

Taxonomists use the term *synonymy* in four different ways (Jeffrey 1989). It can refer to the existence of several names used for the same taxon (the synonyms discussed earlier). For example, the generic name *Orinisobates* Lohmander, 1932 (Diplopoda: Nemasomatidae) has the junior subjective synonyms *Heteroisobates* Verhoeff, 1934, *Tiviulus* Chamberlin, 1941, and *Utoiulus* Chamberlin, 1943. (R. L. Hoffman, personal communication 1997).

It can also describe the relationship between two such names, as in the statement, "Woodman (1996) recently placed the Columbian shrew *Cryptonotis avia* G. M. Allen, 1923, in synonymy with *Cryptonotis thomasi* (Merriam) 1897, on the basis of a principal components analysis of nine cranial measurements."

It is used to speak of all the names, other than the correct name, used for a taxon, as in "The synonymy of some well-studied species can take up several pages of a paper."

Finally, it is used to refer to the synonymy section of a taxonomic description, the bibliographic list of all the names that have been used in published references to that taxon, which precedes the description itself.

The synonymy section of a species description serves two purposes. First, it gives a history of the nomenclature of a species. The information it provides is not only necessary for further taxonomic work, but is often the only approach to the older literature on a species, taxonomic or biologic. If you don't know what a species was called 10 or 50 years ago, you are denied access to all the biological or ecological works in which that species appears, as well as all the taxonomic literature on it. Electronic abstracting services and computerized collection databases may provide some links between synonyms automatically, but even for the more current literature, the more of a species's possible synonyms you are aware of, the more complete your search will be. Second, this section presents the author's conclusions as to taxonomic placement of the species and the validity of the names that have been applied to the species in the past (Schenk and McMasters 1948; Sohn 1994).

New Species

Again, if you are describing a completely new species, you will not have to worry about a synonymy section; you will not have one. However, if you are describing as new, and publishing a name for, a species that has been misidentified in the past, you must include a synonymy citing all the references that you believe refer to that species under another name or names. The following example shows the heading and synonymy for a description of a new species that had been previously studied, but identified incorrectly:

Cheiriphotis williamsoni sp. nov.
Cheiripohotis megacheles, Walker, 1904: 284, pl. 6, figs. 42: Nayar, 1959: 33, pl. 11, figs. 23–25; 1966, pl. 159, fig. 17a. (Not *Melita megacheles* Giles, 1885). (Salman and Jabbar 1990:215)

Ideally, you will have been able to examine the actual specimens on which the publications you cite were based and will be able to list the type locality and type depository for any synonym (which usually is done in the material examined section, but may be made part of the synonymy).

Types of Synonymies

There are two general methods for arranging synonymies: according to chronology of names and according to chronology of bibliographic references.

The first type consists of a chronological list of all the scientific names by which a species has been designated, with bibliographic citations (also in chronological order) after each name. It starts with the original description, as shown by this heading and compactly set up synonymy for a marine alga, *Cladophora vagabunda,* from the Pelican Cays, Belize:

Cladophora vagabunda (Linnaeus) van den Hoek 1963: 144.
Conferva vagabunda Linnaeus 1753: 1167. *Cladophora fasicularis* (Martens *in* C. A. Agardh) Kützing 1843: 268. *C. mauritiana* Kützing 1849: 399. *C. sertularina* (Montagne) Kützing 1849: 396 (see van den Hoek 1982). (Littler and Littler 1997:95)

Or this synonymy of a an Antarctic sea anemone:

Isosicyonis alba (Studer, 1879)
Paractis alba: Studer, 1879, p. 545.–Ridley 1882, p. 101.-Carlgren, 1898, p. 43.
Paractis Studerii: Andres, 1883, p. 271.
Isosicyonis alba: Carlgren, 1927, p. 52.–Carlgren, 1949, p. 42, 57.–Riemann-Zürneck, 1980, p. 19, 24, 32. (Fautin 1984:9)

The bibliographic citation gives the author and year of the reference at minimum. A complete bibliographic reference to each work cited in a synonymy is then given in the references section of the paper. In the interest of brevity, a synonymy reference usually gives just the page number where the species description begins in that citation, not all the pages, but if you want to include all pages, that may be negotiable. It may also give a reference to any figure or plate numbers illustrating that species in the work being cited, as in this fossil diatom synonymy by Patricia Sims (1989:354):

Coscinodiscus solidus Strel'nikova (1971) p. 44, pl. 1, Fig. 8 (Fig 16–21, 69).
Coscinodiscus subtilis Ehrenb., sensu Long, Fuge & Smith (1946) p. 105, pl. 16, Fig. 18.

Synonymies may be further elaborated by the inclusion of information on locality of collection, location of deposited material, or previous taxonomic decisions. For example, John P. Grunshaw (1995:405) gave the following synonymy for an African grasshopper species, in which collection localities and the author's and previous authors' taxonomic decisions were also recorded:

Tylotropidius didymus (Thunberg 1815)
Gryllus didymus THUNBERG 1815: 241 (Rep. South Africa).
Heteracris speciosa WALKER 1870: 661 (Sierra Leon). Syn. by DIRSH 1966: 212.
Tylotropidius didymus var. *voltaensis* SJÖSTED 1931: 34 (Burkina Faso). Syn. by UVAROV 1953: 92.
Tylotropidius didymus var. *citrea* KARNY 1907: 352 (Sudan). Syn. by DIRSH 1966: 212.
Tylotropidius didymus var. *violacea* I. BOLIVAR 1912: 99 (Zaire). N. syn.
Tylotropidius grandis SJÖSTED 1931: 35 (Burkina Faso). Syn. by DIRSH 1966: 210.
Tylotropidius congoensis SJÖSTED 1931: 35 (Congo) N. syn.

In this synonymy, from a paper by Wayne Mathis (1995:628) on the shoreflies of the Galápagos, information is included on type locality, type specimens, where they are deposited, and their catalog numbers:

Atissa luteipes Cresson, 1944a: 2 (Honduras, Puerto Castilla; HT ♀, ANSP [6661]); Wirth, 1968: 4 (Neotropical catalog).

HT stands for *holotype,* which is ♀ (female), *ANSP* stands for the place it is deposited (the Academy of Natural Sciences of Philadelphia), and *6661* is its catalog number.

In the second type of synonymy all published references to a species are arranged chronologically in order of publication date, followed by year, name used, and author bibliographic information. This arrangement has the advantage of clearly displaying the way in which nomenclature of a species has developed or changed over time, but it often requires more space (a commodity jealously guarded by journal editors). The following is an example of a heading and chronological synonymy for a fossil plant:

Brachyphylum crassicaule Fontaine (Figure 17)
Brachyphylum crassicaule Fontaine–Fontaine, 1889, p. 221, Pl. 100, fig. 4; Pl. 109, figs. 1–3; Pl. 111, figs. 6, 7; Pl. 112 figs. 6–8; Pl. 168, fig. 9.
Brachyphylum crassicaule Fontaine–Fontaine, in Ward, 1905, pp. 529, 557, Pl. 113, fig. 6.
Brachyphylum crassicaule Fontaine–Berry, 1911a, p. 393, Pl. 164, figs. 1–6. (Upchurch et al. 1994:18)

In the preceding example, all references to the plant used the same name for it, but any number of names could appear in the synonymy. The point is that the citations are arranged chronologically, rather than grouped by name used. The same information could have been presented in the first style shown, as follows:

Brachyphyllum crassicaule Fontaine (Figure 17)
Brachyphylum crassicaule Fontaine, 1889, p. 221, Pl. 100, fig. 4; Pl. 109, figs. 1–3; Pl. 111, figs. 6, 7; Pl. 112 figs. 6–8; Pl. 168, fig. 9. Fontaine, in Ward, 1905, pp. 529, 557, Pl. 113, fig. 6. Berry 1911a, p. 393, Pl. 164, figs. 1–6.

In some journal synonymy styles (for example, in the publication series of the American Museum of Natural History) the citation for a taxonomic reference to a species is separated from the name of the species by a colon if that author did not originate the name in the paper being cited. The author's name is not separated by punctuation from a name that he or she originated. For example, for their synonymy of the plant bug *Pseudopsallus angularis* (Uhler), Stonedahl and Schwartz (1986:12–13) wrote

Macrotylus angularis Uhler, 1894, pp. 242–274 (for the citation to the original description by Uhler)

but

> *Pseudopsallus angularis:* Van Duzee, 1916a, pp. 224–225 (for an additional citation to the species)

However, other journals use commas, semicolons, or no punctuation to separate species names from authors of citations. In fact, it may seem to the potential author that regardless of certain conventions, every journal uses a slightly different style for synonymies. Because this is the most painstaking and tedious part of the whole description, it is well worth taking a careful look at copies of the journal you hope to publish in and following their style exactly.

Terms Used in Synonymies

Synonymies use many abbreviations, usually in Latin. Some you already know because they are the standard Latin abbreviations used in scholarly publishing (*e.g., i.e., op. cit., loc. cit., etc.*). However, taxonomy also has its own special set of abbreviations. You must become familiar with them in order to understand the information being presented in the synonymy sections of species descriptions. They are listed alphabetically, along with their definitions, in table 7.1.

Different Kinds of Synonymies

Full synonymies are usually given in monographs, revisions, and catalogs. In a full synonymy the author has attempted to find, cite, and evaluate every taxonomic reference to the species; in a thorough work, the author will also have located, borrowed or visited, and checked and verified the actual specimens on which the citations were based. Nontaxonomic (e.g., ecological, physiological) references to a species are usually not included in the synonymy section, although they may be included elsewhere in the species description.

Abbreviated synonymies are used in checklists and faunal studies. These may refer only to records of the species from the area being studied or they may include only major monographic references, leaving out references from checklists or references that cannot be substantiated by voucher specimens.

Table 7.1 **Terms and Abbreviations Used in Synonymies and Taxonomic Description**

Latin Abbreviation	Latin Term	English Term or Definition
aff.	affinis	having affinity with but not identical to
al.	alii, aliorum	others, of others
ap.	apud	with, in the work of
ascr.	ascriptum	ascribed to
auct.	auctorum	of authors
auct., non ..	auctorum, non ..	of authors, not (of), used for a name that has been mistakenly applied
ca.	circa	about (with reference to time)
cf.	confer	to be compared with, compare with
cit.	citatus	cited
comb. nov.	combinatio nova	new combination
cv.	cultivar	agricultural or horticultural variety
descr.	descriptio	description
e.g.	exempli gratia	for example
emend.	emendatio	emended
e.p.	ex parte	in part
ex aff.	ex affinis	of affinity
ex gr.	ex grupo	of the group of
f. forma	form	(a morphologic term to ICZN; the lowest supplementary taxonomic rank to ICBN)
–	fide	on the authority of
gen. et. sp. nov.	genus (novum) et species nova	new genus and new species
gen. nov.	genus novum	new genus
ibid.	ibidem	the same reference
id. ac.	idem ac	the same as
–	in	in
–	incertae sedis	of uncertain taxonomic position
–	infra	below
in litt.	in litteris	in correspondence
–	ipso facto	in the fact itself
–	lapsus calami	a slip of the pen
l.c. or loc. cit.	loco citato	place cited (publication and page)

Table 7.1 Continued

Latin Abbreviation	Latin Term	English Term or Definition
–	mihi	belonging to me (as a new species)
–	nec	and not (of), nor (of)
nob.	nobis	belonging to us (as a new species)
nom. cons.	nomen conservandum	a name that should be preserved (meaning a name that has been preserved by the ICZN or ICBN; see ch. 21)
nom. corr.	nomen correctum	change of spelling, e.g., -somidae to -somatidae because of incorrect original spelling of stem form
–	nomen inquirendum	a name that should be investigated
n.n., n. nov., or nom. nov.	nomen novum	new name, proposed for as a direct substitution for an invalid existing name
–	nomen nudum	a name without a designation (i.e., its description is not acceptable nomenclaturally); singular
–	nomina nuda	names without designation (plural of above)
nom. trans.	nomen translatum	name that has been changed in hierarchical rank with changed suffix (ICZN)
–	not	not
–	non, nec	neither (of), nor (of)
–	non vidi	I have not seen it
–	non viso	not seen
nov.	novum	new
nov. n.	novem nomen	new name
nov. sp.	nova species	new species
–	olim	formerly
op. cit.	opere citato	publication cited (but not page)
orth. mut.	orthgraphia mutata	with an altered spelling (by)
–	partim	part
–	passim	here and there

Table 7.1 Continued

Latin Abbreviation	Latin Term	English Term or Definition
p.p.	pro parte	in part
q.v.	quod vide	which see
–	rite	according to the rules
–	saltem	at least
sc.	scilicet	namely
sec.	secundum	according to
–	sensu	Authorname (e.g., sensu Osburn) in the sense that a particular author understood the taxon
s.a.	sensu amplo	in the broad sense
s.l.	sensu lato	in the broad sense
s.s., s. str.	sensu strictu	in the strict sense
–	seu	either, or
–	sic	thus (to show exact transcription)
–	sin typo	without type
–	sine	without
sp.	species	species (singular) often used when true identity unknown
sphalm.	sphalmate	by mistake, in error
spp.	species	species (plural), used in referring to several species, usually in the same genus
sp. indet. or sp. ind.	species indeterminata	species indeterminate
–	species inquirendae	species of doubtful identity, needs further investigation
sp. nov.	species nova	new species (preferred to nov. sp.)
s-g.	subgenus	subgenus
s-gg.	subgenera	subgenera
stat. n. or stat. nov.	status novus	taxon altered in rank, but retains specific name from former rank, used when the change is made
ssp. or subsp.	subspecies	subspecies
sup.	supra	above
sup. cit.	supra citato	cited above
syn.	synonymum	synonym

Table 7.1 Continued

Latin Abbreviation	Latin Term	English Term or Definition
tab.	tabula plate,	illustration
tax. n. or tax. nov.	taxon novum	new taxon, used when a category name is derived from a preexisting genus for the first time
–	teste	according to (verbal testimony; contrasted to fide, written testimony)
tom. cit.	tomo citato	in the volume cited
trans. nov.	translatio nova	new transfer, taxon altered in position, but keeps species name from former position
typ. cons.	typus conservandus	a type to be conserved
v.	vel	or
–	vice	in place of
–	vide	see
v. et.	vide etiam	see also
–	vidi!	I have seen it (or just "!")

Sources: Compiled from Schenk and McMasters (1948), Hoffman (1980), Jeffrey (1989), and Quicke (1993), with additional assistance from R. L. Hoffman.

References in Headings and Synonymies

Given the value of the literature, especially the older literature, in taxonomic research, it seems that the references used in headings and synonymies would be cited completely and accurately, but this does not always happen. Biologists, especially animal biologists, have not been very consistent about citing authors of taxa in systematic studies (let alone in biological studies), as documented by Kelt and Palma (1992). The references to a species found in a synonymy provide essential information on that species and on taxonomists' interpretation of that species over time. To gain access to that information you must rely on the accuracy of those references. Names found in headings can pose even more difficulties than those in synonymies. Some authors and journal editors do not think it is necessary to include the citations to family and generic names that are found in the headings to species descriptions. However, when the heading reads "*Reptadeonella* Busk," it is not of much use to someone who needs to know how Busk defined that genus. "*Reptadeonella* Busk, 1884" at least gives you a date to start searching for the reference in which the genus was described, but what if Busk wrote several papers in 1884? Only the complete citation, included in the paper's bibliography (e.g., "Busk, G. 1884. Report on the Polyzoa, Part I. The Cheilostomata. *Rep. Sci. Results Voy. Challenger, Zool.* 10(30): 1–216.") satisfies the potential user's needs.

The problem may not be as acute in botany because several compilations of the taxonomic literature exist (e.g., Lawrence et al. 1968; Stafleu and Cowan 1976), but a reader who did not have access to such volumes might waste a considerable amount of time tracking down the correct reference.

Chapter 8

Naming Species: Etymology

Words are in themselves among the most interesting objects of study, and the names of animals and plants are worthy of more consideration than Biologists are inclined to give them.

–Savory (1962:vii)

Words that anesthetize are words of three, four and five syllables, mostly of Latin origin. –Zinsser (1988:130)

Naming Species

Naming your new species is an essential part of describing it. Both codes of nomenclature have the primary purpose of making the names of taxa available for use, thus making subsequent recognition and identification possible. This chapter covers the selection and composition of species names and the etymology section of the species description. Etymology is the study of the origins and meaning of words. The etymology section is the section in which you tell what your new species name means, why you chose it, and how you composed it.

Some people approach naming their new species with trepidation, worrying about stumbling over unfamiliar or half-forgotten Latin and Greek. Others behave like expectant parents, daydreaming about potential names for their offspring as soon as the possibility of its appearance arises. There's no need for fear; the Latin needed is minimal and easy to understand. But there *is* every reason to exercise your creativity to come up with an appropriate name.

As you know by now, the *binomen,* the scientific name of a species, is a *binomial phrase,* made up of two parts: a genus term and a species term. The genus term is the noun of the phrase; it identifies the general kind of plant or animal involved. The species term is the qualifier. If the genus is the kind, the species is what kind (the brown kind, the pointed-toothed kind, the Jamaican kind, etc.). As English speakers we may wonder why the species term follows the genus term–why it is written as "The Lion, pointed-toothed," not the pointed-toothed Lion–but it is simply because in Latin syntax, as in French or Spanish (but not in English), the qualifier or adjective commonly *follows* the noun. The initial capitalization of the genus term but not the species term seems rooted in the classic typographic tradition of capitalizing nouns but not adjectives (as in the extract from the *Systema Naturae* shown in figure 2.1).

The custom of italicizing scientific names (genus and species) seems also to stem from the common typographic custom of setting words in italics or a contrasting typeface for clarity when they are written in a language other than that of the main text.

Grammatically, qualifier terms are of four basic types. Most often, the species name is an adjective or an adjective–adverb or adjective–adjective combination. It can also be a participle in the Latin nominative singular, a noun in apposition to the genus name, or a noun in the genitive (possessive) case.

Zoologists, in particular, have been heard to curse the Latin binomial when they go to name a species because most zoologists today have no more knowledge of Latin than the average educated English speaker (although that English speaker should be reminded that 60 percent of English words are derived from Latin, according to *Wheelock's Latin* [Wheelock 1995]). Botanists, who must use Latin in their original species descriptions, have been forced to continue to develop and maintain their Latin skills. But even if we dislike the extra effort we may have to make to name a new organism, it is probably fortunate for biology that Latin became the standard, rather than some modern language. The use of Latin (which was still the language of science in Linnaeus's time) probably prevented the nationalistic scientific battles that might otherwise have forestalled agreement on worldwide scientific nomenclature. If you are skeptical, read Duane Isely's 1972 article, "The Disappearance," a frighteningly plausible fantasy about what would happen if a system of biological nomenclature had to be reestablished today.

Brief Review of Latin and Greek

The main thing you need to remember about Latin is that, unlike English but like a number of modern languages such as German and Russian, Latin is an inflected language. In English, we depend on word order to determine the meaning of sentences. In an inflected language, meaning depends on word endings, and word order is flexible. Latin nouns, pronouns, and adjectives are *declined.* Latin words have an unvarying *base* (also called the *root* or *stem*) and a variable *ending.* Word endings change depending on their case, the way they are used in a phrase or sentence (e.g., as subject, object, or after a preposition), their gender (masculine, feminine, neuter), the gender of the word they modify, and their number (whether they are singular or plural). For example, instead of using *of* or an apostrophe to show possession, Latin changes the word ending to the *genitive* (possessive) case, singular or plural. To use a zoological example, **lumbricus** (the earthworm) becomes **lumbrici** (of the earthworm, the earthworm's) and **lumbricorum** (of earthworms, the earthworms') (Yancey 1946; Brown 1956; Savory 1962).

There are six cases in Latin. Luckily, for us, though, only two Latin cases–the *nominative* and *genitive* (possessive)–are used in constructing scientific names. Unluckily for us, Latin nouns and adjectives are put into groups called declensions (five for nouns, three for adjectives) on the basis of their gender and endings. The nominative case (the one to be used when it is the subject of a sentence) of the word is just the word as it appears in the dictionary. Genitive endings vary. The genitive singular is the most important case in Latin because it supplies the base of the noun for all the endings except the nominative singular. Table 8.1 gives some examples of nominative and genitive of each of the five declensions. However, some words do not fit the regular patterns, so you will need to consult a Latin grammar, dictionary, or specialized sourcebook to find the proper genitive singular ending for a word. These guides will identify what declension a word belongs to and give its gender and genitive singular (and hence its base). They also have tables with sets of endings for each declension.

First declension (sometimes called a-declension) nouns have **-ae** as their genitive singular ending (Table 8.1). The nominative singular (and most other cases) end in **-a.** They are feminine, with the exception of words that refer to a male (e.g., farmer = **agricola**). Second declension (also called

Table 8.1 Declension Endings of Latin Nouns in the Nominative and Genitive Cases

Declension		Case	Singular Ending	Plural Ending	Examples
GENDER: ALL FEMININE, EXCEPT FOR WORDS THAT INDICATE A MALE (E.G., *AGRICOLA* [FARMER])					
I		Nominative	**-a**	**-ae**	**rosa, rosae** = rose, roses
					agricola, agricolae = farmer, farmers
		Genitive	**-ae**	**-arum**	**rosae, rosarum** = of the rose, of the roses
					agricolae, agricolarum = of the farmer, of the farmers
GENDER: MASCULINE OR NEUTER					
II	m	Nominative	**-us**	**-i**	**locus, loci** = place, places
		Genitive	**-i**	**-orum**	**loci, locorum** = of the place, of the places
	n	Nominative	**-um**	**-a**	**cilium, cilia** = eyelash, eyelashes
		Genitive	**-i**	**-orum**	**cilii, ciliorum** = of the eyelash, of the eyelashes
GENDER: MASCULINE, FEMININE OR NEUTER					
III	m, f	Nominative	Varies	**-es**	**homo, homines** = man, men
					dens, dentes = tooth, teeth
		Genitive	**-is**	**-um**	**hominis, hominum** = of the man, of the men
					dentis, dentium = of the tooth, of the teeth
	n	Nominative	Varies	**-a**	**caput, capita** = head, heads
				-ia	**animal, animalia** = animal, animals
		Genitive	**-is**	**-um**	**capitis, capitum** = of the head, heads
			-is	**-ium**	**animalis, animalium** = of the animal, of the animals

GENDER: MASCULINE AND NEUTER					
IV	m	Nominative	**-us**	**-us**	**quercus, quercus** = oak, oaks
		Genitive	**-us**	**-uum**	**quercus, quercuum** = of oak, of oaks
	n	Nominative	**-u**	**-ua**	**cornu, cornua** = horn, horns
		Genitive	**-us**	**-uum**	**cornus, cornuum** = of a horn, of horns
GENDER: ALL FEMININE, EXCEPT *DIES*, DAY					
V		Nominative	**-es**	**-es**	**facies, facies** = face, faces
		Genitive	**-ei**	**-erum**	**faciei, facierum** = of the face, of faces

Sources: Based on Yancey (1946), Brown (1956), and Hendricks (1992).

o-declension) nouns have **-i** as the ending of the genitive singular. Those whose nominative singular ends in **-us**, **-er**, or **-ir** are masculine. Those whose nominative singular ending is **-um** are neuter. Third declension (i-declension) nouns are the largest group of Latin nouns. They may be masculine, feminine, or neuter, and their endings in the nominative singular are quite variable, making it most important to look at the genitive singular for the base, if you are going to use it as a plural or possessive form. Third declension masculine, feminine, and neuter nouns all have **-is** for the genitive singular. Dropping that ending gives the base for the other cases. If the irregularity of third declension nouns irritates you, just consider how much more irregular many English nouns are in their singular and plural forms (e.g., some of those in table 8.1). Fourth declension (u-declension) nouns ending in **-us** in the nominative singular are masculine; those ending in **-u** are neuter, with a few exceptions. Both gender nouns end in **-us** in the genitive singular. Fifth declension (e-declension) nouns usually end in **-es** in the nominative singular and **-ei** in the genitive singular; except for **dies**, which is generally masculine, they are all feminine in gender.

Although it is possible to use a general English–Latin dictionary to search for descriptive terms to compose a new scientific name, several books have been compiled especially to help biologists in naming organisms. One of the most useful is *The Composition of Scientific Words*, by Roland W. Brown (1956), which, in addition to an introductory section that discusses Greek and Latin as languages, and the formation of scientific words, has a cross-referenced dictionary section, so that you can find a Latin or Greek equivalent for an English term or find the English definition and derivation of a Greek or Latin term.

A Source-Book of Biological Names and Terms by Edmund C. Jaeger (1955) is also helpful. It contains a dictionary of about 14,000 elements (words, stems, prefixes, and suffixes) used in biological nomenclature, with their Greek, Latin, or other origins and their English definitions. This book has the advantage of containing more terms, but it is harder to use because it works in only one direction.

Botanical Latin, 4th ed., by W. T. Stearn (1992), is invaluable for botanists, who must still compose part of their original descriptions in Latin. It is very useful for animal biologists as well. The book includes a Latin/English–English/Latin dictionary of terms used in describing and naming plants. Of course, many of these terms are also applicable to animals.

In the Ancient Language wars, the Greeks seem to have lost the battle against the Romans, at least when it comes to biological nomenclature. Whereas ancient (classical), medieval, and modern scientific Latin may be used in constructing scientific names, only ancient Greek words may be used and they must be transliterated and treated as Latin words (ICZN, Article 11b, ii).

Table 8.2 shows the regular endings of the three Greek declensions: first declension or a-nouns, second declension or o-nouns, and third declension or consonant-nouns. They roughly correspond with the first three Latin declensions. The transliterated Latin endings are given in parentheses. Those who know the Greek alphabet can look up terms directly in a Classical Greek dictionary, and transliterate them into Latin (using a source such as the transliteration and Latinization table given in Appendix B of the ICZN). The rest of us must use one of the sourcebooks mentioned here or in this chapter's *Sources.* Of course many Greek nouns were taken over by Latin and received Latin endings, just as, over time, English speakers have taken over many Latin and Greek words and given them English endings.

Basic Rules of Species Names

The codes of nomenclature give some basic rules about the formation of species names. For a name to be considered properly published and acceptable (*validly published* according to the ICBN and *available* according to the ICZN), it must be constructed according to these rules.

A species name must be spelled in Latin letters (meaning in the alphabet we use today for English, French, etc., and not in Arabic, Chinese, Cyrillic, Greek, Hebrew, Greek, or other alphabets or characters). The Latin alphabet the Romans used was similar to ours, but lacked the letters *j, v,* and *w.* However, in scientific Latin we are allowed to construct names from words with those letters.

A species name begins with a lowercase letter, even when formed from a personal name (ICZN, Article 28). For example, a species of harvestman named after its collector is called *Acuclavella merickeli,* not *Acuclavella Merickeli* (Shear 1986).

A species name can contain no diacritical marks (accent marks, apostrophes, etc.) (ICZN, Article 27). For example, when *Cyrioctea calderoni,* a Chilean spider, was also named in honor of the collector of the holotype,

Table 8.2 **Declension Endings of Greek Nouns in the Nominative and Genitive Cases, with Latinized Endings in Parentheses**

Declension		Case	Singular Ending (Latinized Ending)	Plural Indent (Latinized Ending)	Examples
I		Nominative	*-a*	**-ai (-ae)**	**cardia, cardiai** = heart, hearts
			-e	**-ai (-ae)**	**cephale, cephalai** = head, heads
		Genitive	**-as (-ae)**	**-on (-arum)**	**cardias, cardion** = of the heart, of the hearts
			-es (ae)	**-on (-arum)**	**cephales, cephalon** = of the head, of the heads
II	(1)	Nominative	**-os (-us)**	**-oi (-i)**	**bios, bioi** = life, lives
		Genitive	**-ou (-i)**	**-on (-orum)**	**biou, biorum** = of the life, of the lives
	(2)	Nominative	**-on (-um)**	**-a**	**ganglion, ganglia** = swelling, swellings
		Genitive	**-ou (-i)**	**-on (-orum)**	**gangliou, ganglion** = of the swelling, of the swellings
III	(1)	Nominative	**-ps**	**-es**	**phelps, phlebes** = vein, veins
			-x	**-es**	**pharynx, pharynges** = throat, throats
			-is	**-es**	**ophis, ophies** = snake, snakes
			-as	**-es**	**gigas, gigantes**= giant, giants
			-on	**-es**	**geron, gerontes** = old man, old men
		Genitive	**-os (-is)**	**-on (-um)**	**phlebos, phlebon** = of the vein, of the veins
			-os (-is)	**-on (-um)**	**pharyngos, pharynon** = of the throat, of throats
			-os (-is)	**-on (-um)**	**ophios, ophion** = of the snake, of snakes
			-os (-is)	**-on (-um)**	**gigantos, giganton** = of the giant, of the giants
			-os (-is)	**-on (-um)**	**gerontos, geronton** = of the old man, of old men

(2)	Nominative	**-a**	**-a**	**chiasma, chiasmata** = crossing, crossings
		-ar	**-a**	**hepar, hepata** = liver, livers
		-as	**-a**	**keras, kerata** = horn, horns
	Genitive	**-os (-is)**	**-on (-um)**	**chiasmatos, chiasmaton** = of the crossing, of the crossings
		-os (-is)	**-on (-um)**	**hepatos, hepaton** = of the liver, of the livers
		-os (-is)	**-on (-um)**	**keratos, keraton** = of horn, of the horns
(3)	Nominative	**-er**	**-es**	**ther, theres** = beast, beasts
		-er	**-es**	**gaster, gastres** = belly, bellies
		-er	**-es**	**ander, andres** = man, men
		-is	**-es**	**rhis, rhines** = nose, noses
		-is	**-es**	**ornis, ornithes** = bird, birds
		-on	**-es**	**axon, axones** = axle, axles
	Genitive	**-os (-is)**	**-on (-um)**	**theros, theron** = of the beast, of the beasts
		-os (-is)	**-on (-um)**	**gastros, gastron** = of the belly, of the bellies
		-os (-is)	**-on (-um)**	**andros, andron** = of the man, of men
		-os (-is)	**-on (-um)**	**rhinos, rhinon** = of the nose, of noses
		-os (-is)	**-on (-um)**	**ornithos, ornithon** = of the bird, of birds
		-os (-is)	**-on (-um)**	**axonos, axonon** = of the axle, of the axles
(4)	Nominative	**-ys**	**-es**	**ichthys, ichthyes** = fish, fishes
	Genitive	**-os**	**-on**	**echthyos, ichthyon** = of the fish, of fishes

Sources: Based on Yancey (1946) and Brown (1956).
Note: All words transliterated into the Latin alphabet.

R. Calderón G. (Platnick 1986), the acute accent found in the collector's name was not used in the species name.

A species name must be more than one letter long (ICZN, Article 11n, i). The name itself may be formed from a Latin word, a Latinized word (one that has been given a Latin ending), a word in another language that uses the Latin alphabet (whether or not the word ending is a Latin ending), a word in a language using another alphabet that has been transliterated into Latin (ICZN Article 11b, i, ii), or even a nonsense word composed from an arbitrary combination of letters (ICZN, Article 11b, iii).

There are a few differences between the botanical and zoological codes with regard to constructing species names. Until fairly recently botanists capitalized species names based on personal names, feeling that not to do so was to show disrespect to the person they were intended to honor (McClintock 1969), but although the ICBN (rec. 60F, p. 78) *strongly recommends* that specific names begin with lowercase letters, it does allow capitalization of species names when they come directly from a person's name (human or mythical), are vernacular (non-Latin) names, or are names that were once generic names.

The zoological code does not allow hyphens between the two or more terms of a compound species name; the ICBN states that they must be "united or hyphenated" (ICBN 23.1). However hyphenated compounds are allowed only when "an epithet is formed of words that usually stand independently, or if the letters before and after the hyphen are the same" (ICBN 60.9).

Both sets of nomenclatural codes state merely that a binomial name is acceptable if the combination of genus and species is *unique,* not used for any other organism covered by that code, and meets the preceding guidelines. However, the best names are those that are in some way descriptive of the organisms, euphonious (or at least pronounceable), and unlikely to cause later nomenclatural problems. The next sections describe the morphology and construction of the various categories of species names: *descriptive, geographic, commemorative,* and *nonsense,* giving the advantages and disadvantages of each.

Descriptive Species Names

Descriptive names point out some characteristic or combination of characteristics of the species:

Mytilus edulis: the edible mussel

Lobelia cardinalis: the bright red cardinal flower, the flower that is the color of a Cardinal's robes

Ophiomyxa flaccida: the soft (skinned) brittlestar

Asteroporpa annulata: the brittlestar with annulated arms

Savory (1962) lists some of the characteristics that can be used: size, shape, proportion, color, appearance, behavior, habitat, environment, locality, pattern rarity, peculiarity, resemblance (e.g., earthworm-like, *lumbricoides*). Often the characters used are the diagnostic characters, those that distinguish the new species from other related species. Descriptive names can be *simple* or *compound,* composed of two or more word stems.

Simple Descriptive Names

Simple names have the advantage of being short. But unless a species is very distinctive it may not be possible to characterize it in a single word. And words characteristic of a whole group tend to have been used for the first described members of that group, and so are unavailable for naming later discoveries.

Grammatically, the simple descriptive species name is most often an adjective. An adjective is a word that modifies a noun (or pronoun), describing it or telling something about it:

The beautiful *Xxxxxus*

The blue *Xxxxxus*

Simple adjective names can sometimes be extended in meaning by being made comparative (= more something, somethinger). The comparative adds the endings **-ior** (masculine and feminine singular), **-iores** (masculine and feminine plural), and **-ius** (**-iora**) for the neuter singular and plural. More commonly, the superlative (the most something, the somethingest) is created by adding **-issumus, -a, -um,** or **-imus, -a, -um.** Apparently we'd rather think of our new species as the most something than merely more something:

Conopeum tenuissimum: the thinnest *Conopeum* bryozoan

Ophioderma squamosissimum: the scaliest *Ophioderma* brittlestar

Rattus villosissimus: the hairiest rat

Latin adjectives are grouped into three declensions. They must agree in gender with the noun (for scientific names, the genus term) they modify. If the noun is feminine, it must be followed by the feminine form of the adjective, and so on. This is easy enough to do when the gender of the noun is clear from its ending (as with most first and second declension nouns), but this is not always true. You must check a Latin dictionary or sourcebook, which lists the gender of each noun. For example, *Felis* (cat) is a feminine noun, so adjective species names for members of the cat genus must be feminine. The house cat is *Felis domestica,* not *Felis domesticus.* But *Canis* (dog), which also ends in **-is,** is masculine in gender, so we have *Canis familiaris,* not *Canis familiare* (the neuter adjective ending), and *Canis lupus,* not *Canis lupa* (the feminine ending). Note that matching in gender doesn't necessarily mean having identical endings, as that depends on the declensions of the nouns and adjectives involved.

Although matching nouns and adjectives may sound tricky, it is made easier by the fact that, with a very few exceptions, the only cases you must be concerned with for scientific names are the nominative and genitive singular. With the nominative singular to use for simple descriptive species terms and the genitive singular to give you possessive terms and to provide the base for the combining form to use in compound species terms, you will be pretty well equipped as a wordsmith. Table 8.3 shows the nominative and genitive singular endings for Latin adjectives. First and second declension (-a and -o declension) adjectives have three forms: masculine, feminine, and neuter. Adjectives that end in **-er** in the masculine singular may either keep or drop the **-e-** (for each adjective of this type, the genitive singular ending given in the dictionary will show you which). A few adjectives in this group are declined in the regular way except that they have **-ius** for the genitive singular for all three genders. Those commonly used in scientific names are included in the table. Third declension (or -i declension) adjectives occur in three groups, according to the number of endings they have in the nominative singular: one ending for all three genders, one ending for masculine and feminine and one for neuter, and one ending for each gender.

Present and past participles that function as adjectives may also be used. A present participle is the form of a verb that ends in *-ing.* Examples of descriptive names using present participles are as follows:

The startling *Xxxxxus*

The confusing *Xxxxxus*

Table 8.3 Groups of Latin Adjectives Showing Nominative and Genitive Singular Endings

Declension/Case	Gender	Singular Ending	Examples	
I and II				
Adjectives that end in **-us, -a, -um.**				
Nominative	m	**-us**	**bonus**	**puer bonus** = good boy
	f	**-a**	**bona**	**casa bona** = good house
	n	**-um**	**bonum**	**frumentum bonum** = good grain
Genitive	m	**-i**	**boni**	**pueri boni** = of the good boy
	f	**-ae**	**bona**	**casae bonae** = of a good house
	n	**-i**	**boni**	**frumenti boni** = of good grain
Some adjectives that end in **-er** drop the **-e;** some keep it.				
Nominative	m	**-er**	**miser**	**puer miser** = wretched boy
		-r	**pulcher**	**liber pulcher** = pretty book
	f	**-era**	**misera**	**puella misera** = wretched girl
		-ra	**pulchra**	**puella pulchra** = pretty girl
	n	**-erum**	**miserum**	**bellum miserum** = wretched war
		-rum	**pulchri**	**oppidum pulchrum** = pretty town
Genitive	m	**-eri**	**miseri**	**pueri miseri** = of the wretched boy
		-ri	**pulchri**	**libri pulchri**= of the pretty book
	f	**-erae**	**miserae**	**puellae miserae** = of the wretched girl
		-rae	**pulchrae**	**puellae pulchrae** = of the pretty girl
	n	**-eri**	**miseri**	**belli miseri** = of the wretched war
		-ri	**pulchri**	**oppidi pulchri** = of the pretty town

A few adjectives in this group have **-ius** for the genitive singular for all three genders, but other endings as in rest of the group. Those useful in creating names include **unus, una, unum** = one; **alter, altera, alterum** = the one, the other; **solus, sola, solum** = only, alone; **totus, tota, totum** = whole, all.

Continued

Table 8.3 Continued

Nominative	m	**-us**	**solus**	**puer solus** = the only boy
	f	**-a**	**sola**	**puella sola** = the only girl
	n	**-um**	**unum**	**frumentum unum** = one grain
Genitive	m, f, n	**-ius**	**solius**	**pueri solius** = of the only boy; **puellae solius** = of the only girl
		unius	**frumenti**	**unius** = of one grain
III				
Some adjectives in this group have one spelling for the nominative and genitive singular for all three genders (e.g., **audax**).				
Nominative		**audax**		**puer audax** = the bold boy; **puella audax** = the bold girl **iudicium audax** = the bold judgment
Genitive		**audacis**		**pueri audacis** = of the bold boy; **puellae audacis** = of the bold girl **iudicii audacis** = of the bold judgment
Some adjectives in this group have one spelling for the masculine and feminine nominative singular, and another for the neuter nominative singular (e.g., **omnis**).				
Nominative	m, f	**omnis**		**puer omnis** = every boy; **casa omnis** = every house
	n	**omne**		**bellum omne** = every war
Genitive	m, f, n	**omnis**		**pueri omnis** = of every boy; **casae omnis** = of every house **belli omnis** = of every war
Some adjectives in this group have three endings in the nominative singular, one for each gender (e.g., **celer**).				
Nominative	m	**celer**		**puer celer** = the swift boy
	f	**celeris**		**puella celeris** = the swift girl
	n	**celere**		**iudicium celere** = the swift judgment
Genitive	m, f, n	**celeris**		**pueri celeris** = of the swift boy; **puellae celeris** = of the swift girl **iudicii celeris** = of the swift judgment

Sources: Based on Yancey (1946) and Hendricks (1992).

The past participle is the verb form that ends in *-ed* for regular English verbs (and in several other ways for irregular verbs):

The confused *Xxxxxus*

The frozen *Xxxxxus*

Simple descriptive names may also be nouns used *in apposition,* that is, placed beside another noun to identify, explain, or emphasize it. For example, the scientific name of the African Lion, *Felis leo,* is composed of a species noun, **leo,** in apposition to the genus noun, **Felis.** This construction sounds awkward when translated into English, something like "Cat, the Lion-one." Others, if translated, sound like baby-talk or pidgin-English:

Ficus carica: the Fig-fig

Mus rattus: the Mouse-rat

Equus caballus: the Horse-horse

But the construction is perfectly acceptable in Latin, just as it is in many other languages. Those who do not want to worry about matching endings will see the advantage of using this construction; each term can be used just as it is found in the dictionary. No endings need to be changed.

In naming animals (ICZN) the identical word can be used for both genus and species terms, making even sillier sounding (but easy to remember) names. Such names are common in ornithology and malacology and as names for pelagic animals. They seem to be used in two different senses, to show that this is the "real one" or that this is the only such animal in the world:

Buteo buteo: buzzard

Pica pica: black-billed magpie

Oliva oliva: olive snail

Gemma gemma: amethyst gem clam

Velella velella: by-the-wind-sailor

Mola mola: ocean sunfish

These names, called *tautonyms,* are not allowed in botanical nomenclature.

Compound Descriptive Names

Coining a species name by combining two terms makes a more accurate description possible. When two terms are combined they become one word with no hyphenation:

Uvularia grandiflora: bellwort, a U.S. wildflower with a large drooping flower

To compose a compound adjective name, the stem of the first term is joined to the second adjective by a linking vowel (**a, o, i**). Only the ending of the second adjective is affected by the gender of the genus term. For example to name a species "yellowish-red," you could combine the adjective *yellow* (**flavus**) with the adjective *red* (**ruber**), giving you three possible names depending on the gender of the genus name:

Xxxxus flavoruber (m)

Xxxxa flavorubra (f)

Xxxxum flavorubrum (n)

In general, for Latin adjectives, the combining form is found by adding **-i** to the masculine genitive singular base. For Greek adjectives **-o-** is used, except for adjectives ending in **-ys.** Their combining form is the masculine singular without the **-s** (e.g., for **polys** = many, the combining form is **poly-**) (Brown 1956).

Some actual examples of compound species names:

Corbus cryptoleucus: a raven characterized by a hidden white patch on its neck

Macropodus viridiauratus: a green-golden paradise fish

An adverb modifies verbs, adjectives, and other adverbs. A compound descriptive name may be an adverb–adjective combination:

The seven-spined *Xxxxus*

The perfectly round *Xxxxus*

To make things even more complicated (or even more descriptive, if you prefer), present and past participles can be combined with adjectives and adverbs:

The completely frozen seven-spined *Xxxxus*

But that's where you can *really* get into names that are unpronounceable, so if you decide to use a long descriptive name, test it verbally first.

A compound descriptive name can also be an adjective–noun combination. This is probably even more common than adjective—adjective combinations:

Narrow-leaf: *angustifolium*

Long-leg: *longitarsus*

Short-tail: *brevicauda*

Different lobe: *diversiloba*

The noun term can be treated as a noun in apposition (Savory 1962). In that case the ending does not change; for example, **cauda** (tail) is a feminine noun, so even if the genus name is masculine the species name, *short-tail,* would remain *brevicauda,* as in *Xxxxus brevicauda.* But it is also permissible (and seems more common) to make the second noun into an adjective in form and function; for example, *long-leg* becomes *long-legged* (**longitarsus -a -um** or **longitarsis -e**). Or if *short-tail* is made into *short-tailed,* it must then match the genus name in gender:

Ophioderma (a neuter genus name) *brevicaudum*

Two nouns can also be joined together to form a species name. In this case the second noun is treated as an adjective and its ending must match the genus name in gender. They can also be made up of a noun and a verb. In that case the verb ending does not change: From the Latin **arena** (= sand) and **colo** (= I inhabit) comes the name **arenicola** (inhabits sand). Note that in this case the second **a** becomes a linking **-i-**, apparently just because it looks and sounds better (Savory 1962).

Compound descriptive names are probably also best in the sense that they are less likely to lead to a nomenclatural problem if a species later is transferred to a different genus. For example, suppose you named a species *Xxxxus minutus* (having checked the literature first, of course, to make sure that *Xxxxus minutus* was not already being used as a species name in that genus). However, *minutus* is a pretty commonly used descriptive adjective. If someone later revising the group on the basis of new information decides that your species really belongs in the genus *Yyyyus,* a genus that already contains a species *Yyyyus minutus,* there would be suddenly be two different species named *Yyyyus minutus.* The more recently described of the two species would have to be given a new name (see chapter 21).

The only disadvantage to the use of compound descriptive names is that it can be hard to keep them from becoming "complex, long, and difficult to pronounce" (Mayr 1969:354), the kinds of tongue-defeaters that should be avoided.

Geographic Species Names

Geographic names are derived from the habitat or locality in which the organism was first found:

Didelphis virginiana: the Virginia opossum

Lepus californicus: the black-tailed jackrabbit

Lilium canadense: the Canada lily

The advantage to giving a species a geographic name is that it provides information about a species's first known locality. However, giving a geographic name to a species you know little about has a certain risk because the name may become less and less appropriate as more information is gathered, as happened with all three examples above. Although most common in the southeastern U.S., the Virginia opossum has been expanding its range northward and has even been introduced in several areas of the western United States. *Lepus californicus,* the most common North American jackrabbit, is found throughout most of the west, and the Canada lily, *Lilium canadense,* grows not only in southeastern Canada, but in the northeastern United States and on down into Carolina mountain habitats. Once in a while, an even worse mistake can occur with a geographic name; for example, the plant *Simmondsia chinensis* is actually endemic to California, but its author mistakenly thought the type had been collected in China.

Geographic names may be treated as adjectives with the following endings:

-ensis, -iensis (this is preferred by the ICZN)

or a species name may be formed from a locality name by adding an adjectival ending:

-ensis, -(a)nus, inus, or **-icus** (this is preferred by the ICBN)

Or, like personal names, as possessive (in the genitive case) nouns:

sancti pauli
romae

Note that although the ICZN recommends that if the places were given Latin names in classical or medieval times those should be preferred–for example, *viendobonensis* (for Viennese) rather than *viennensis*–there is so little recognition of those names among present-day scientists that the modern name might be the best choice. If you do want to put place names into the Latin form, those versions can be found in Grasse (1971) or Stearn (1992).

Commemorative Species Names

Commemorative names honor a person or event (such as an expedition). Commemorative names were perhaps in greater favor in the nineteenth century, when scientists depended on patrons who could be rewarded by having a species named after them, but they are still constructed today. They may honor the person who first collected specimens of the organism, a person who supported the research project or expedition, or an author's friend or colleague:

Cleigona margarita, a milliped, was named for Margaret Cannon Howell of Atlanta, Ga., and Highlands, N.C., patron of the Highlands Biological Station.

Astarte claytonrayi, a fossil bivalve mollusk, was named "in honor of Dr. Clayton E. Ray of the National Museum of Natural History, a longtime advisor, cohort, and friend" (Ward 1992:86).

These days, it would be considered pretty tacky, in most scientific circles, to name a species after yourself. But species are still named after teachers, celebrities, friends, lovers, spouses, children, and grandchildren, as well as in honor of other workers on the group of organisms involved. If you choose to name your species after a living person, be sure to get his or her permission.

Most people are probably pleased to be thus immortalized, but a professional systematist may not be flattered by the request. As a colleague once told me, "I don't feel very happy to hear that someone wants to name a species after me, because usually that means I saw it, or even published on it, but failed to recognize it as new." For example, in 1873 Swedish bryozoologist F. A. Smitt identified a species in a collection from Florida as *Membranipora irregularis.* R. C. Osburn, a later worker who had spent much of his career studying tropical bryozoan faunas, realized that the species Smitt had been given was not *M. irregularis* (a European species now called *Alderina irregularis*), but a new species, one that Smitt (who did describe a number of new species in his study) had missed. So Osburn (1950) described and named it *Alderina smitti.* The new species name honors Smitt as the person who first saw it, but it also points out the fact that Smitt missed recognizing it as new. However, if a systematist has given you a lot of help with the process of describing your new species but doesn't want to share authorship of the description, a commemorative name would be a very nice way of saying thank you. Just be sure to check with the person you want to honor first.

A species name based on a personal name may be formed from the first name (e.g., *Perophora regina*), the last name (called a patronym, e.g., *Necturus lewisi*) or a combination of first and last names (e.g., *Strigiphilus garylarsoni*).

Species names formed from personal names may be nouns in the Latin genitive (possessive) case (ICZN, Art. 11h):

> *Membranipora annae* (= Anna's *Membranipora,* or the *Membranipora* of Anna)

If a species name constructed as a possessive is formed directly from a modern personal name, the ending depends on gender and number. A name based on a surname (considered a masculine name) ends in **-i**. So a species name honoring Hoffman would become *hoffmani* (one **i**). A surname ending in **-i**, such as Julliani, would be *jullianii* (two **iis**) according to the ICZN.

Botanists favor using Latin terminations for personal names whether genitive or adjectival in form (ICBN Art. 23A.3(a) and Rec. 60C). For genitive endings, names ending in a vowel or **-er** add **-i** (e.g., *scopolii* for Scopoli, *glazioui* for Glaziou), except names ending in **-a,** which add **-e** to become **-ae.** If the personal name ends in a consonant, botanists add an extra **-i-** (called stem augmentation) plus the genitive ending appropriate to the sex of the person, as in *lecardii* for Lecard (a male botanist) and *wilsoniae* for Wilson (a female botanist). For adjectival endings, if the personal name ends in a vowel, adjectival names are made by adding **-an-** plus the gender-appropriate (to the genus term) nominative singular adjectival ending (e.g., *Cyperus heyneanus* for Heyne, *Vanda lindelyana* for

Lindley), except when the ending vowel is an **-a**, in which case **-n-** plus the appropriate ending is added (e.g., *balansanus, -a, -um,* for the name Balansa). If the personal name ends in a consonant, adjectival species names are formed by adding **-i-** (stem augmentation), plus **-an-** (adjectival suffix stem), plus the gender- appropriate (to the genus term) nominative singular adjectival ending (e.g., *Desmodium griffithianum* for Griffith, *Rosa webbiana* for Webb).

The author of the name has a little latitude in choosing the stem to which the Latin ending is attached. Generally the entire name is used, but a name that ends in *e,* such as Eldredge, could yield either *eldredgei* or *eldredgi.*

A name based on a woman's personal name ends in **-ae** For example, Anna = *annae.* Many women's names end in *a* or *e.* If that final vowel gives an awkward sound when made into species name, it can be dropped (e.g., *ernestinae* rather than *ernestineae*).

Commemorative species names constructed in the genitive can also be plural:

of men: **-orum**

of women: **-arum**

Greek personal names such as those from mythology are treated like other Greek words and Latinized. It is also recommended that a first name of classical origin be used in its Latin form.

Prefixes to surnames present another problem. They are run together with the name and lose their punctuation:

Le, la l', el, il, lo, du, de, la, des del, della (e.g., *laclairi*)

Mc or *M'* becomes *mac* (e.g., *macdonnelli*)

O' loses the apostrophe (e.g., *obrieni*)

With the German and Dutch prefixes (*van, von, van der,* and *zur*), composition depends on usage. If the prefix is normally united with the noun then the prefix is retained in the species name

Vanderbilt = *vanderbilti*

von Letkemann = *letkemanni*

Besides possessive species names constructed in the Latin genitive, commemorative species names may be in the nominative singular, as substantives, or nouns in apposition:

Perophora regina

Contyla melinda

Platygobiopsis akihito, a Indonesian goby fish, which was named after Japanese emperor Akihito, a marine biologist with a long-standing interest in gobies (Springer 1992).

A commemorative surname may also be a noun used in apposition, that is, a personal noun placed beside another noun to identify or explain it:

Xxxxxus cuvier

Again, this construction sounds awkward translated into English ("Xxxxxus, the Cuvier-one") but it works in Latin. However, the ICZN recommends against using surnames of well-known biologists such as Cuvier this way because they can be too easily mistaken for the authority citation of the genus name.

Botanists follow the general rule that if the person being commemorated was involved in the discovery of a new species, whether by collecting the type or by drawing it to the author's attention, then the name is treated as a substantive and composed according to the guidelines given in Recommendation 60C 1a and b of the ICBN. If the commemoration is being made for scientific achievement in the area, or to honor donors, then the adjective forms are used, as in Rec. 60C.1c and d, hence *C. noritagii* (for the collector) but *C. noretagiana* to honor the Noretagi's scientific achievements (D. W. Stevenson, personal communication 1997).

Species names based on personal names show how applying the rules of Latin grammar can yield a variety of names even for one stem and one case (possessive or genitive). For example a personal name may be Latinized (e.g., Cuvier becoming *cuvierius*), treated as a Latin word (if it has an ending like one), or treated as modern and non-Latin (even though from its ending it could be Latin). For example, Fabricius yields *fabricii* when treated as a Latin word, but *fabriciusi* when treated as a modern word. However, the ICZN now recommends against treating modern personal names as Latin by adding **-ius** to the name.

In the case of a well-known worker in a field, an increased variety of species names can be created from the person's name by using both first name and surname and both modern and classical forms. For example, Pierre Strinati is a Swiss speleobiologist for whom many new species had already been named ("strinatii"), so when Richard Hoffman wanted to name a cave-dwelling milliped in his honor he decided to call it

Lophodesmus petrinus, Pierre being the French version of Peter, the Latin **Petrus,** which becomes **Petrinus** when used in the genitive case.

One special case is that of species names that stem from the name of an organism on which the new species is an associate or parasite. For example, *Lernaea lusca* is a copepod parasite on *Gadus luscus.* This name is a substantive in the genitive case treated as possessive, the parasite belonging to that species of fish. Other names of parasites are in the genitive plural case. For example, *Syndesmis echinorum* is a flatworm parasite of "spiny things," or echinoderms.

The practical advantage of the commemorative name is that it offers an almost endless variety to choose from. The commemorative name you come up with is more likely to be completely unique, and therefore unlikely to cause nomenclatural problems in the case of subsequent changes in the taxonomic position of the species.

Nonsense Species Names

In groups such as insects in which a large number of species may have to be named in one publication, systematists sometimes give up trying to find descriptive names and make up species names from random combinations of letters. This is acceptable as long as the nonsense word thus produced is pronounceable (has some vowels). For example, *Meringa hinaka* as a name for a New Zealand spider is acceptable (Forster et al. 1990), but a *Meringa hnk* would not be accepted.

Students of reading development have tested the ease with which children and adults remember nonsense words versus meaningful words. They have concluded that nonwords that sound like real words in the speaker's language (because they have vowel and consonant patterns common in the language) are more readily pronounced and remembered (Treiman et al. 1990), a factor to keep in mind if you decide to make up a nonsense name.

Names based on arbitrary combinations of letters are left exactly as they are no matter how they are used grammatically. They are not declined.

The Etymology Section

Each description of a new species should contain an etymology section, giving the meaning and origin of the species name. The ICZN makes no provisions about this, but many journals require it. For example, according to the *American Museum of Natural History Style Manual* (1998:10), "Authors publishing new genus group and species group names are required to include pertinent etymologies."

Even if it is not required by the journal in which you publish your description, it is good practice to include this section because it tells others why and how you composed the word, a step that may eliminate confusion if the species name must later be corrected because its generic placement has been changed. For example, a species term ending in **-fer** or **-ger** could be either a masculine adjective or a noun in apposition. If the species is placed in a new genus that is feminine in Latin, the adjective would change to the feminine form to agree with the noun, but the noun in apposition would not. This has happened often enough that the ICZN has ruled that if the author did not make the word's construction clear it should be treated as a noun in apposition.

The following examples illustrate etymology sections from new species descriptions. They may be detailed and literary, like this etymology for *Conus kahiko:*

> *Etymology:* The species name *kahiko* derives from the rich and poetic Hawaiian language. Primarily, *Kahiko* means ancient, the remote past, or an ancient person. *Kahiko* was the name of the first native Hawaiian man. When the initial syllable is stressed and long, *kahiko* means ornamented or dressed in fine clothes (see Malo, 1951; Pukui and Elbert, 1971). The word thus appropriately describes the only extinct, endemic Hawaiian species of *Conus,* a shell with a richly ornamented color pattern. (Kohn 1980:537)

Or brief and utilitarian, as in this entry for a new gastropod, *Perotrochus charlestonensis:*

> **Etymology:** Named after the type locality, commonly referred to as the Charleston Lumps. (Askew 1988:89)

Authors may explain how the name chosen illustrates some characteristic of the organism, as with this new species of frog from New Guinea, *Cophixalus verecundus*:

> ETYMOLOGY: The Latin *verecundus* (shy) refers to the males' habit of calling softly from concealment. (Zweifel and Parker 1989:16)

Or, with this species of species of sipunculan from Florida, *Phascolion psammophilum*:

> *Etymology:* The name *psammophilum* refers to the sand habitat in which this species is found. (Gr. psammos, sand; Gr. philos, having affinity for.) (Rice 1993:599)

Or, this southwestern Virginia milliped, *Rhysodesmus restans:*

> NAME: Derived from the Latin stem *resto-* with the implied meaning "one who remains behind." (Hoffman 1998:78)

used for the only species of the genus *Rhysodesmus* remaining in what is considered to be its biogeographic area of origin.

The author may use this section to give the reason for selecting a commemorative name, as for *Sphaerolana karenae,* a new isopod crustacean from Mexico:

> *Etymology:* Named for the first author's daughter Ana Karen. (Rodríguez-Almaraz and Bowman 1995:209)

Or for a new parasitic bee, *Triepeolus loomisorum:*

> ETYMOLOGY: I take pleasure in naming this species for Mr. and Mrs. Alfred Lee Loomis, Jr., in recognition of their interest in and concern for entomological field research and the American Museum of Natural History. (Rozen 1989:17)

Finally, as mentioned earlier it offers a chance to explain how the name was constructed grammatically, as with this new species of spider from Copiapó, Chile, *Segestrioides copiapo:*

> ETYMOLOGY: The specific name is a noun in apposition taken from the type locality. (Platnick 1989:6)

Sources

Sources for Constructing Scientific Names

Blinderman, C. 1990. *Biolexicon: A Guide to the Language of Biology.* Springfield, Ill., Thomas.

Borror, D. J. 1960. *Dictionary of Word Roots and Combining Forms Compiled from the Greek, Latin, and Other Languages, with Special Reference to Biological Terms and Scientific Names.* Mountain View, Calif., Mayfield.

Brown, R. W. 1956. *Composition of Scientific Words,* rev. ed. Washington, D.C., Smithsonian Institution Press.

Jaeger, E. C. 1955. *A Source-Book of Biological Names and Terms,* 3d ed. Springfield, Ill., Charles C. Thomas.

Savory, T. 1962. *Naming the Living World: An Introduction to the Principles of Biological Nomenclature.* New York, Wiley.

Stearn, W. T. 1992. *Botanical Latin: History, Grammar, Syntax, Terminology, and Vocabulary,* 4th rev. ed. Portland, Oreg., Timber Press.

Yancey, P. H. 1946. *Introduction to Biological Latin and Greek,* 3d ed., rev. Mt. Vernon, Iowa, F.G. Brooks.

Latin Dictionaries and Grammar Books

Collins Latin Dictionary plus Grammar. 1997. London, HarperCollins.

Gildersleeve, B. L. 1963. *Gildersleeve's Latin Grammar,* 3d ed., rev. and enlarged by B. L. Gildersleeve and G. Lodge. London, Macmillan.

Hendricks, R. A. 1992. *Latin Made Simple.* Rev. by L. Padol. New York, Doubleday.

Humez, A. 1976. *Latin for People–Latina pro Populo.* Boston, Little, Brown.

Kennedy, B. H. 1958. *The Shorter Latin Primer,* new ed., rev. by J. W. Bartram. London, Longmans, Green.

Simpson, D. P. 1968. *Cassell's Latin Dictionary,* 5th ed. New York, Macmillan.

Wheelock, F. M. 1995. *Wheelock's Latin,* 5th ed., R. A. LaFleur, revision ed. New York, Harper Perennial.

Other Etymological Reading

Bailey, R. M. 1988. Changes in North American fish names, especially as related to the International Code of Zoological Nomenclature, 1985. *Bulletin of Zoological Nomenclature* 45: 92–103.

Isely, D. 1972. The disappearance. *Taxon* 21: 3–12.

McClintock, D. 1969. *A Guide to the Naming of Plants, with Special Reference to Heathers.* n.p., The Heather Society.

McDowall, R. M. 1991. The "ii's" may have it at the end: Patronyms should be amended only if demonstrably incorrect. *New Zealand Natural Sciences* 18: 25–29.

Wells, D. 1997. *100 Flowers and How They Got Their Names.* Chapel Hill, N.C., Algonquin Books.

Chapter 9

Type and Voucher Material

Holotypes, syntypes, lectotypes, and neotypes are the bearers of the scientific names of all animal taxa. They are the international standards of reference that provide objectivity in zoological nomenclature. They are held in trust for science by all zoologists and by persons responsible for their safe keeping.

–ICZN, ART. 72(g)

The purpose of typification is to fix permanently the application of names of all ranks governed by the Code so as to preclude the possibility of the same name being used in different senses; i.e., for different plants. –HAWKSWORTH (1974:127)

Rationale for Types and Vouchers

An original species description should include a section on type material and its deposition. The different kinds of type material you might encounter in museum research were covered in chapter 5, but the basic concepts are reviewed briefly here.

The type is the name-bearer, the specimen associated with a name by description and publication. Its use provides an objective base for our system of biological nomenclature. The type as a nomenclatural object has nothing to do with the common idea of "typical" as expressing some kind of average, or with the pre-Darwinian typological species concept. A type specimen may or may not be typical of the species; the important point is that it provides a fixed reference for the use of the name (Jeffrey 1989). Most ecologists, environmental biologists, molecular biologists, and natural products chemists are familiar with voucher specimens. In order to maximize the potential value of a collection made for ecological and pharmacological reasons, it is important to save a specimen or some part of the material of each taxon collected and prepare it as a *reference sample,* or *voucher.* This step is particularly important when the rest of the sample is being processed destructively (e.g., for chemical or molecular genetic analysis). This voucher specimen has two purposes: to enable the taxon to be identified (or to have its preliminary identification verified by a systematist) and to enable the researchers (or later workers) to collect the same taxon again if necessary (Pomponi 1988).

In a sense, the type is a super voucher specimen, providing a fixed reference to the named taxon for all time.

Holotype

The *holotype* is a single specimen chosen as the nomenclature type. The Botanical Code allows an illustration of a specimen as the type element: "A holotype . . . is the one specimen or illustration used by the author, or designated by the author as the nomenclatural type" (ICBN, Article 9.1).

Type Series

The type series includes all specimens on which a description of a new species is based, except those that have been purposely and specifically excluded from type status by the author of the description (e.g., by the author calling them variants or questionable in the description). *Syntypes* are two or more specimens included in the type series.

A *lectotype* is a single specimen selected from syntypes of a previously described species to serve as the equivalent of the holotype.

Paratypes

Paratypes are additional specimens selected from the material the author had on hand at the time at the time of writing the description. Usually they are distributed to other institutions to serve as vouchers for that taxon in their collections.

Isotypes

In botany an isotype is a duplicate of the holotype, part of a single collection of the species. The difference between an isotype and a syntype is that an isotype is really a *duplicate,* not just another specimen in a type series. In practice, because a single genetic individual of a plant species might yield several dried specimens, isotypes may not be as likely to cause later problems than zoological syntypes or type series. In botany there may also be *isolectotypes.*

Rules of Nomenclature Regarding Types

Along with publication and priority, *typification,* the designation of a type of a scientific name, is one of the chief principles of the international codes of nomenclature (Jeffrey 1989). Both current codes for plants and animals have sections with rules concerning types. The ICBN requires that a type be indicated in order for a new plant species name to be accepted as validly published. In fact, not only must a type be indicated, but the institution in which it is deposited must be specified in some way (ICBN, Article 37). In zoology, the designation of a type specimen was only a recommended, not required part of the species description up through the third edition of the ICZN: "Designation of holotype.–An author who establishes a new nominal species group taxon should clearly designate its holotype" (ICZN 1985, Recommendation 73 A). The fourth edition of the ICZN finally does make the fixation of a name-bearing type and its deposition in an institutional collection mandatory. All descriptions of new species published after 1999 must include fixation of a name-bearing type and must state the institution in which it has been deposited.

Even before the fourth edition, however, the designation of a type was a requirement of responsible taxonomic practice in zoology as well as botany. Reviewers and editors of refereed journals almost always required its inclusion in any description of a new animal. For example, the "Instructions for Contributors" of the Biological Society of Washington stated, "When appropriate accounts of new taxa must cite a type specimen deposited in an institutional collection" (inside back cover of journal, v. 109(3), 1996).

Selection of Types and Vouchers

Although the botanical and zoological codes deal in detail with the kinds of nomenclatural problems that can arise when the author of a species fails to designate a type, they give few guidelines for type selection. The best information comes from practicing taxonomists and the books and papers they have published (e.g., Mayr et al. 1953; Mayr 1969; Lee et al. 1978; Pomponi 1988).

The type specimen should be selected before the description is published. The description should be written with the type specimen clearly in mind, even though you are describing characteristics of the population as a whole, not just of that one specimen. Not only is that specimen going to be the reference standard against which other specimens will be checked, but later workers who read the description and look at the type will judge how good your original description was on that basis (Blackwelder 1967). Once a type is selected and the description published, that type cannot be changed, even by the author, without going through a good deal of red tape (see chapter 21).

If an illustration is used as a botanic type, botanists consider the *illustration* to be the type of that name. However, use of an illustration as a type is rare in modern botany. Although article 8.3 of the ICBN allows an illustration to be used "if it is impossible to preserve a specimen," illustrations are generally used only to deal with historic names, names of species that were described before the type concept was in use, and in cases where later workers can find no specimen that the author reliably saw.

Although the ICZN allows illustrations to be designated as the type, the zoological type is still considered to be the *specimen* illustrated, not the picture or photograph. It is very rare in modern zoology to see an illustration cited as the type. But it does happen, and it may become more common for very rare or threatened species (e.g., birds) when the researcher does not want to harm even one of the remaining population of the species. In these cases DNA (from a blood sample) may be taken and sequenced (Quicke 1993 and references therein) to create a genetic record.

In surveys and systematic work, the new species are often the rare ones. Many times there is no problem deciding which specimen should be made the type because you only have one specimen of the taxon. (Murphy's Law of Types dictates that when this is the case, you will probably have done something to that specimen–coated it for scanning electron microscopy, dissected it, etc.–that you would not have done had you realized it was going to be the only specimen of that taxon found.) Ecologists and paleoecologists are less likely than museum systematists to encounter this problem because they can't carry out ecological studies on species that rare in an environment.

If you have more than one specimen available, there are some guidelines for type selection. Select a specimen that is in good condition, has been well preserved, and clearly shows as many diagnostic characters (those distinguishing it from other taxa) as possible. In other words, use common sense and pick a specimen that is representative of the species, not a bizarre variant. Although a single specimen can't represent the whole range of variation of a species, a representative specimen is obviously going to be more useful as a standard of comparison than an extreme or deformed specimen. If your description is not a very good one, a later worker may need to examine your type specimen and use it to improve or add to the description. Even if your description is the best that it was possible to write at the time it is published, it still might not be sufficient to distinguish your species from a new relative discovered at some future time, without restudy of your type specimen (Mayr 1969; Blackwelder 1967, Quicke 1993).

When you are dealing with fragmentary specimens, as with vertebrate fossil material, or with dredged specimens of soft-bodied or colonial invertebrates, and any time there is any doubt about whether all the pieces belonged to one individual, try to pick the most diagnostic piece to be the type (Mayr 1969).

Type Locality

It has also been suggested that, when possible, you should choose a specimen from a place where there is a fairly large population of the species. Specimens from a large population are most likely to show the range of variation possible within the species. The type locality should also be a place for which you have clear and accurate locality data. Both of these guidelines will make it easier for later workers to collect *topotypic material* or specimens from the same locality as the type (unless the site has been paved over or plowed under) (Mayr et al. 1953; Blackwelder 1967).

Composition of Type Material

For plants, the type can be a dried or otherwise preserved specimen. For animals, type material can be a preserved specimen of the entire animal or its parts, a colony or part of a colony. There are a number of useful books on methods of collecting and preserving plants and animals (see this chapter's *Sources*). These references explain how best to collect and preserve specimens of different groups of organisms. Examining taxonomic papers in which previous new species of that particular group have been described will tell you what is usually considered adequate material for type purposes.

For fossils, the type can be the fossilized organism or its parts or a natural impression, mold, or cast of an animal or colony. Artificial molds or copies cannot be used. Scientific names and type specimens can no longer be created for the work of an animal (an *trace fossil* or *ichnofossil*) according to the current codes, although names established before the rules were changed (1931) are still good (ICZN, Art. 72 C i, p. 143).

For protozoans, if the species can't be recognized by a single individual, the type can consist of a number of individuals mounted in slide preparations, or by a series of preparations showing different stages of the life cycle. This is called a *hapantotype* (ICZN, Art. 72 C iv, p. 143).

Documentation of Type Material

The following information should be included somewhere in the published description, as well as on the label of any type specimen. Ecologists should be familiar with these requirements if they have dealt with voucher specimens because it is similar to the information required for voucher specimens (see Pomponi 1988). Of course, having the information available to begin with depends on good field and good collecting techniques and records, including field notes, photographs, or even videotapes (see chapter 13).

Exact collecting locality, including city, county, map coordinates, latitude and longitude, GPS data, loran coordinates, height above sea level, depth below it, substratum type, habitat, collecting gear used, and so on. For fossil specimens information on the geological horizon should be included.

Date of collection. There are different systems of writing out dates. The important thing is to use a method that is clear and consistent throughout your work.

Name of collector(s), written out in full, not abbreviated.

Sex and relevant life history information, such as developmental stage or form (if other than adult), host species in the case of parasites.

Name of institution where the specimen has been deposited and the *catalog number* assigned to the specimen.

Deposition of Types

The quotation from the ICZN that began this chapter emphasized the importance of types as international standards of reference, held in trust for safe keeping. This is just as true for plants as it is for animals. This is why both codes now specify that the name of the herbarium, museum, or other institution in which the type is conserved must be given for any new species, clearly indicating the importance attached to such institutional safe keeping. Letting a type specimen languish in your personal collection is the biggest mistake. If you die before you can transfer it to an institutional collection it easily just be dumped in the trash by someone eager to acquire the space you have vacated.

The type specimen or specimens should be deposited in a reputable scientific institution. An accredited museum or herbarium is preferable to a university departmental collection unless the university has expressed long-term commitment to collections. Many university departments and research laboratories that used to keep collections have gotten rid of them. Some years ago, personnel at a well-known university literally defenestrated its natural history collections, tossing them out the second-story windows of the building in which they had been housed (although administrators had second thoughts and the collections were not discarded, but put in storage).

Although natural history museums and herbaria are perpetually short of space, they will probably be willing to accept your material if it is properly documented. You should call or write the collection management or curatorial staff at the institution or its appropriate department well ahead of time to request permission. At the very latest, this should be done when you submit your paper for publication. Institutions most likely to accept your specimens are those that have regional collections for the area in which your species occurs (e.g., some state museums are even mandated by state law to accept specimens from within the state), major research museums or herbaria housing collections that are international in scope, and smaller museums or herbaria that have a good collection or an active worker on that particular group. By the time you have done your research and written your description, you will know an appropriate institution to contact.

More and more often, museum staff are having to consider the cost of taking in and housing your specimen before deciding whether to accept it. It has been calculated that the average cost of processing a specimen (unpacking, accessioning, cataloging, storing, and updating the institutional database to include it) is about $5.00 per specimen for an insect and as much as $10–15 for a mammal (P. Cato, personal communication 1997). Of course, this is also true of library collections. If you have ever lost a library book you may have found to your sorrow that the library charged you not only the replacement cost of the book, but also a processing fee, which could be higher than the cost of the book. The cost of processing and storing an item (in terms of staff time, materials, and overhead) is a very real cost that those institutions must meet in order to stay in business. Obviously, if you hadn't planned on finding a new species during your project, there isn't much you can do except to make whatever financial contribution you can afford to help defray the costs. However, any new research proposal for which you anticipate collecting voucher and other specimens to be deposited in an institutional collections should include processing and storage fees for the receiving institution in the budget.

Paratypes

You must designate one specimen to serve as the holotype of the name. When you have a fair number of specimens available, it is a good idea to make one of them the holotype and to pick out good specimens from the rest of the material to serve as paratypes. Often paratypes are distributed to several geographically distant institutions so that they can be examined more conveniently by researchers who live far from the institution where the type was deposited. Although they are not considered nomenclatural types, paratypes are given special care and labeling in many collections, as are types, although restrictions on borrowing them may be less stringent than those on types. You may contact the collection personnel at additional museums yourself, or discuss possibilities with the collection staff at the institution where you are depositing the holotype.

Voucher Material

It is good practice to deposit voucher specimens of any organism or organisms you have studied and published research papers on into an institutional collection, even if they are not new species. This may be unnecessary in the case of well-studied vertebrates, but for most invertebrates and plants it can make all the difference between your published work being useful to others or just wasted effort. This is especially true if you did not find a systematist to verify your identifications. If you have deposited a voucher, even if you did make a mistake, it can later be rectified by study of the voucher specimen. For example, certain bryozoan species are so distinctive that they can be easily identified by anyone who has the right literature at hand. Unfortunately, in other common genera and families species can be very difficult to sort out, even for specialists. A paper was recently published on the toxicity of certain pollutants to one species of this latter group. Unfortunately, the chemist and ecologist who carried out the work did not consult a specialist to have their identification verified. The name they gave the species in their publication was that of a species that does not even occur in the geographic area in which the work was carried out. Three related species are found in the region, and any or all of them could be expected to occur at their study site. Unfortunately, the researchers kept no vouchers, so their work is completely useless. It does not apply to the species in their title, and there is no way to find out to what species it does apply. Their samples may have included a mixture of species, an even worse scenario.

By the time you have selected a type, you will have learned how to fix, preserve, dry, press, or otherwise properly prepare that specimen according to the standards for that group. This chapter's *Sources* list includes references on collection, preservation, and processing for identification of animals, plants, and fossil material. Remember that many field guides, keys (such as the *Pictured Key Nature Series*), and taxonomic works also include this information for the group they cover.

If you can't deliver the specimens to their new home in person, pack them well and send them by the safest means possible. It is important to use a strong container for transport. Use sturdy, heat-sealed plastic bags or jars with nonleaking lids, wrap specimens well in bubble-wrap, and secure them in sturdy boxes, well padded with foam or other packing material. The goal is to immobilize the specimen during shipping and prevent it from being damaged or broken in transit. If the specimen is in a jar, vial, or other container, the label should be inside, not outside. Only pencil or indelible India ink (never ball-point pen) should be used on labels. Photographs or slides of living specimens (invaluable for material that fades and loses its natural color when preserved) should be sent with the specimens. Copies of field notes and collecting permits should be sent, too.

Old guides to taxonomic procedure used to recommend against sending types to a museum or herbarium until the paper on the new taxon was published, on the grounds that once the specimens have been deposited in the institution, other workers would have access to them and might beat you to the draw on publication. Of course, doing so would certainly be a violation of taxonomic (and general scientific) ethics. Also, most invertebrate groups have so few people currently working on them that the idea of some underemployed systematist lurking around a museum, hoping to scoop a competitor on a new species, seems ridiculous. In my experience, collections personnel who receive a type specimen of a new species treat it with care and full regard for its special value. They would never allow another scientist to make use of it before the paper describing it was published. Scientific staff at museums and herbaria that list their collections online have developed policies to make sure that a new taxon is not listed until the name and description are officially published.

Some institutions or institutional departments still assign catalog numbers to your type specimens, or other material on which you are publishing, before actually receiving the specimens. Usually, this is done when you call to tell them the paper is in press, or that you have proofs in hand. Other institutions, having been burned too many times by giving numbers to specimens that they never receive (because the researcher never bothered to send them in), now insist on having the specimens in their possession before they will assign them catalog numbers. So you must either send the material to them before you need the number or put only the name of the institution in your paper.

Once the specimens have been received, the museum will assign catalog numbers to any type or other material you are depositing and you will be able to add them to your paper when you correct the proofs. Once deposited, type material is specially marked or labeled and is often segregated from the reference collections.

■ Type Section

In an original description you may record information on the type material in a special section. For example, in this description of a new poison frog, the type section, with information on holotype and paratypes, immediately follows the heading with the species name, and the numbers of the figures illustrating the new species:

Dendrobates castaneoticus, new species

Figures 1–4, 5A, 6A, 8A

HOLOTYPE: MZUZP 64775 (field no. JPC 7115), an adult female caught by Laurie J. Vitt on February 24, 1987, in primary lowland rain forest near Cachoeira Juruá, Rio Xingu, State of Pará, Brazil. The type locality is approximately 03°22'S, 51°51'W on the international millionth map (sheet SA-22); the site is situated within a loop of the Rio Xingu, about 220 km S of its junction with the Rio Amazonas.

PARATYPES: A total of 18 specimens, all from northern Pará, as follows: MZUSP 30767, 35681, from Taperinha (02°32'S, 54°17'W), about 48 km ESE Santarém, collected by H. Reichardt., February 1–11, 1968, and by P. E. Vanzolini, October 22, 1970, respectively. MZUSP 64763–64771, 64777, AMNH 133451–133455, from the type locality, and MZUSP 64776 from 13 km NW of the type locality; preceding specimens collected January 18–March 5, 1987, by Janalee P. Caldwell, Gaurino Rinaldi Colli, Jeffrey M. Howland, Pamela T. Lopez, João Silva de Oliveira, and Laurie J. Vitt. (Caldwell and Myers 1990:3)

Note that when the names of institutional collections are abbreviated, as they are here, the full names of the institutions are given somewhere else in the paper.

You may also include the type information as part of your section on material examined, as shown here:

Eudendrium bermudense, sp. nov.

Figs. 27–29

Material Examined

Holotype: Sailor's Choice Cave, Hamilton Parish, on ledge at entrance, −1.5 m, 30 June 1983, one colony, 2.0 cm high, with female gonophores, ROMIZ B333. Paratypes: Sailor's Choice Cave, Hamilton Parish, on ledge and survey line at entrance, −1 to −2 m, 20 June 1983, two colonies, 3.2 and 3.8 cm high, with male gonophores; one colony 2.0 cm high, with female gonophores; ROMIZ B 334. . . . (Calder 1988:39)

Botanists generally list the type information after the diagnosis and description sections. Here is an example for a new red algal species, *Rhodogorgon carriebowensis:*

> *Type*–Carrie Bow Cay, Belizean Barrier Reef, Belize, spur and groove zone, 4.6–12.2 m depth, 1 May 1979, K. E. Bucher, JN-7520 (holotype: Alg. Coll. #US-098360).
> *Distribution*–Caribbean Sea: Bahamas, St. Croix, Belize, Martinique, Panama.
> *Paratypes*–Caribbean Sea: BAHAMAS.–Chub Cay, 4 m depth, among corals, 12 Jun 1989, R. Sims, s. n. (US). U.S. VIRGIN ISLANDS).–St. Croix: Boiler Bay, growing on fore-pavement in front of boiler, 3–4 m depth, 19 Aug 1978, W. Adey, s. n. (US).; Tague Bay, patch reef, 2.4 m depth, 6 Jan 1973, C. Bowman, IAA-11592 (MELU), 5 Apr 1973, P. Adey, IAA-11447a & b (BISH, US), and patch reef, 25 Jan 1974, R. Burke & R. Steneck, IAA-11783 (MICH, US). . . . (Norris and Bucher 1989:1057)

Some journals have a fixed style for such material; others are more flexible. Before writing this section you should examine several articles in recent issues of the journal to which you are submitting the description, and decide which style to follow.

You may also use a type section in a revisionary study, when you are redescribing a known species and want to note where the type is located and give information concerning the type.

This example is from "Revision of *Zygothrica* (Diptera: Drosophilidae), Part II. The first African species, two new Indo-Pacific groups, and the *bilineata* and *samoaensis* species groups" by David Grimaldi (1990):

> LECTOTYPE: Female, with only the following labels on the specimen: "Pres. [presented] by P. A. Buxton/BM 1934–97." Specimen examined, but genitalia not dissected. In the BM(NH). Malloch's description includes the following type locality in the collection data: **Samoa:** Upolu, Malololelei, 2000', 30/XI/24, Buxton and Hopkins. Measurements: HD = 0.77; ED = 0.63; PL = 0.73; ThL = 1.33; WL = 2.39; C.I. = 3.28; 4-V = 1.77. (Grimaldi 1990:9)

Giving the type location and the institutional collection holding the type material can be useful to others even if you were not able to see that specimen yourself:

> HOLOTYPE: **Fiji:** Levuka, Draiba, VII/29/73, H. Takada. In the Biological Laboratory, Sapporo University (BLSU), Hokkaido, Japan. Type not examined. (Grimaldi 1990:17)

As these examples show, not all type specimens have special catalog numbers. Also, almost every institution has a different system for cataloging specimens, including their type specimens. Most institutions separate types from the regular collection in some way. But particularly with older material, this information may not have been marked on the specimen, so some types may not have been separated from the regular collection because they weren't recognized to be types. In those cases, it is especially important to include in your paper any additional information that might help someone else locate the specimen in the institutional collection.

Sources

General

ATTC Guide to Packaging and Shipping of Biological Materials. 1994. Rockville, Md., American Type Culture Collection.

Brown, P. A. 1997. *A Review of Techniques Used in the Preparation, Curation and Conservation of Microscope Slides at the Natural History Museum, London.* n.p., Biology Curator's Group.

Goldblatt, P., P. C. Hoch, and L. M. McCook. 1992. Documenting scientific data: The need for voucher specimens. *Annals of the Missouri Botanical Garden* 79: 969–970.

Hangay, G. 1985. *Biological Museum Methods. Volume 2. Plants, Invertebrates and Techniques.* New York, Academic Press.

Knudsen, J. W. 1966. *Biological Techniques: Collecting, Preserving, and Illustrating Plants and Animals.* New York, Harper & Row.

Simpkins, J. 1974. *Techniques of Biological Preparation.* Glasgow, Blackie.

SPNHC Leaflets. 1996–. (Various aspects of preservation and management of natural history collections.)

Valdecasas, A. G., J. M. Becerra, and D. M. Marshall. 1997. Extending the availability of microscopic type material for taxonomy and research. *Trends in Ecology and Evolution* 12: 211–212. (Digital enhancement of microscopic images.)

Wagstaffe, R., ed. 1968. *The Preservation of Natural History Specimens. Zoology: Vertebrates. Botany. Geology.* New York, Philosophical Library.

Animals

American Society of Ichthyologists and Herpetologists 1987. *Guidelines for the Use of Live Amphibians and Reptiles in Field Research.* Lawrence, Kansas, American

Society of Ichthyologists and Herpetologists, Herpetologist's League, and the Society for the Study of Amphibians and Reptiles.

American Society of Mammalogists 1987. Acceptable field methods in mammalogy: preliminary guidelines approved by the American Society of Mammalogists. *Journal of Mammalogy* v. 68, no. 4, Supplement: 1–18.

Anderson, R. M. 1977. *Methods of Collecting and Preserving Vertebrate Animals,* 4th ed. rev. Ottawa, Department of the Secretary of State (*National Museum of Canada Bulletin* 69).

Blake, E. R. 1949. *Preserving Birds for Study.* Chicago, Chicago Natural History Museum.

Bub, Hans 1991. *Bird Trapping and Bird Banding, a Handbook for Trapping Methods All Over the World.* Ithaca, New York, Cornell University Press.

Chapin, J. P. 1946. *The Preparation of Birds for Study.* New York, American Museum of Natural History.

Coad, B. 1995. *Fishes,* 2d ed. London, Expedition Advisory Centre.

Crossman, E. J. 1980 The role of reference collections in the biomonitoring of fishes, pp. 357-378 in C. H. Hocutt and J. R. Stauffer, Jr., eds. *Biological Monitoring of Fishes.* Lexington, England, Lexington Books.

Dindal, D. L. 1990. *Soil Biology Guide.* New York, Wiley. (Includes information on collecting and preserving soil-dwelling organisms.)

Foster, M. S. and P. F. Cannell 1990. Bird specimens and documentation: critical data for a critical resource. *Condor* 92: 277–283.

Genoways, H. H., C. Jones, and O. L. Rossino, eds. 1987. *Mammal Collection Management.* Lubbock, Texas Tech University Press.

Haedrich, R. L. 1983. Reference collections from faunal surveys., pp. 275-282 In L. A. Nielson, D. L. Johnson and S. S. Lampton, eds. *Fisheries Techniques.* Bethesda, Maryland, American Fisheries Society.

Hall, G. R. 1962. *Collecting and Preparing Study Specimens of Vertebrates.* Lawrence, Kansas, University of Kansas Museum of Natural History.

Higgins, R. P. and H. Thiel, eds. 1988. *Introduction to the Study of Meiofauna.* Washington, D.C., Smithsonian Institution Press.

Measuring and Monitoring Biological Diversity. Standard Methods for Mammals. 1996. Washington, D.C., Smithsonian Institution Press.

Notes on Methods for the Narcotization, Killing, Fixation, and Preservation of Marine Organisms. Compiled by H. D. Russell. Woods Hole, Mass., Systematics Ecology Program.

Peden, A. E. 1974. *Collecting and Preserving Fishes.* Victoria, British Columbia Provincial Museum.

Peterson, A. 1970. *Entomological Field Techniques: How to Work with Insects,* 10th ed. Ann Arbor, Mich., Edwards Bros.

Pisani, G. R. 1973. *A Guide to Preservation Techniques for Amphibians and Reptiles.* Lawrence, Kansas, University of Kansas Museum of Natural History.

Rose, C. and A. deTorres, eds. 1995. *Storage of Natural History Collections,* 2 vols. Washington, D.C., Society for the Preservation of Natural History Collections.

Rudloe, J. 1984. *The Erotic Ocean: A Handbook for Beachcombers and Marine Naturalists.* New York, E.P. Dutton. (Includes methods of collecting, narcotizing, and preserving different marine invertebrate groups.)

Simmons, J. E. 1987. *Herpetological Collecting and Collections Management.* Oxford, Ohio, Society for the Study of Amphibians and Reptiles.

Stoddart, D. R. and R. E. Johannes, eds. 1978. *Coral Reefs: Research Methods.* Paris, UNESCO. (Includes methods for sampling and processing coral reef animals for identification.)

Thorp, J. H. and A. P. Covich. 1991. *Ecology and Classification of North American Freshwater Invertebrates.* New York, Academic Press. (Includes group-by-group information on collecting, raising, preserving, and preparing for identification.)

Williams, S. L., R. Laubach, and C. M. Laubach. 1979. *A Guide to the Literature Concerning the Management of Recent Mammal Collections.* Lubbock, Texas Tech Museum Science Program, pub. no. 5. (Bibliography on collecting and preserving mammals, organized by mammal group.)

Plants

Bridson, D. M. and L. Forman. 1992. *The Herbarium Handbook,* rev. ed. Kew, U.K., Royal Botanic Gardens. (Includes section on what material to collect for herbarium specimens of different families of plants, guides to preparation of specimens, and bibliographies.)

Carroll, L. E. 1975. *Herbarium Procedures for Collections of Fungi, Lichens and Mosses.* Lubbock, Texas Tech University, Museum Science Dept.

Ferren, W. R., D. L. Magney, and T. A. Sholars. 1995. The future of California floristics and sytematics: Collecting guidelines and documentation techniques. *Madroño* 42: 197–210. (Recommendations for making voucher collections of plants.)

Hawksworth, D. L. 1974. *Mycologist's Handbook. An Introduction to the Principles of Taxonomy and Nomenclature in the Fungi and Lichens.* Kew, U.K., Commonwealth Mycological Institute.

Hendricksen, L. 1975. *Herbarium Procedures: Vascular Plants.* Lubbock, Texas Tech University, Museum Science Dept.

Stuessy, T. F. and S. H. Sohmer. 1996. *Sampling the Green World: Innovative Concepts of Collection, Preservation, and Storage of Plant Diversity.* New York, Columbia University Press. (Contributions include methods for anatomical, ultrastructural, tissue and DNA sampling, as well as collecting for voucher material.)

Vogel, E. F. 1987. *Manual of Herbarium Taxonomy, Theory and Practice.* Jakarta, UNESCO.

Fossils

Collins, C. , ed. 1995. *The Care and Conservation of Palaeontological Material.* Oxford, U.K., Butterworth-Heinemann.

Converse, H. H. 1984. *Handbook of Paleo-Preparation Techniques.* Gainesville, Florida State Museum.

DiMichele, W. A. and S. L. Wing. 1988. *Methods and Applications of Plant Paleoecology.* The Paleontological Society and the University of Tennessee, Special Publication no. 3.

Feldmann, R. M., R. E. Chapman, and J. T. Hannibal. 1989. *Paleotechniques.* The Paleontological Society Special Publication no. 4.

Kummel, B. and D. Raup, eds. 1965. *Handbook of Paleontological Techniques.* San Francisco, Freeman.

Leiggi, P. and P. May, eds. 1994. *Vertebrate Paleontological Techniques,* vol. 1. Cambridge, U.K., Cambridge University Press.

Rixon, A. E. 1976. *Fossil Animal Remains: Their Preparation and Conservation.* London, Athlone Press.

Chapter 10

Diagnosis

"Diagnosis" is ultimately from Greek διαγιγνώσκω, *"distinguish between [two] things," and "definition" from Latin* definio, *"enclose within limits." The concepts are importantly different, but in fact taxonomists seldom use the terms consistently and commonly assume that they are synonymous.*

–Simpson (1961:138)

What Is a Diagnosis?

In the chapter in which he gave these definitions, George Gaylord Simpson was actually discussing the difference between "splitters" and "lumpers" among taxonomists, between those who look primarily for dissimilarities and ways to tell taxa apart and those who look for shared similarities that limit collective boundaries. There are still splitters and lumpers among taxonomists, but it is not quite fair to claim that neither group can tell diagnosis from description. However, when it comes to writing taxonomic papers, it does seem to be true that most biologists, including some taxonomists, are still confused about the difference between a *description* and a *diagnosis* and when asked to write a diagnosis just write a shorter version of their description.

But diagnosis and description are not the same, although diagnosis may be a part of description. Whereas the number of taxonomic characters that could be mentioned in a description may indeed be limited "only by the patience of the investigator" (Mayr et al. 1953:106), a *diagnosis* includes just the characters that *distinguish* that species from others, in the same way that the disease identification you receive when you visit the doctor is called the *diagnosis* because the doctor has *distinguished* your illness from all other possibilities on the basis of your symptoms and tests.

You can distinguish a taxon from similar or related species by *comparison,* by *contrast,* or both. Comparison uses ways in which the species differ in shared character states; for example, "Species 1 has avicularium length less than one-third of autozooid length; Species 2 has avicularium length greater than one-half of autozooid length." Contrast shows ways in which they differ completely (e.g., in the presence or absence of a character or character state). For example, "Species 1 has avicularia; Species 2 does not have avicularia."

A *differential diagnosis* specifically shows the ways in which the taxon differs from other taxa that you identify by name (Mayr et al. 1953; Wiley 1981). In the example that follows, Daphne Fautin gives a differential diagnosis for a new species of shell-encrusting anemone, pointing out the ways in which the new species, *Stylobates loisetteae,* differs from *Stylobates aeneus* and *Stylobates cancrisocia,* the two other species of the genus.

Differential Diagnosis

Comparisons with the other species of *Stylobates–S. aeneus* and *S. cancrisocia*–are based on data in Dunn, Devaney, and Roth (1980). Carcinoecia of *S. loisetteae* vary more than those of *S. aeneus.* This may be related to the nature of the original gastropod shell, which in *S. aeneus* is always very small, so that the carcinoecium is almost entirely a production of the anemone. By contrast, a high proportion of the *S. loisetteae* specimens examined (and the only one of *S. cancrisocia*) occur on large shells, which the anemone coats with chitinous material. . . . Growth lines of *S. loisetteae* carcinoecia . . . may be wavy, in contrast to their concentricity in *S. aeneus.* The carcinoecium of *S. aeneas* has a deep umbilicus, unlike that of *S. loisetteae.* . . .

Orientation of the actinian on the carcinoecium differs as well. . . .

Nematocysts also differentiate *Stylobates loisetteae* from the two previously described species. . . . Spirocysts (Fig. 6a) are broader in both other species, particularly *S. aeneas.* Large tentacle basitrichs (Fig. 6b) are smaller in *S. cancrisocia.* Small tentacle basitrichs (Fig. 6c) were not found in either of the other species. . . . (Fautin 1987:4–5)

In contrast to a description, which may be as long as necessary, a diagnosis should be as concise as possible–from one line or sentence to a few sentences in length, although a differential diagnosis may be longer if there are a number of other species to which comparisons must be made.

It should not repeat characters that are found in all members of the group, but identify only those that make it possible tell the taxon in question from the others. In a sense, it is a key to help readers identify species in a group or determine whether a taxon they have in hand is identical to one of them or possibly something undescribed (Wiley 1981).

Diagnosis in Zoological Taxonomy

In zoological taxonomy, diagnosis is sometimes called *definition.* You may see this in older papers, but right now about the only zoological literature that still uses that term is the ICZN itself (Art. 12 (a)): "To be available . . . every new name published before 1931 . . . must be accompanied by the description or a definition of the taxon that it denotes." For names published after 1930, "To be available . . . every new species-group epithet published after 1930, must satisfy the provisions of Article 11, and must be (i) accompanied by a description or definition that states in words characters that are purported to differentiate the taxon, or (ii) accompanied by a bibliographic reference to such a published statement." (Art. 13 (a)).

Diagnosis in Botanical Taxonomy

The biggest hurdle most biologists face when confronted with the need to describe a new plant species is the fact that the original description must be written in Latin to be *validly published* according to the ICBN (Article 36.1, 36.2). But even this requirement may not be as difficult as it sounds; although the entire description may be written in Latin, the Latin portion may also be limited to the diagnosis, the characters separating the new taxon from other species. Again, the diagnosis is supposed to be concise, consisting of a single phrase to a few sentences. Hawksworth (1974), writing for mycologists, made several points on the writing of diagnoses that are useful to all writers of botanical descriptions:

Botanical Latin has diverged from classical Latin for many years. It has newer technical vocabulary and different usage, so, in constructing your diagnosis, it is important to consult a dictionary of botanical Latin (e.g., Stearn 1992) as well as a general Latin dictionary.

You should compare your diagnosis with published examples by others working in that or related groups to make sure that the way you are using terms is consistent with accepted usage in that group. (This is also true for diagnoses of animal taxa, of course).

If possible, you should show the draft of your diagnosis to a specialist in systematic botany who is knowledgeable about the group of plants to which your new species belongs. Lacking that, at least show it to a botanist familiar with botanical Latin and have that person check meaning and grammar before you submit the manuscript for publication.

What Is a Diagnostic Character?

Naturally, it follows that a *diagnostic character* is one that is useful or necessary to distinguish a species or other taxon from its relatives (Stace 1989). In general, a diagnostic character should be easily recognizable. Mayr (1942) claims that the most practical diagnostic characters are often superficial ones and compares the diagnosis with directions you'd give someone locating a certain house. For example, your visitors will find it easy to follow your directions if you can tell them that your house is the only red house on the street. There may be other, more fundamental architectural differences between the houses on that street, but if your house is the only red one, then color is probably the most easily recognized character you can use. In other words, the characters you choose to distinguish your species in the diagnosis may not reflect the most basic biological differences, but they should be differences that can be quickly recognized by others.

Ideally, a diagnostic character should be constant for all members of the taxon in question and should not appear in the other related taxa. This may not always be possible. Sometimes you must use quantitative and variable characters. However, there has to be at least a statistically testable difference between the frequency of that character present in the selected group and in those to which it is being compared.

It is important to remember that a character that is useful for diagnosis in one group may be useless for diagnosis in another, even within the genera of one family (Mayr et al. 1953). Or as botanists have put it, "The exact nature of the diagnosis is not stated in the Code, nor can it easily be specified because of the enormous diversity of organisms covered by the Code" (Kirk and Cannon 1991:281).

In a diagnosis you usually compare a new taxon to its close relatives, which, in a new species description, would be the other species in the genus. When close relatives are poorly known, it may be helpful to compare your new species to a well-known but not so closely related species (Mayr et al. 1953).

The Diagnosis Section: Animals

The position of the diagnosis section in an original description varies depending on the style requirements of the journal and the choice of the author. Sometimes it is placed immediately after the headings and before the etymology and type material sections, as is shown in this excerpt from a description of a new bryozoan from Florida:

Drepanophora torquata, new species

Figures 93–96

DIAGNOSIS: Uniserial *Drepanophora* with tubular peristome and curved orificial process, but no peristomial avicularium. Ovicells globose, lateral pores large.
HOLOTYPE: AMNH 699.
PARATYPES: AMNH 700, 701, 702.
ETYMOLOGY: From the Latin *torquatus,* meaning adorned with a necklace or collar.
DESCRIPTION: Colonies are primarily uniserial, rarely loosely pluriserial, arising from an 8?-spined tatiform ancestrula, with a gymnocytal area as large as the opesia. Zooids are ovoid, the frontal surface convex proximally, and rising distally into a thick-walled tubular peristome. . . . (Winston and Håkansson 1986:36–37)

It is also sometimes placed at the very end of the description section, where it provides a bridge between the more detailed descriptive section and the taxonomic discussion that follows, as in this example from a paper on a new Norwegian polychaete, *Ophryotrocha cosmetandra,* by Eivind Oug (1990):

Diagnosis
Body with 28–32 setigers. Prostomium with a pair of papilliform antennae, no palps. Parapods simple, uniramous, without dorsal and ventral cirri. Males with conspicuous dorsolateral processes from setiger 15–17 to posterior end, on setigers 26–29 a posterior single funnel-shaped mid-dorsal protuberance. Two anal cirri. Compound setae scarce, with pointed blades. Mandibles with three strong teeth. K-type maxillae present in males only.
Remarks
Most *Ophryotrocha* are small and morphologically similar. Distinguishing characters include presence/absence of palps, presence/absence of dorsal and ventral cirri. . . . (Oug 1990:196).

Often diagnostic material is found not only in a diagnosis section, but is also incorporated into the taxonomic discussion section, as in the preceding example or in the following example. In this description of a new olive tree root-knot nematode species, the diagnosis section explains briefly how the new species differs from the other members of the genus; the relationships (discussion) section that follows elaborates on differences between the new species and its relatives.

Diagnosis
Meloidogyne lusitanica n. sp. can be distinguished from other species in the genus by the following characteristics: The perineal pattern has a medium to high trapezoidal dorsal arch and distinct punctuations in the tail terminus region. Distance from the excretory pore to head end of the female is 44.1 (22.0–60.0). Head cap in the male is rounded and extends posteriorly into the head region. This male characteristic is confirmed by SEM observations which show that labial disc and median lips are fused to form elongate lip structures. Head region is often marked by one broken annulation. Second-stage juvenile head region is smooth or with one or two short broken annules. Tail shape is conoid with rounded unstriated terminus.
Relationships
Meloidogyne lusitanica n. sp. is most similar to *M. megatyla* Baldwin & Sasser, 1979 and *M. partityla* Kleynhans, 1986. However, *Meloidogyne lusitanica* n. sp. differs from these species by the presence of distinct punctuations above the anus in perineal patterns, by the absence of deep indentations in the stylet knobs of the female, male, and second-stage

juvenile; by the more posterior position of the excretory pore in the female. . . . (Abrantes and Santos 1991:220–221)

Occasionally the diagnosis may be the entire Remarks section, as in this description of a new southwest Pacific sponge, *Tethya pellis,* by Patricia Bergquist and Michelle Kelly-Borges:

> **Remarks.** *Tethya pellis* is comparable to *T. australis* and *T. popae* but differs in a number of points of general skeletal arrangement and morphology. *Tethya pellis* has no interstitial megascleres as do *T. australis* and *T. popae,* and oxyspherasters are common throughout the entire width of the cortex rather than being restricted in distribution to the upper and lower regions of the cortex, as in *T. australis.* The choanosome of *T. pellis* contains only scattered micrasters and oxyasters, unlike *T. australis* which has all microsclere forms in abundance in the choanosome. In such general features as uniform oxyspheraster distribution throughout the cortex, an absence of interstitial megascleres and the low microsclere density in the choanosome, *T. pellis* approaches *T. robusta.* The prominent diagnostic feature of *T. pellis* is the extremely thick cortex. (Bergquist and Kelly-Borges 1991:51)

It may also be included in an otherwise labeled section that points out ways in which the new species differs from similar species. In this original description of *Goniopora cellulosa* by J. E. N. Veron, the diagnostic material is placed in a section labeled "Similar species," essentially a differential diagnosis:

> Similar species:
>
> The cellular appearance of the corallites is unlike that of any other *Goniopora* known to the author. Corallites are larger than those of *G. minor* Crossland, 1952 and *G. burgosi* Nemenzo, 1955, but smaller than those of the other massive species except *G. tenuidens* Quelch, 1886. Of other nominal species, *G. pendunculata* Quoy & Gaimard, 1833 from New Guinea and *G. gracilis* (Edwards & Haime, 1860) from the Red Sea have corallites of similar size and wall structure, but the septation of the former consists of a neat pattern of primary septa, while the latter consists of [a] small number of irregularly fused spines. Both are unlike the present species. (Veron 1990:118)

Diagnostic information can also be included in a key to species in a genus. For some groups of insects, for example, morphological differences may be very slight and keys may be needed to diagnose species. In the neotropical caddis-fly genus *Eosericostoma,* for example, Oliver Flint (1992:495) wrote in the introduction to his generic description, "The specific diagnoses will contain the genitalic descriptions, the only structure by which I can discriminate between the species." He also gave a key and provided illustrations of diagnostic characters immediately before the species descriptions. Then, in the description of the first species, he stated once more that "The species are only to be told apart with certainty by a careful study of the genitalia, either male or female, as outlined in the key" (Flint 1992:500).

Key to Species, Adults

1. Apex of abdomen with claspers, tenth tertiges, phallus extending freely: males 2

– Apex of abdomen almost semicircular in outline, no free processes: females 3

2. Clasper in ventral aspect with inner margin curving irregularly toward lateral margin at its apex; dorsal surface with its apical process near inner margin, slender (several times as wide as long) . . . . *inaequispina*

– Clasper in ventral aspect with inner margin sharply angled near apex, then extending nearly straight to lateral margin; dorsal surface with apical process near center, broader (barely longer than broad) . . . . *aequispina*

3. Vaginal sclerite with its apicolateral, bandlike supports attached near posterior margin and running more or less directly to a bilobate structure almost attaining apex of terminal segment . . . . . *inaequispina*

– Vaginal sclerite with its supports attached midlaterally, then curving distinctly mesad, ending before a knobbed, internal plate attached to a bilobate structure only extending half length of terminal segment . . . . *aequispina* (Flint 1992:500)

Some form of diagnosis is usually expected for a formal species description, although it is less commonly used by workers on some groups of organisms than on others. It is not necessary when the species diagnosis is identical to the diagnosis for the genus, which happens when there is only one species known in that genus (see chapter 18). Therefore, in their description of the Transylvanian fossil turtle *Kallokibotion bajazidi,* Eugene Gaffney and Peter Meylan had only to write:

> *Kallokibotion bajazidi* Nopcsa 1923a
>
> TYPE SPECIMEN: Designated lectotype BMNH R4916. Consists of most of carapace, posterior lobe of plastron, pelvis, and a caudal vertebra. DIAGNOSIS: Same as for genus. (Gaffney and Meylan 1992:3)

The Diagnosis Section: Plants

As previously mentioned, in botanical taxonomy, either the diagnosis or the description (or both) must be in Latin. Although experienced botanists may become knowledgeable enough to write their diagnosis directly in Latin, most would figure out what they wanted to say first in English, and then make the translation, getting a Latin expert or another botanist to check the translation for accuracy.

Sometimes the Latin diagnosis may lead off the description, as in this paper on a new red algal species from the Dry Tortugas, a species discovered during studies of the recolonization of an area of the marine reserve after a ship grounding:

> Diagnosis
>
> ***Laurencia coelenterata*** **sp. nov. Figs. 1–8.**
> Plantae e crusta crescentes, usque ad 2 mm crassae, usque ad 1.0 cm latae; axes cylindrici 0.6–1.0 mm diametro aut leviter clavati ad apices, erecti aut patentes, raro ramosi, usque ad 12 mm alti; omne segmentum axiale ramuli com quattuor cellulis pericentralibus; ramuli clavati plerumque ad apices; cellulae corticales in sectione transversali lateriores quam longi (23–31 × 25–42 μm); multi foveocolligationes secondarii inter cellulas; cellulae medullosae parietibus crassis, 130–350 μm; tetrasporangia abaxialiter abscissa, usque ad 114 × 140 μm, perpendiculares ad axem longum ramuli; cystocarpia apicaliter posita, in textura gametophyti profunde inclusa, vix procurrentia carpostomate priminente. (Ballantine and Aponte 1995:417–418)

In this paper the Latin diagnosis was followed (after information on the etymology, type material, and other material examined) by an

English-language diagnosis identical to the Latin version and then a much longer section called "Observations" in which more detailed descriptive information was presented, with illustrations of living specimens, photomicrographs and drawings of sections, and so on, followed by a discussion section.

Another botanical example follows. In this paper, Dieter Wasshausen was describing three new species of the Acanthus family from Brazil in *Proceedings of the Biological Society of Washington.* He used a format standard for plant descriptions in that particular journal: heading, followed by the diagnosis (in Latin), followed by description–in this case an expansion of the diagnosis, rather than just a translation. The descriptive sections were then followed by type, distribution, and a final discussion section.

Justicia (sect. *Chaetothylax*) *andersonii* Wasshausen, sp. nov.
Figs. 1A-C, 2A-D

> Herba, caulibus erectis vel ascendentibus, bifariam puberulis; foliorum lamina oblonga-ovata, obtusa vel acuta et aristata, basi cuneata in petiolum decurrens, aliquanto firma, utrinque glabra vel subglabra, costa et venis lateralibus parce hirtella; petioli hirtelli; paniculae axillares et terminales, subsessiles, ex spicis densis floribus; bracteae oblanceolatae, gradatim angustatae et cuspitatae, reticulatae, virides, parce pilosae et glanduloso-punctatae; . . .
>
> Herbaceous, erect or ascending plants 15–25 cm tall; stems bifariously pubescent with sordid, recurved eglandular trichomes; leaves petiolate, the blades oblong–ovate, 7–10 cm long and 2.7–3.7 cm wide, obtuse to acute and often aristate at the apex, narrowed at the base and briefly decurrent on the petiole, rather firm, drying dark green, both surfaces glabrous except the costa and lateral veins (4–5 pairs) sparingly hirtellous, the trichomes sordid and recurved, the cystoliths conspicuous under a lens. . . . (Wasshausen 1992:666)

Additional Uses for Diagnoses

A diagnosis is not usually found in descriptions in faunal studies and surveys, but it is often made part of species descriptions in revisionary studies. However, even when a formal diagnosis is unnecessary, it is a good idea to keep the diagnostic characters in mind when writing your descriptions, for the benefit of those who may use your article for identification purposes. If you have a taxonomic discussion section in your species descriptions, distinguishing characters can be repeated or included there (as in the examples in this chapter and in chapter 12).

In addition to their role in taxonomic discussion, diagnostic characters are the essentials of any key. They are considered again in chapter 19.

Chapter 11

Description Section

Selecting the characters necessary to give a good idea of a species in a short description requires a thorough knowledge of the subject and a methodical mind. –BENTHAM (1874:48)

The preferred styles in all my surveys used, where possible, short words rather than long, ordinary words rather than grand, familiar rather than unfamiliar, non-technical rather than technical, and concrete rather than abstract.

–KIRKMAN (1992:19)

While skill at drawing, or a steady hand, may help a person create an illustration, his skill will be of little help if he lacks the quality of careful observation, of attention to every detail of his subject; accuracy is the most sought after virtue in biological illustration. –DOWNEY AND KELLY (1982:viii)

Descriptive Writing

Whereas the diagnosis consists only of the characters that distinguish the new species from its relatives, the description section should include all the information necessary to provide a clear and accurate mental image of the organism. Ideally, the writing should be so vivid that a reader could recognize that species from the written description alone. But producing a taxonomic description that good is not easy, even for the expert with the "thorough knowledge of the subject" and "methodical mind" of the introductory quotation. In a sense, a species or any other taxonomic description combines two of the most difficult types of writing:

the technical description and the poem. Like a technical description of a piece of machinery and its operation, for example, it demands all the skill you can muster in terms of logic, conciseness, and precision of language. But, like a poem, its success depends also on the right word in the right position, on connotation and association as well as the dictionary definition of a word, and on the imagination of the author in making words work in a visual way. Even professional systematists admit that most descriptions must be supplemented with illustrations to be truly useful (Mayr 1969:267). And, as you have probably already discovered from your literature search, bad unillustrated descriptions are useless unless you can get access to the specimens on which they were based.

But don't be too discouraged. Most of us are aren't poets, and most of us aren't great writers, but with practice we can become good enough. When the morphology and functioning of a bryozoan species I'm working on seems impossible to describe in plain English, I remember the lesson of the Caterpillar Tractor Company (Kirkman 1992). Faced with the need to train non–English-speaking operators and mechanics all over the world, a potentially immense translation problem, they created a 784-word vocabulary and a restricted grammatical structure. They used this controlled English to produce manuals describing the operation and maintenance of their equipment. After only 30–60 hours of training, their employees could learn to operate and repair Caterpillar equipment using these manuals. Their program was so successful that it was expanded and marketed as ILSAM (International Language for Service and Maintenance) and BASIC 800. A Caterpillar tractor seems a lot harder to describe than a bryozoan. Remembering this story puts my own problem into perspective: If they could accomplish all that with 800 words, then surely, with all the words and grammar I have available, I can finish my description.

The basic principles used in developing such restricted technical languages are important to remember for another reason. Like the technical writers at Caterpillar, you are writing for an international audience. Some of the people who need to know what is in your description will not be able to read it with ease. They will be sitting at their desks, dictionary in hand, deciphering your description line by line. In Restricted English the variety of words used must be strictly limited, each word used must have only one meaning (synonyms are avoided), the number of verbs is reduced, statements are as short and positive as possible, and explanations are carefully sequenced in steps (Kirkman 1992). These rules make it easier for readers with only a limited grasp of the language to use the technical description. They are good guidelines to remember as you write taxonomic descriptions, for the same reason. Of course, each group of organisms has its own specialized terminology, but this technical vocabulary is not a problem. The terms are often similar in many languages. Technical dictionaries and glossaries explain them to novice taxonomists, and specialists are already familiar with most of them.

Information Used in the Description Section

Whatever the organism, the species description should provide the essential taxonomic characters along with other relevant information. The two opposing pitfalls of the description section are leaving out information that is crucial and including so much information that the reader becomes confused. The line between good description and bad lies in putting in everything that is relevant, yet not cluttering the description by throwing in everything but the kitchen sink. You can't study everything about the organism before you describe it, and you can't put all you know in the description. You must choose what is most important biologically and what is most diagnostic taxonomically. In doing so, you will be led (although not blindly) by what is traditionally included in descriptions of that group, in terms of both taxonomic characters and general biological information.

Taxonomic Characters

Taxonomic characters, as you recall from chapter 3, are all the features of an taxon by which its members actually or potentially differ from those of another taxon. They are the characters that can provide phylogenetic information. These characters are genetically controlled and represent homologies, deep similarities based on descent from a common ancestor.

These characters and their distinctions form the basis of all phylogenetic work, starting with the basic species description, but in a species description their function is limited to the characters useful in distinguishing the species from others at the same level. You do not need to include in a species description characters that are taxonomically useful for higher categories. For example, a species description does not need to include a character that occurs in all the species in a family unless there is some problem with the classification of the family or the assignment of that species to that family is doubtful. Morphological characters are always included. They still form the basis for most descriptions, but other kinds of characters (chemical, genetic, genomic, etc.) have becoming increasingly important in many groups as we learn more about those groups' biology.

Life History Characters

Care should be taken to include the pertinent aspects of the organism's life history. For example, for organisms with two sexes, both should be described. If differences are complex, this can be done by describing each sex differently, as in this description of a new species of cumacean crustacean, *Oxyurostylis lecroyae:*

> *Description of the Marsupial Female–*
> Length: 2.9 mm to 5.9 mm.
> Carapace (Figs. 1, 23–25, 37, 38): width exceeding depth, 0.73–0.85 times length. Carinae variably developed, acute and distinct or blunt and perceptible only by rotating the specimen. . . .
>
> *Description of the Adult Male–* In addition to the sexual differences in the development of the second antennae, exopods, and pleopods, the adult male differs from the adult female in the following characters:
> Length: 4.7 mm to 5.7 mm.
> Carapace (Fig. 31): width 0.66–0.78 times length, dorsal outline almost straight. Oblique carina weak, blunt, perceptible only by rotating the specimen. . . . (Roccataglia and Heard 1995:604, 607–608)

Or it can be handled by pointing out the differences between the sexes in a single description. In this description of a new amphipod, *Leucothoe laurensi,* the description is based on the female holotype; the differences between male and female are pointed out in a separate paragraph at the end of the description section.

> *Description–* Female, 2.1 mm, USNM 266424. Article 3 of antenna 1 about one-half as long as article 1; rostrum small, anteroventral margin of head rounded; eye with 19 compact ommatidia; coxa 1 slightly smaller than 2–4, broadly rounded anteriorly, coxae 2–3 subtruncate, ventral margins with few short setae, posteroventral margin of coxa 4 produced into broad lobe, excavate dorsally, coxae 5–6 bilobed, 5 larger than 6, posterior lobe of 6 the deepest, coxa 7 small, evenly rounded. . . .

Description of Male: Similar to female except for article 6 of gnathopod 2, palm slightly more oblique, reaching only 64 percent along anterior margin of propodus (versus 82 percent in female). (Thomas and Ortiz 1995:613, 616)

For organisms with polymorphic individuals or structures, all polymorphs should be discussed. For example, many hydroids have both a polyp stage and a free-swimming medusa stage. Cheilostome bryozoan colonies have feeding individuals (autozooids); they may also have female zooids with brood chambers (ovicells) and defensive or cleaning zooids (avicularia and vibracula). All these polymorphs may contain important taxonomic characters.

Life History Stages

All life history stages, immature and mature, should be covered, although not necessarily in great detail. It may be enough to state that juvenile forms of species in the study group cannot be distinguished taxonomically. For example, in the cumacean *Oxyurostylis lecroyae* described earlier, immature instars were not distinguishable from those of related species, so descriptions were based on adults of both sexes (Roccatagliata and Heard 1995). On the other hand, in whiteflies the adults can be indistinguishable, and it is the larval stages that are diagnostic, whereas in caddisflies, for example, larvae, pupae, cases, and adult characters are all useful taxonomically (Flint 1992).

For colonial organisms, differences resulting from astogeny (colony development) as well as ontogeny (individual development) must be considered. For example, in encrusting colonies of cheilostome bryozoans there may be a zone of astogenetic change consisting of the first few generations outward from the metamorphosed larva, in which zooids gradually increase in size and lack polymorphs and reproductive structures. The peripheral regions of a mature colony have a zone of astogenetic repetition, zooids are much the same in size, and polymorphism is expressed. Obviously, comparing young zooids in one colony with old zooids in another (even from the same population of the species) would be misleading.

Color

The value of color in a description varies from group to group and species to species. In some cases environment or diet has a controlling influence on coloration. For example, captive flamingos must be fed a diet rich in carotenoids if their plumage is to remain pink. Other groups show strong genetically controlled variation with species. In some cases variation in coloration is linked with population or subspecific variation, as in the fox squirrel, *Sciurus niger* (figure 11.1), or the gray squirrel, *Sciurus carolinensis* (Moncrief 1993), or in birds such as the northern junco, *Junco hyemalis*. In other cases, as in some colonial ascidians (figure 11.2), no such correlation exists and differently colored or patterned individuals occur even within populations.

But coloration (of plumage, pelage, flowers, larvae, etc.) is important in adequately describing many species. Unfortunately, there is still no entirely satisfactory method of standardizing descriptions of colors. A

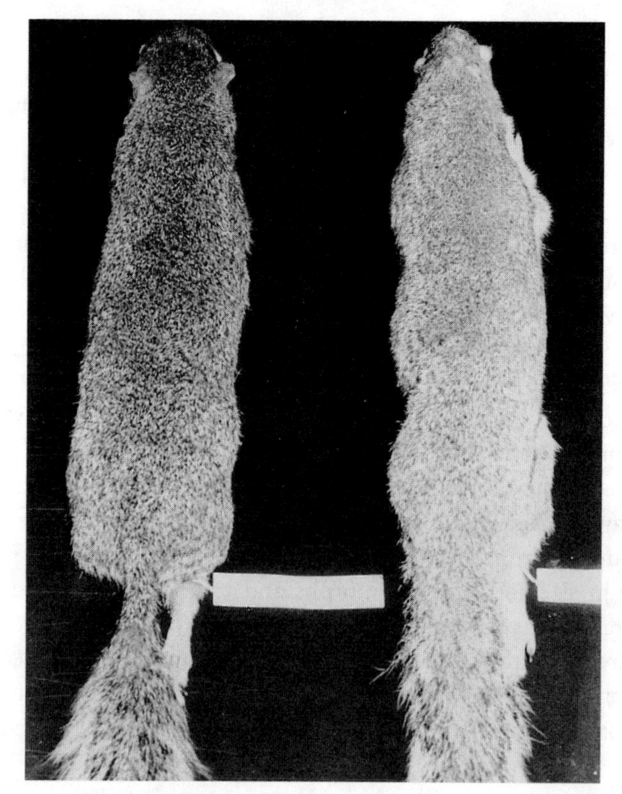

Figure 11.1 Regional population variation in fox squirrel, *Sciurus niger,* pelage color. **Left,** Specimen from mainland population with dark orange-brown coloration. **Right,** A Delmarva specimen, showing the pale gray pelage characteristic of Eastern Shore populations. *–Photo: J. E. Winston.*

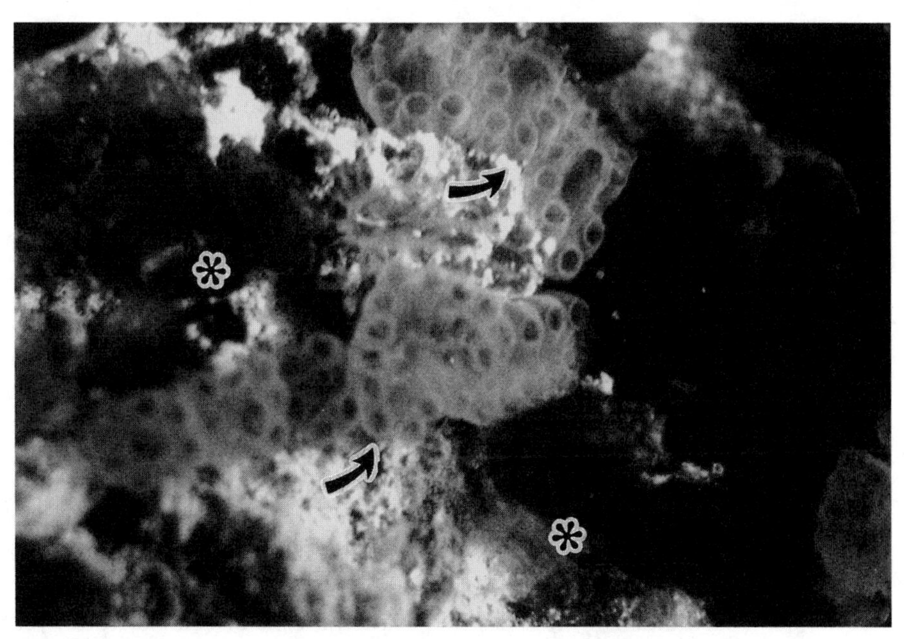

Figure 11.2 Both orange (arrows) and dark purple (asterisks) colonies of the colonial ascidian *Distaplia corolla* occur within the same habitat, as shown here on a mangrove root at Twin Cays, Belize. –*Photo: J. E. Winston.*

number of color standard books and charts have been published over the years (see this chapter's *Sources*), but none of the older ones are in print, and copies are scarce even in libraries. Tucker et al. (1991) produced an excellent review of the biological charts produced up to that time, with remarks on how particular colors in them may have faded or changed over the years.

Biological color charts currently available include the *Royal Horticultural Society Colour Chart* (actually a sturdy plastic box of color swatches) put out by the Royal Horticultural Society, London. Although this chart was developed for botanists, it includes a complete spectrum of colors and is useful to all biologists. Many ornithologists use Smithe's (1974, 1975, 1981) *Naturalist's Color Guide.* For paleontologists, two rock and sediment color charts, the Geological Society of America's *Rock Color Chart* (1991) and the *Munsell Soil Color Charts* (1994), may be useful (see this chapter's *Sources*).

Books of color standards for artists and printers (such as the Pantone Color Books series) may also be useful to biologists. If you cannot find or afford one of these, probably the best practical solution is to use traditional and translatable artists' terms for colors (e.g., the difference between scarlet, a bright red, and crimson, a deep purplish red) are translatable and recognizable worldwide. You should also be careful to standardize your own use of color terms at least within a work (e.g., if you call a structure peach-colored in one description and a structure in another species has an similar color, call that one peach, also, not pinkish-orange).

In some groups, such as mammals, in which coloration is important in taxonomic description, there are already terms in use such as *agouti* (a variegated pelage coloration in which each hair is barred in dark and lighter colors) that are understood by those who study the groups and are explained in field guides and glossaries for novices.

Portable spectrophotometers are increasingly used by those who work on birds and mammals. As satisfying as it may be to match the color we see with a standard, it does not mean that it is the color the animals themselves (or their predators) see. Color vision is variable among groups and species. For example, stomatopods have 12 classes of color receptors (vs. 3 for humans). *Sepia* has just 1 spectral class of color receptor, yet both can camouflage themselves effectively in the visually similar shallow underwater habitats in which they live (Osorio and Vorobyev 1997). Colors of marine animals seen at the surface show up very differently under water. Red and orange colors are quickly absorbed as light penetrates water at increasing depths. A sponge or brittle star whose water surface color is red looks black at 100 feet. Living colonies of the Caribbean reef bryozoan *Trematooecia aviculifera* look pink in very shallow water, but a striking fluorescent green in deeper water. We also know that the color patterns of mammals change with environmental conditions, and these changes may be biologically significant (Endler 1990).

In the past, museum taxonomists often preferred to ignore color because the dried or preserved material available to them had often lost the colors it might have had in life (e.g., "Color: white in alcohol") Today, the thorough taxonomist includes the color of the preserved material in the description, along with any information available on living colors, as in this brittle star description:

> Large individuals are entirely scarlet red or orange-hued vermilion; the brilliant pigment quickly and completely bleaches in alcohol-preserved specimens. Small individuals have pale orange patches on the disk. (Hendler et al. 1995:141)

But there are exceptions; some organisms change rather than lose their color when preserved or dried. For instance, Hawksworth (1974) mentions lichens, grayish green in life, which become bright yellow when dried.

Incorporating more information on coloration of living organisms and its role in their ecology is one important contribution ecologists can make to improving taxonomic descriptions. In turn, their observations of living coloration and markings detectable in the field will help other ecologists, enabling them to identify their organisms in the field without destructive sampling.

Quantitative Characters

Measurements should be included either as part of the description or in a separate section. When only one specimen is available, all you can do is measure that holotype specimen. If a number of specimens are available, then you should present ranges and average values for your measurements. You may need only a little basic information on size to adequately characterize some organisms. But in other groups, quantitative characters such as size differences are essential in delimiting or describing species. Thinking two things are similar on the basis of descriptions that lack measurements and, on finding the specimens, discovering that one is twice the size of the other is not uncommon but very annoying. Moreover, differences in size of a whole organism or its parts may have important ecological implications. For example, different beak sizes and shapes in Galapagos finches collected by Charles Darwin provided a first clue, later confirmed by years of field study, that rather than just filling a tree-dwelling, seed- and fruit-eating niche (like an English finch), the Galapagos finches had radiated into different niches (e.g., insect-eating, ground-dwelling) filled in mainland communities by many different kinds of birds (Lack 1947; Grant 1986).

The use of continuously varying characters may present problems for phylogenetic analysis (see chapter 22), but often they must be accommodated in a description, In fact, for some groups (fossil hermatypic corals or fenestrate bryozoans, for example) quantitative characters may be the only ones that work to separate related species (Foster 1984; Hageman 1991).

Scanning the major systematic works for a group will tell you the measurements most commonly used and the methods used to make them. It is then up to you to carry out the measurements and make the way you used them clear to your reader. You can do this either by reference to a standard work or by using diagrams or definitions in the methods section of your paper.

Much has been written about analysis of variation in ecology and taxonomy (see chapter 13's *Sources*), and most of you will already be familiar with statistics and their use. If you have resorted to statistical analysis in determining that your species is new, the results should be incorporated into the description, but only to the extent necessary to recognize the species and distinguish it from others. Details of the analysis can be given in the discussion section.

Behavioral and Ecological Characters

Brief summaries of behavioral and ecological characteristics should be included in the description, along with morphological data, particularly if they are of diagnostic importance. Detailed results are best handled in discussion or ecology sections (chapters 12 and 13).

The description should cover the range of variation both in all the material available and in the holotype you've selected. You can do this either by basing the description on all available material and noting any differences of the type or by basing the description on the holotype and then describing any variations in the rest of the material.

Finally, don't forget that what is unknown about a species may be as important as what is known. If you haven't found a male or observed a reproductive structure at the time of year when related species have them, this may be significant, and should be noted in the description.

Writing the Description

The most exhaustive description of a species is usually the original description, in which the species is named and described for the first time. Other types of description vary in completeness depending on use and intended audience. Consider the following points before writing any description.

The most important point is to *work from the material.* Use the specimens, measurements, field notes, photographs, and drawings you have made. Pay attention to what others have written on the subject (and make the most of it in the discussion section or mention it in the description section), but depend primarily on and write from what you have in front of you, and from what *you* can determine from that material. Do not rely on what you have read about the species, or what someone else has told you about it (even if the person telling you is the world's authority). Find a specialist to help you with your description, if at all possible, and learn as much as you can from that person, but if an expert or book claims something is true, be sure you see it for yourself. Don't rely in your description on what you have read or a picture in a book (although you may want to incorporate such information in your discussion or point out any differences with published descriptions, either in the discussion or the description section itself).

In addition to a standard vocabulary, a standard order of description is usually followed in descriptions of each group of organisms. It is important to follow that standard sequence because it makes it easier to compare your description with other descriptions. It will also help convince others that you have done your homework and that you understand the terminology, anatomy, and biology of the organism you are writing about.

Like any good description, a taxonomic description works from the familiar or more general to the specific (what Mayr called following the natural order). For example, a description of a colonial invertebrate (e.g., a bryozoan species) would include, first, appearance of the colony, then individual (zooid) characters, polymorph characters, reproductive information, and any other biological information. A description of a solitary ascidian or sea squirt would include body surface texture, size, color, siphon morphology (external anatomy), then spicules, brachial tentacles, stigmata, brachial sac, alimentary canal, anus (internal anatomy), then gonads and larvae (reproductive biology) (Millar and Goodbody 1974). A description of a fish would cover, first, a description of its external morphology (with table of measurements), then descriptions of teeth, gill rakers, scales, bones (osteology), and coloration (Stiassny and Reinthal 1992). A description of a mammal also works from external morphology, including fur, skull, and teeth (which provide many useful characters) to other skeletal characters, and finally any other biological information.

Telegraphic Style

Many taxonomists and journal editors favor the *telegraphic style* of taxonomic description, in which adjectives and verbs are eliminated, over the *telephonic style*, in which complete sentences are used. Telegraphic style *can* save space. In a long monograph, a few lines per species can add up to pages saved. As any editor will tell you, journal space gets more expensive all the time, a compelling reason for presenting descriptions as concisely as possible. In telegraphic description the verb *to be* is eliminated and remaining verb forms are limited pretty much to passive participles. Abbreviations (previously defined by the author in a methods section) are used as much as possible. For example, this description of a new spider, *Mangua gunni,* from New Zealand, shows the compression of information that can be achieved with telegraphic style:

> MALE: Total length 3.39. Carapace 1.31 long, 1.08 wide. Abdomen 2.16 long, 1.16 wide, 1.31 high. Leg lengths: I 10.55, II 7.39, III 5.16, IV 7.78. Eye sizes and interdistances: AME 0.04, ALE 0.07, PME 0.07, PLE 0.07; AME–AME 0.01, AME–ALE 0.03, PME–PME 0.07, PME–PLE 0.07, ALE–PLE 0.01–1 MOQ length 0.17, front width 0.09, back width 0.21. Clypeal height over four times AME diameter. Chelicerae 1.08 long, vertical, divergent, with six evenly spaced promarginal teeth; fang strong, sinuous. Carapace dark with weak pattern; abdomen pyriform, patterned as in figures 256, 258. Palp 1.09 long; patella twice length of tibia, with strong distodorsal bristle; tibia with strong macrosetae (figs. 268, 269); distal processes of bulb complex (figs. 270, 271).
>
> FEMALE: Total length 3.45. Carapace 1.31 long, 1. 16 wide. Abdomen 2.31 long, 1.46 wide, 1.42 high. Leg lengths: I 9.78, II 7.08, III 8.62, IV 7.24. Eyes as in male except PME more widely separated. Clypeal height three times AME diameter. Chelicerae shorter (0.62). Coloration as in male. Internal genitalia (fig. 257) with paired receptacula fused but lumen of each distinct. (Forster et al. 1990:78)

Another advantage of telegraphic style is that, when used in a standardized format and with a standardized vocabulary, it becomes similar to the restricted technical languages described earlier in this chapter, making it more useful to an international audience. Once you master the vocabulary of a field and understand how passive participial grammatical constructions are formed in the target language, you can sit down with the proper dictionaries and consistently and fairly accurately translate species description after species description, even though you might still have a tough time understanding the discussion section of the same publication.

On the other hand, the chief disadvantage of the telegraphic style lies in its being, like poetry, very dependent on exact placement of punctuation, parallel order of presentation, and selection of exactly the right word to preserve the intended meaning.

Styles and length of species descriptions can vary tremendously from group to group. Mayr (1969) made the generalization that a description should be longer and more detailed for species in poorly known groups than in those better studied ones because for the poorly studied groups it might be impossible to know what characters would be most useful to future workers. In practice, however, descriptions tend to be longer in well-studied vertebrate groups, in which only a few to a few dozen species are described per year. In groups such as birds, mammals, fish, or even fossil vertebrates, a description of a single new species can fill a paper. The description section alone may run several pages and is often written in telephonic rather than telegraphic style. In contrast, in entomology, where millions of species remain undescribed and where hundreds or thousands are described in a year, a species description, almost by necessity, consists of a very short telegraphic paragraph covering only the most crucial characters. There may be dozens of such descriptions in each paper. It is not surprising that entomologists were the first to come up with computerized formats for species description and the composition of keys (see chapter 19). To most entomologists the idea of a paper in which only one species is described is almost unthinkable, unless that species represents a new genus as well.

Although some of these stylistic differences may be ascribed to differing degrees of knowledge in a field and some to time constraints, others may really have to do with the difficulties of writing about complex anatomy and functional morphology. A good deal of functional morphology (along the line of "the hipbone is connected to the thigh bone") is included in vertebrate descriptions. Complete sentences are clearer for describing complex structures and operations, as in this extract from a description of *Crunomys* (a rodent):

> On the side of each braincase posterior to the orbit and above the pterygoid ridge, the alisphenoid region is without a lateral strut of alisphenoid bone. The alisphenoid canal is an open channel in this area, its entry into the sphenoidal fissure can be seen and the foramen ovale is conspicuous in side view. The alisphenoid canal transmits the internal maxillary artery and masticatory nerve, which pass along the bottom of the channel and through the anterior opening of the canal into the sphenoidal fissure and then the orbit. Without the strut of alisphenoid bone, the buccinator–masticatory foramina are functionally absent and the foramen ovale accessorius lacks an anterior boundary (fig. 23). The masticatory and buccinator branches of the maxillary nerve emerge directly from

> the foramen ovale. The buccinator nerve runs anteriorly, most of the masticatory courses onto the lateral surface of the alisphenoid bone in a shallow channel. . . . (Musser 1982b:23)

With the help of the author's figure 23 (a plate comparing the relevant parts of *Crunomys* and *Rattus* skulls), even a nonmammalogist can understand the anatomy being described. Descriptions such as this show that it is possible to use a clear and active writing style even in a format as technical and restrictive as that of a taxonomic description.

In choosing a style, you will need to look carefully at the style allowed by the journal in which you hope to publish. Some journals allow flexibility (e.g., the description may be written telegraphically and the discussion telephonically); others even allow you to alternate between the two types within the description section. Still others may have a more restrictive format.

The Description Section: Animals (Examples of Styles for Different Groups)

The description section usually follows the diagnosis section when one is used, or the types and material examined section when the diagnostic information is included in a taxonomic discussion section that follows the description section.

By the time you come to write a description yourself, you will have looked at many others during your literature searches. You will have come to understand the terminology used for your group of interest, and you will probably have made some decisions about what information to include and what to avoid in your own description. The following examples are just samples. They show some of the ways in which style and content differ for descriptions of organisms from different taxonomic groups and also according to the kind of paper (whether a composite paper, combining taxonomic description with other research results, a single-species description, or a taxonomic revision).

Descriptions of Invertebrates

A NEW HISTOPHAGOUS CILIATE, *MESANOPHRYS CHESAPEAKENSIS*, PARASITIC IN BLUE CRAB HEMOLYMPH This description is from a composite paper in which the systematic section describing the new parasite prefaces results on its culture, prevalence in blue crab populations, and histopathology. The description mixes telegraphic and telephonic styles:

> **Description.** Body fusiform, 37.6 μm long (range 28.0–47.6 μm) by 13.4 μm wide (range 11.0–18.3 μm) (n = 23) (Table 1). Macronucleus just posterior to mid-body and slightly wider than long. Rounded micronucleus on surface of, or, at times, within an indentation in the macronuclear envelope (Fig. 1). Ten equally spaced bipolar ciliary rows, or kineties, covering the ciliate. First oral polykinetid (OPK1) composed of a file of up to 6 ciliated kinetosomes; the posterior 4 were paired and the anterior 2 were single. Second oral polykinetid (OPK2) pyriform and composed of up to 7 kinetosomes long by 3 kinetosomes wide. Third oral polykinetid (OPK3) short and wide. The oral dikinetid (ODK) apparently lacked an **a** portion. Oral dikinetid **b** (ODKb) originated between OPK2 and OPK3. Oral dikinetid **c** segment kinetosomes were arranged in a "Y" formation (Figs. 1, 2). (Messick and Small 1996:3)

In this case, the brief, tightly worded description was supplemented by a table of measurements and illustrations and was followed, after sections on type, host location, specimens, and etymology, by expanded descriptions titled "Morphology of live and MHE-stained specimens" and "Morphology of prototargol-stained specimens."

A New Hydroid from Florida, ***Zyzzyzus floridanus*** This species was described as part of a phylogenetic analysis of capitate hydroids and medusae. Telegraphic style was used throughout.

> Medusa reduced to cryptomedusoid gonophore without distal processes. Hydroid solitary; total length about 3.5 mm, hydrocaulus embedded in sponge *Callyspongia vaginalis* (Lam.); hydranth slimly vasiform 0.7 m high, with 14–16 fairly long, round oral tentacles arranged in two to three closely set whorls, and 16–20 long aboral tentacles; oral tentacles adnate to hypostome forming longitudinal ridges, of circular cross-section, with concentration of nematocysts at tip and along aboral side; aboral tentacles laterally flattened, rectangular in cross-section, about three times as long as hydranth, with nematocysts concentrated along aboral surface; neck region slimly cylindrical, as long as hydrocaulus proper, thin-walled, covered by closely fitting filmy perisarc originating in circular groove immediately under hydranth; transition between neck and hydrocaulus

marked by ring-shaped swelling, hydrocaulus below this area covered by somewhat thicker perisarc; hydrocaulus slim, cylindrical with about six longitudinal peripheral canals of equal size; hydrorhiza developed at base of hydrocaulus as four closely-set, rounded bulbs covered by heavier perisarc; six gonophores developed singly on short pedicels in circle just oral to aboral tentacles; gonophores cryptomedusoid, without radial and ring canal, ovoid, mature female gonophore enlarged, transparent, lacking collar around orifice, as long as hydranth, reaching well over oral tentacles, containing one large actinula with circle of 12–14 long aboral tentacles. (Petersen 1990:178–179)

Although this description is clearly written, like all descriptions it greatly benefits from illustration, in this case a clear line drawing (see figure 11.3) accompanying the text.

A POLYCHAETE WORM, *LUMBRINERIS FAUCHALDI,* A NEW ARCTIC SPECIES FROM DAVIS STRAIT, COLLECTED DURING AN OCEANOGRAPHIC CRUISE

Description: The material consists of 7 specimens of which only one is complete (Holotype).

The holotype is coiled, contains approximately 80 setigerous segments and measures about 10 mm in length. The longest anterior fragments measure 12 mm (44 setigers) and 15 mm (65 setigers) in length. The specimens are light tan in color with no body pigmentation.

The prostomium is conical and slightly longer than wide (Fig. 1a). There are no visible nuchal organs or eyes. The two peristomial segments are achaetous and similar in size. There are paired anterior prolongations of the first peristomial segment.

The body is rather slender, tapering at the posterior end and terminating in two anal cirri (Fig. 1b). One specimen had three anal cirri (Fig. 1c).

Setigers throughout are wider than long. The parapodia contain short rounded presetal and postsetal lobes. Setigers 1–10 contain fascicles of limbate setae. Thereafter, the dorsalmost setae in the fascicle exhibit blunted ends. These blunt setae grade into hooded hooks over succeeding setigers. The fully developed hooks do not occur until setigers 25–35. Some of the limbate setae of setigers 20–50 contain greatly elongated capillary tips (Fig. 1d). The hooded hooks completely replace the limbate setae in the last 1/3 of the body. Hooded hooks are multidentate with about 10 small teeth, of which the lowermost tooth is only slightly thicker than the rest (Fig. 1e). The acicula are yellow, straight, and may number two per parapodium in posterior setigers.

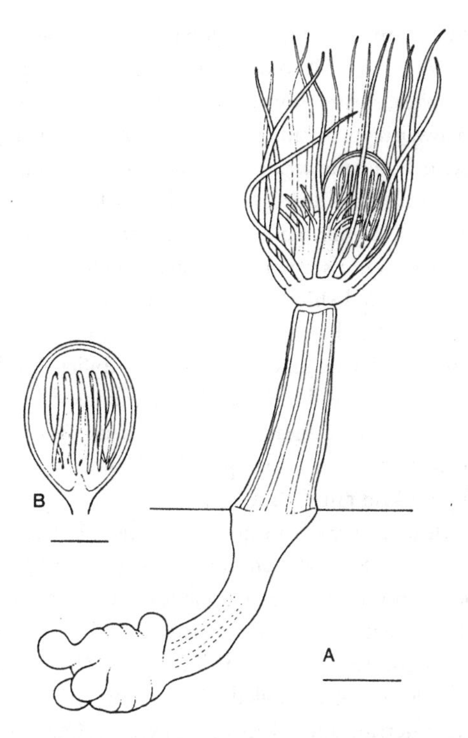

Figure 11.3 *Zyzzyzus floridanus.* **A,** Type specimen growing in *Callyspongia vaginalis* (Lamarck), Miami, Florida, U.S.A., 27 January 1965, C. Nielsen collection; scale bar = 0.5 mm. **B,** Detail of female gonophore with large actinula; scale bar = 0.25 mm. *–Source: Figure 29 of Petersen (1990).* Use of a clear line drawing to illustrate a species description.

> The pharyngeal apparatus is as follows (Fig. 1f): the maxillary carriers are short and sharply pointed, with a weak lateral notch; maxilla I has a weakly curved tip with no distinct teeth along the inner margin; each maxilla II has five teeth on the inner margin and a posteriorly directed lateral prolongation; each maxilla III and IV has one tooth; the mandibles were not observed. (Blake 1972:128)

A NEW TERRESTRIAL SNAIL, *MICRARIONTA OPUNTIA,* FROM SAN NICOLAS ISLAND, ONE OF THE CHANNEL ISLANDS OF CALIFORNIA. Shell characters figure prominently in gastropod taxonomy. In this example, the description of shell characters immediately follows the diagnosis. Further on, after discussion of type locality, type material and other material, and before the discussion section, comes a section called "Anatomy," which covers nonshell morphology.

Description: Shell, small, depressed-globose, openly but narrowly umbilicate, of 4 3/4 to 5 1/4 whorls, moderately thin; spire low-conic, whorl profile convex, sutures impressed. Surface glossy to silky, very finely granular, with low, irregular, radial wrinkles more or less papillose on early whorls. Protoconch radially wrinkled. Whorls of spire punctate; punctations arranged in diagonal lines, less regular on later whorls, becoming spirally elongate on penultimate and body whorl, sparse on base and around umbilicus. Whorls enlarging slowly; body whorl slightly constricted and descending behind aperture. Persitome subcircular, its ends strongly convergent; lip sharply turned outwards, not greatly thickened, moderately reflected at base, encroaching on umbilicus for less than 1/4 of its diameter. Parietal wall convex, lightly calloused between ends of peristome. Color, under a light tan periostracum, pale brown with a purplish tinge, with chestnut-brown peripheral band bordered with white; base lighter; lip pinkish tan within. . . .
Anatomy: Mantle pale gray with extensive patches of black; mantle collar grayish white. Sole of foot whitish; dorsal integument light grayish tan, darker in head region. Right ocular apparatus passing between male and female genitalial systems. Atrium short, penis capacious, containing short, spherical verge, separated from epiphallus by a slight constriction. Epihallic caecum long, complexly coiled. Dart sac absent. Single mucus glad present, its duct inserting about half way up vagina. Duct of spermotheca stout; no spermothecal diverticulum present. . . . (Roth 1975:94–95)

A NEW CRETACEOUS AMMONOID MOLLUSK FROM SOUTH DAKOTA, *CALYOCERAS BOREALE*, FROM A PAPER PRESENTING NEW RECORDS OF BIOGEOGRAPHIC IMPORTANCE

DESCRIPTION: The dimensions of the holotype and three of the paratypes are listed in table 1. The holotype, AMNH 44332 (fig. 51, J), is completely septate and shows depressed costal and intercostal sections (fig. 8). Ten or eleven massive primary ribs bear strong umbilical and inner ventrolateral tubercles and give rise to one or two secondary ribs; there are occasional intercalated ribs; altogether there are 22 ribs per whorl. All of these ribs are straight and coarse and bear strong, blunt outer ventrolateral and somewhat weaker, siphonal clavi.
Paratype 44329 (fig. 5E-H) is also completely septate. Coiling is moderately involuted with a deep umbilicus compressing approximately 30 percent of the diameter. The intercostal whorl section is very depressed and reniform. . . .

> *Paratype AMNH 44330* (fig. 5-A-D is completely septate and partially crushed. Coiling is moderately involute. . . .
> *Paratype USNM 441255* (fig. 6) is a large, completely septate fragment of the outer whorls, with a maximum whorl height of approximately 80 mm. . . . (Kennedy et al. 1996:5)

Note the inclusion of information on holotype and paratype specimens in the description.

A MIOCENE BIVALVE MOLLUSK, *ASTARTE CLAYTONRAYI,* **FROM A WORK ON THE MOLLUSCAN BIOSTRATIGRAPHY OF THE MIDDLE ATLANTIC COASTAL PLAIN OF NORTH AMERICA** This telegraphic description includes measurements on the type specimens.

> Description.–Very small, regular, evenly spaced, narrow undulations originating from the earliest stages of shell development and persisting over one-half of the disk. Undulations grading to a lower amplitude until reduced to faintly impressed concentric lines, giving the ventral portion of the shell a smooth, polished appearance.
>
> Beak prominent and placed well forward. Lunule narrow, lanceolate, and deeply concave. Escutcheon slightly longer and narrower than the lunule and deeply excavated. Posterior dorsal slope elongate and gradual, anterior dorsal slope short and abrupt. Ventral margin broadly rounded.
>
> Hinge plate strong, moderately arched, with a deeply incised ligamental suture and smooth ligamental nymph. Left valve with long, shallow socket to house posterior lateral tooth of right valve. Left valve with a posterior lateral tooth crenulated on the anterior margin and an anterior cardinal with crenulations on anterior and posterior margins. A socket between the anterior and posterior cardinals houses the large crenulated cardinal tooth of the right valve. A very small, shallow socket located just anterior to the anterior cardinal and posterior to the lunule in the left valve houses an equally diminutive anterior cardinal on the right valve. Anterior-dorsal margin of left valve, just below lunule, produced to form long projecting anterior lateral tooth. Right valve with corresponding socket.
>
> Anterior adductor muscle scar lunate, posterior adductor scar subcircular, both well impressed. Anterior pedal retractor scar very deep, small, and located several millimeters above the anterior adductor. Posterior pedal retractor scar less well defined, small, and located adjacent to and dorsal to posterior adductor scar. Three smaller, very deeply impressed

scars located below and under the hinge plate probably represent the pedal elevator muscle scars.

The pallial line is simple, entire, clearly defined, and remote from the ventral margin. Ventral margin not crenulated in any of the studied specimens.

Measurements **(in millimeters):**

Measured Specimen	Valve	Height	Length	Width
Holotype (USNM 405241)	LV	15.9	16.1	4.3
Paratype (USNM 405242)	RV	14.0	16.4	3.8
Paratype (USNM 405243)	RV	17.5	20.1	4.3
Paratype (USNM 405244)	LV	19.4	22.0	5.3
Paratype (USNM 405245)	LV	16.0	17.5	4.2
Paratype (USNM 405246)	LV	14.3	15.2	3.6

From Ward 1992:85–96.

AN INSECT, *TROCHOIDEUS VENEZUELENSIS*, A NEW SPECIES OF ENDOMYCHID BEETLE FROM VENEZUELA Note the concise telegraphic style.

DESCRIPTION: *Male:* very similar to *T. boliviensis* from which it can be separated by the following characters: Body brown; head and antennae darker; legs and palpi lighter. Prothorax transverse, 1.6 times wider than long; anterior third 1.1 times wider than base; pronotal anterior margin concave, when viewed perpendicular to vertex; lateral borders almost straight, and slightly converging to base on posterior two-thirds. Elytra 1.3 times as long as their combined maximum width (at the end of the anterior third); 3.0 times longer than prothorax. Anterior femora of males with tooth in the middle of infero-anterior border, lacking bristles on its internal border.
Aedagus symmetrical (figs. 17–19).
FEMALE: Unknown
MEASUREMENTS: (mm, male): Total length 4.00; length of prothorax 0.72; maximum prothoracic width 1.14; basal prothoracic width 1.07; eletral length 2.14; maximum eletral width 1.72. . . . (Joly and Bordon 1996:6–7)

A FOSSIL INSECT, A NEW WOODGNAT, *VALESGUYA DISJUNCTA*, FROM OLIGO-MIOCENE AMBER. Note that although this is a much longer description, it is still telegraphic in style.

DESCRIPTION:

Head: Antenna strongly pectinate; with 12 flagellomeres, each approximately triangular in lateral (broadest) view, becoming slightly shorter apicad; apical flagellomere almost oval, with tip narrowed. Dorsal and ventral lobes of flagellomeres not symmetrical: ventral lobe slightly longer than dorsal lobe. Scape and pedicel each with ring of fine setulae in a single row. All three ocelli large, of equal size; diameter approximately equal to that of scape. Vertex covered with short, stiff, black setulae. Eye bare, not holoptic in male, but dorsally separated by approximately width of ocellus. Eye large, occupying almost entire lateral part of head; frontal margin concave, extended around base of antenna. Ventral margin of eye extended close to lateral oral margin and separated by distance approximately equal to width of palpus. Facets on frontal margin of eye more distant from each other than elsewhere. Proboscis vestigial; palpus black, protruding well beyond oral margin. Labellum present, but very tiny, at least half the size of palpus. No remnants of clypeus apparent. Frons with median suture or furrow running from anterior ocellus to slightly beyond dorsal margin of scape.

Thorax: Mesonotum and scutellum with even covering of acrostichal setulae, irregularly arranged. Postpronotal lobe dorsoventrally flat, long, almost flangelike, with row of fine setulae on ridge. Mesonotum and pleuron dark to light brown. Coxa long, slightly longer than 1/2 length of femur. Femur slightly shorter than tibia. All legs with five tarsomeres. Relative lengths of tarsomeres as follows: tarsomere 1 about 3 × length of t-2; t-3 slightly longer than t-2; t-4 and t-5 equal, slightly shorter than t-3 (see table 2). No apical spurs on tibiae. Legs without large, bristlelike setae, microtrichiae only, arranged randomly and evenly, not in rows. Pleuron and coxae entirely devoid of setulae.

Wing: Color slightly dusky, with dense, minute microtrichia only (no macrotrichia), arranged randomly. Wing slightly longer than body. Venation unique in Diptera: Costa ending midway between apices of R_{2+3} and R_{4+5}; Sc short, ending free; vein R_1 long, reaching to 2/3 or slightly more the wing length; R_{2+3} and R_{4+5} forked, base slightly longer than 1/2 length of fork. Vein R_1 about twice as thick as posterior veins, with fine setulae along 4/5 its length on dorsal surface, becoming more dense and numerous apicad. Fine setulae either in one row (rarely) or randomly arranged. Costa slightly thicker than R_1, with dense fine setulae randomly arranged. Large b-r-m cell present, M_1, M_2, CuA_1, and CuA_2 connected directly to cell, no forks. Anal vein present as fold only; anal lobe absent. Halter long, with stem about same length as knob, evenly covered with stiff, short black setulae, particularly on knob. Knob dark brown, stem light.

Abdomen: Uniformly light brown, with even covering of short black setulae. Setulae present on pleural membrane, but apparently light-colored and finer. Female genitalia with segment (VIII?) tubular, protuberant from segment (VII?). Arising out of apex of segment (VIII?) is a long, thin "ovipositor." Base of "ovipositor" is bulbous, with microtrichia, abruptly tapered to stylelike portion slightly longer than twice the length of basal bulb. Style consists of two "valves," apparently separated proximally to basal bulb. Each valve with a minute apical bifurcation; tips of dorsal bifurcation lobate and setose, ventral one pointed. Opposing surfaces of valves with longitudinal furrow. Fine stylet apparently occurs within ventral valve. Ventral valve apparently an elongate hypopygnial valve (sensu Peterson, 1981); dorsal valve is cercus. Male genitalia simple and plesiomorphic: with flat, lobate cerci; cylindrical, simple gonocoxites; membranous, trowel-like aedeagus/aedeagal guide; pair of sclerotized, flat, dentate lobes on anteromesal margin of gonocoxites (gonostyli?).

MEASUREMENTS: Body length (head to apex, in mm): males, mean of 5.0 (3.57–6.25, N = 8); females, mean of 4.24 (4.39, 4.09). Wing length: males, mean of 4.90 (4.39–5.59, N = 8); females, mean of 5.50 (5.14–5.84, N = 4). See table 2. (Grimaldi 1991:5–11)

A CORAL REEF TUNICATE FROM JAMAICA, *BATHYPERA GOREAUI,* **FROM A PAPER ON NEW ASCIDIANS FROM THE WEST INDIES** Note that this description is tightly written, but in telephonic style.

The body is oval in outline, forming a low dome with the siphons placed toward each end of the upper side. The length varies from 5 to 11 mm. (Fig. 111a, b). In life the animal is deep red, but preserved specimens tend to be orange or colorless.

The test has superficial spicules arranged in rows as in other species of the genus. Each spicule has a flattened bilobed base and a short stalk surmounted by a spherical knob which bears numerous spinous processes (Fig. 111c, d).

The siphons are slit-like, at least in the preserved specimens, and bear no trace of lobes. Inside each siphon is a ring of spicules similar to those on the surface of the body.

The mantle musculature consists of strong strands radiating from the siphons, passing down across the body, and crossed by more delicate

> circular strands. Circular muscles are also present round the siphons (Fig. 111e). The ganglion is visible as an elongated white body about half way between the siphons.
>
> 12–14 branchial tentacles are present, each long and thin with only 2 or 3 short simple side branches (Fig. 111f). The dorsal tubercle is oval with a simple longitudinal or slightly curved slit (Fig. 111f), and the neural gland forms a compact mass close to the tubercle.
>
> There are six conspicuous folds on each side of the branchial sac, and in one specimen examined the arrangement of longitudinal bars was: mid dorsal line 0 (10) 2 (12) 2 (14) 1 (14) 2 (13) 2 (12) 2 endostyle.
>
> A series of pointed languets occupies the mid dorsal line of the branchial sac. The stigmata tend to form infundibula on the branchial folds, and there, as well as between folds, they generally appear to be essentially spiral in arrangement (Fig. 111h).
>
> The alimentary canal forms a simple loop in the left side of the body (Fig. 111i). The oesophagus is short and sharply bent and the diverticula are either symmetrically placed or one slightly further forward than the other (Fig. 111j). The remainder of the alimentary canal forms a simple open loop and the anus is either plain-edged or has two shallow lips each bearing a small rounded lobe.
>
> There is one gonad on each side, that of the left being enclosed in the intestinal loop. Each gonad consists of an elongate ovary and a series of irregular and slightly lobed male follicles adjacent to the body wall. Both oviducts and common sperm duct are short (Fig. 111k). (Millar and Goodbody 1974:156, 158)

Figures accompanying this description include the whole animal, its spicules, and the taxonomically relevant details of its anatomy (figure 11.4).

Descriptions of Vertebrates

Note the use of telephonic style in these descriptions.

A NEW SPECIES OF FRESHWATER FISH, *BARBUS CARCHARHINOIDES,* **FROM WEST AFRICA** Note the use of a table of specimen measurements.

> DESCRIPTION: Based on the holotype and six paratypes. Morphometric measurements and meristic counts are given in table 1. See also figure 1.

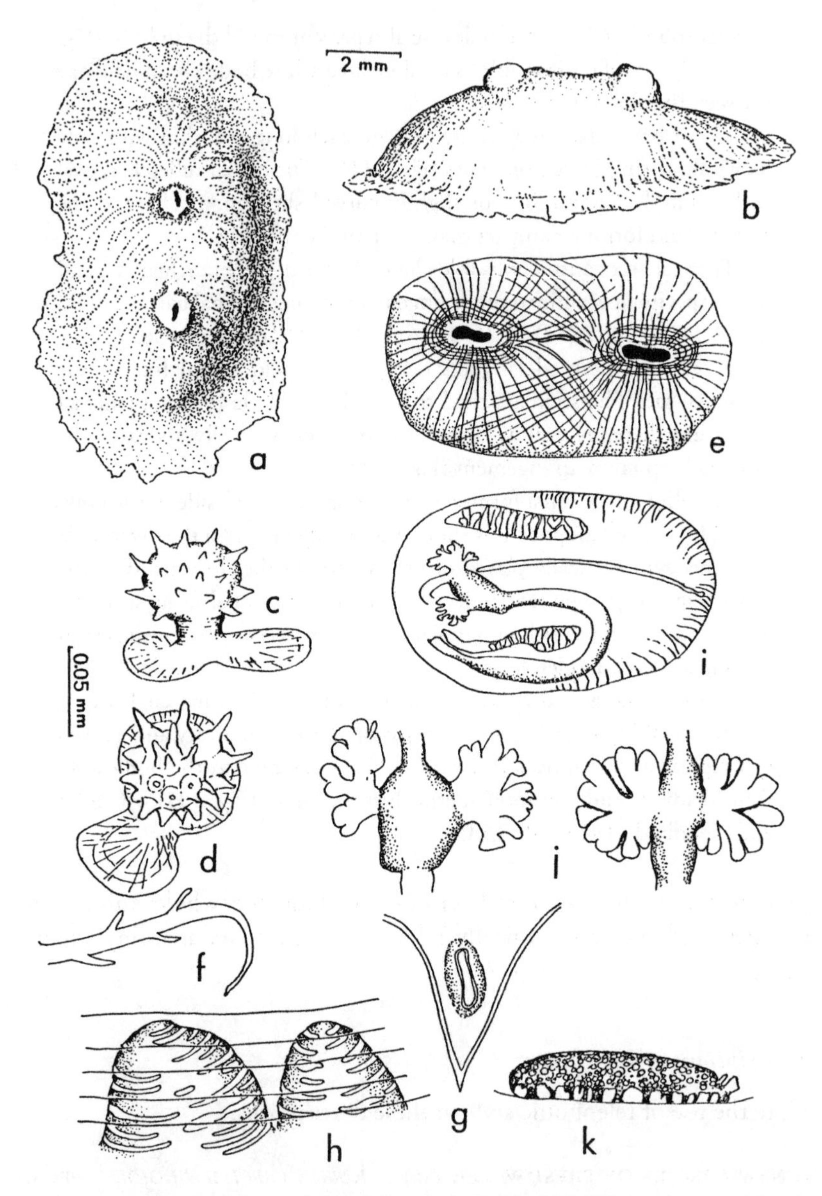

Figure 11.4 *Bathypera goreaui.* **A, B,** whole animals; **C, D,** spicules; **E,** mantle musculature in dorsal view; **F,** branchial tentacle; **G,** dorsal tubercle; **H,** portion of branchial wall; **I,** ventral view of animal showing alimentary canal and gonad; **J,** stomach and diverticula; **K,** gonad. *–Source: Millar and Goodbody (1974),* their figure 111.

The body of *Barbus carcharhinoides* is moderately compressed and attenuate. The predorsal body depth is relatively high, but tapers sharply behind the dorsal fin spines lending a somewhat sway-backed appearance to the body of the fish. The caudal peduncle is extremely elongate (LPC/LT 85–87%). The snout is bluntly pointed in profile, and the subterminally positioned mouth is moderate in size. Both pairs of barbels are relatively short: the anterior barbels extend only to the anterior margin of the orbit, and the posterior pair extend to the posterior border of the pupil. Cephalic pit lines are well developed and present in a series of rows on the cheek, and on the dorsal surface of the head between the orbits.

The pectoral fins are long and the tips overlap the origin of the pelvic fins. The prominent dorsal fin is considerably longer than the head and is composed of four unbranched spinous rays and eight branched rays. The last unbranched spinous ray is strongly ossified basally with a serrated posterior border and weakly segmented tip (fig. 2). The dorsal fin is expansive with a strongly falcate posterior margin. The anal fin bears three unbranched spinous rays and five branched rays. The caudal fin is strongly forked with pointed lobes.

The body is covered with large regularly imbricate cycloid scales. Striations on the scales are few in number and diverge from the central focus of each scale. Predorsal scales range between 9 and 10 scales along the midline. There are 28 or 29 scales in the longitudinal series, and the lateral line is more or less straight along the longitudinal series. Additional scale counts are given in table 1.

Gill rakers are weakly developed on the first gill arch. There are two or three small, nublike rakers on the ceratobranchial, one in the angle of the arch, and one on the epibranchial. All specimens have a vertebral formula of 16 abdominal and 17 caudal elements. The first infraorbital (= lachrymal) is elongate, but the remaining series is unexpanded (fig. 3). The pharyngeal bone is robust with an invariant tooth formula of 5–3–2:2–3–5.

The holotype and largest paratype (81.3 mm LS) are both sexually mature individuals with enlarged gonads.

COLORATION: In preserved specimens the dorsal body surface is darkly pigmented. Each scale bears numerous fine melanophores that are somewhat less densely scattered peripherally leaving a narrow paler zone at each scale margin. The lateral line scale pockets are marked with a dark, crescentic patch bisected horizontally by the lateral line pore. The dorsum of the head, the dorsolateral aspect of the snout, the upper jaw, and infraorbital region are also darkly pigmented. Three circular, black, midlateral spots are present. The first lies a little above the lateral line canals of the 6th and 7th scales. The second spot overlies the canals of the 14th and 15th lateral line scales, a little behind the vertical through the insertion of the last dorsal fin ray. A third spot is invariably present at the base of the caudal peduncle. In most individuals a fourth, smaller spot is also present between the second and third spots. This additional spot is not always present on both sides of the fish. The dorsal, caudal, and pectoral fins are slightly pigmented and a dusky gray in color, while the pelvic and anal fins are clear. Juvenile specimens lack much of the dorsal pigmentation of the adults.

Based on color photographs of freshly preserved material, *Barbus carcharhinoides* is silvery in life. The dorsum is a dark greenish brown, the midlateral spots are jet black, and the venter is silvery white. The fins are clear with a pale greenish yellow tinge. (Stiassny 1991:3–5)

A NEW BIRD, ***STACHYRIS LATISTRIATA,*** **THE PANAY STRIPED-BABBLER, FROM THE PHILIPPINES** Note the importance of color in this bird description.

DESCRIPTION OF HOLOTYPE:–Lores pale Cream Color 54 (capitalized color names and numbers are from Smithe 1975, 1981) grading into cream white in lower superciliary, auricular, and malar region; band of black on the forehead extending into the forecrown, upper superciliary and looping behind and below the auriculars; eye-ring cream white, crown and nape Greenish Olive 49 a shade darker than the back, which becomes lighter toward the rump; upper tail coverts rusty olive; tail dark Greenish Olive with a tinge of rust; edges of the rectrices have the same color as back; primaries Sepia 119 with pale yellow outer edges; secondaries dark Olive Green 48 with outer edges same color as the back and inner edges pale yellow; spot distal to bend of wing light Sulphur-Yellow 57; chin and upper throat cream white; lower throat light Sulphur Yellow becoming Buff-Yellow 53 on breast and belly; flanks including shank features Greenish Olive; broad black streaks beginning on throat, most prominent on breast, fading to dark Olive Green on belly and under tail coverts. Soft part colors of the living holotype 1 h after capture as described by artist John Ruthven, "Eye bright rust, outer edge lighter grading to bright rust toward pupil; eye-ring pale horn; bill with upper mandible dark horn, deeper at base and pale horn at cutting edge; lower mandible dark horn at the tip grading to light horn toward the cutting edge; gape pale horn with greenish overtone grading posterior to dull ochre then pale flesh; tongue pale horn at tip grading posteriorly to pinkish white; under-

tongue pale flesh; legs and feet bluish olive; toe pads medium chrome yellow fading to lighter on edges: nails horn, darker on ridge."
PARATYPIC VARIATION:–Overall colors vary from slightly darker to slightly paler than the holotype. There is variation in the intensity of color of the black streaks, which normally are confined to the breast but may occasionally (especially in males NMP 16358, 11673 and CMNH 34211, and female CMNH 34226) extend onto the belly. The black patch on the forehead and forecrown also varies in some specimens (particularly males NMP 16664, 16677, and female NMP 16675) with black feathers extending onto the crown just posterior to the eyes. Plumage variations seem to be individual and do not reflect the sex or age of the specimens in the series. The sexes are different in size, with males significantly (two sample t-test, $p < 0.05$) larger than females in weight, length of culmen, bill width and tarsus length, and highly significant ($p < 0.01$) in wing chord. Differences in tail length and distance from the bill to p to the nostril were not significant. (Gonzales and Kennedy 1990:368–369)

A FOSSIL LIZARD, *PROGLYPTOSAURUS HUERFANENSIS,* FROM THE EARLY EOCENE OF COLORADO

DESCRIPTION: The holotype skull (AMNH 7431) is slightly distorted. The posterior right inferior temporal region is represented by articulated osteoderms, and is folded inward toward the sagittal plane. The skull roof is flat. The skull itself is elongate, measuring 71 mm from tip of the premaxilla to the posterior nuchal osteoderms along the sagittal plane. The maxilla is slightly convex and bears (or has spaces for) 16 maxillary teeth. Premaxillary tooth count is 9. The teeth are obtuse, with blunt crowns and are essentially homodont along the length of the maxillae.

Dorsally, the cephalic osteoderms are sub-conical and bear a concentric ring pattern of tubercles around their periphery. Five longitudinal rows of osteoderms occur between the orbits. Five supraocular osteoderms are preserved in articulation in the right orbit. Three supraocular osteoderms (the two posteriormost and anteriormost) as preserved in the left orbit. The osteoderms along the left and right supratemporal region are folded downward off the parietal table forming a parallel ridge. The parietal foramen region bears minute cephalic osteoderms that become larger outward. The labial surface of the maxilla above the alveolar margin, and the entire premaxilla, are devoid of osteoderms as in other glyptosaurines.

Ventrally, the paired vomers are preserved in articulation along with the premaxilla, both maxillae, and the extreme anterior ends of both pterygoids are present. The vomers are toothless, complex, and measure 22 mm in length. The ventral surface of the parietal and left postorbital, left supratemporal, left jugal, and posterior part of the left frontal are preserved in articulation (fig. 3). Seen in interior view, the descending jugal margin is curved as in other glyptosaurs and lacks the jugal process found in melanosaurs. In ventral view, the supratemporal fenestrae are open but covered dorsally by cephalic osteoderms. (Sullivan 1989:4–5)

The Description Section: Plants (Examples of Styles for Different Groups)

Like original descriptions of animals, original descriptions of plants should be long enough to cover the essential points and short enough not to obscure them (Davis and Heywood 1965). Style should be concise, and the description should follow a conventional sequence that works from general to specific. For example, in angiosperm taxonomy the plant is described from below upward and the flower and floral parts from outside inward. So the sequence is underground parts (roots, tubers), its habit or growth form (size, shape, branching, leaf outline, bark, etc.) and then leaves (arrangement, shape, size, etc.) and epidermal characters of stems and leaves. Flowers are first described in general (size, position, color, scent), then the organs of the flower (calyx, corolla, stamens, staminodes, ovary), with fruit and seeds being described last (Davis and Heywood 1965).

Measurements may be included and are often important. As with animals, the measurements used in the description should be limited to those useful in distinguishing the species and in comparing it with relatives or similar species.

As with animals, attention should be paid to all life history stages and to nonmorphological characters. For example, although angiosperm taxonomy is based on the plant at a mature and reproductive stage, important taxonomic differences can be found in juvenile plants and in embryos and seedlings (Davis and Heywood 1965). In addition, ultrastructural, physiological, genetic, and phytochemical characters have now been incorporated into the systematics of many groups of plants and have been useful in clarifying their phylogenetic relationships (Stace 1989; Stuessy 1990). Only research into the literature on the groups of plants to which your species belongs will show which of these characters you should include in your description.

If the diagnosis has been given in Latin, the description section does not have to be written in Latin, although some botanical taxonomists recommend it. They do so not to torment nonsystematists, but for the same practical reason the Caterpillar tractor people developed their Restricted English system: communication. Botanical Latin is a standardized code that can be deciphered by the international audience of botanical systematists, the readers most likely to need the information. In fact, they will probably find it easier to translate the telegraphic Latin diagnosis section than the remainder of a paper in an unfamiliar language.

As with animal groups, style and terms used may be different for each group of plants. If your work is to be accepted by other students of the group, it is important that you know how to use the terms in accordance with current practice in the field and follow accepted style. The following are examples of description sections for new species from groups covered by the ICBN.

A NEW DIATOM, *COCCONEIS ANDERSONII,* FROM THE GULF OF EILAT, ISRAEL

Valve elliptical or almost circular to linear-elliptical. **Raphe valve** with narrow axial area and quite narrow hyaline marginal area. Central area lacking. Raphe filiform with proximal raphe ends only slightly enlarged, distal ends appearing straight. Striae slightly radiate, composed of single rows of poroid puncta (SEM). **Pseudoraphe valve** with narrow axial area. Striae slightly radiate, covered by a perforate membrane. Perforations (puncta) indistinct, seemingly in double rows (SEM). Interstriae (costae) with one or two narrow rows of longitudinal connecting costae (internally). Externally, the interstriae occasionally bifurcate near the margin (SEM). Striae 28–30 in 10 μm. Length 3.5–7 μm. Width 3–4.5 μm. (Reimer and Lee 1988:340)

A NEW FERN, *POLYPODIUM ALANSMITHII,* FROM MESOAMERICA

Epiphytic, rarely epipetric or terrestrial; rhizome 2–4 mm wide, long-creeping, the scales 2–4 mm, lanceolate, bicolorous, black medially with pale brown margins, clathrate or subclathrate medially, appressed, denticulate to erose; petiola ca. equaling the lamina, dark brown, nonalate; lamina (7-)10–18(-22) × 4.5–15 cm, broadly deltate to broadly oblong, pinnatisect, sparsely scaly on both surfaces, the scales usually circular or ovate, occasionally with an acicular apex, subentire to denticulate; pinnae 2.5–8 × 0.3–0.6(-8) cm, pairs 8–16, entire to crenate, the distal ones ascending; veins free, obscure; sori round, not embossed adaxially. (Moran 1990:845)

A NEW ANGIOSPERM, ***CYANEA DOLICHOPODA,*** **MEMBER OF A GENUS OF TREES AND SHRUBS ENDEMIC TO THE HAWAIIAN ISLANDS**

Shrub, 1 m tall; stem unbranched, erect, unarmed, glabrous. Lamina ovate or oblong, 6.5–10 cm long, 3.7–5.8 cm wide; upper surface green, glabrous; lower surface light green, with scattered hairs along the midrib; margin callose-serrulate; apex acute or obtuse; base cordate, often markedly asymmetric; petiole 9–16 cm long, slender, glabrous. Inflorescence (not fully expanded) spreading or pendent, 8–11–flowered, sparsely pubescent; peduncle 4–4.5 cm long, slender; rachis 2–2.7 cm long; bracts linear, 5–7 mm long; pedicels 12–15 mm long, spreading, bibracteolate at the middle; bracteoles linear, 7.5–8 mm long. Corolla (in bud) pinkish, glabrous. Fully expanded flowers, fruits, and seeds not seen. (Lammers and Lorence 1993:432)

A FOSSIL PLANT, ***SAPINDOPSIS MINUTIFOLIA,*** **FROM THE LOWER CRETACEOUS OF VIRGINIA**

Description: Pinnately compound leaf with at least 5 leaflets preserved. Leaf > 2.8 cm long by about 2.5 cm wide, distance between the terminal leaflet and the adjacent pair of lateral leaflets less than the distance between the two pairs of lateral leaflets; rachis without a distinct wing of laminar tissue. Terminal leaflet with a symmetric base, contracted at rachis, rachis prolonged 1 to 2 mm beyond lateral leaflets; lateral leaflets sessile, each with an asymmetric base that narrows at its attachment to the rachis; basal pair of lateral leaflets incompletely preserved; terminal and lateral leaflets otherwise similar in size and shape. Leaflets approximately 15 mm long by 3–4 mm wide, exact length difficult to determine because the apex of most is missing, estimated L/W 3.5 to 4; apex of one leaflet almost complete, probably acute; base of leaflets acute to decurrent, without a joint at the point of attachment to the rachis; margin serrate, serrations straight to concave on the apical side and convex to straight on the basal side, strongly asymmetric, minute, 1 to 2 mm long by < 1 mm tall. Primary venation of leaflets pinnate; primary vein massive. Secondary venation of leaflets poorly preserved, appearing to be semicraspedodromous; secondary veins very thin relative to primary vein, opposite, > 10 pairs (estimated), irregularly spaced, diverging at moderate acute angles, often sinuous to zig-zag in course, showing little apical curvature until approaching margin, each secondary vein typically branching near margin to form two veins, with the admedial branch connecting with the superadjacent secondary vein to form a brochidodromous loop and the exmedial branch terminating at the apex of a tooth; intercostal regions of variable size and shape, looping angular. Tertiary venation poorly preserved, appearing to be irregularly reticulate; tertiary veins appearing to originate at acute to obtuse angles. Teeth with sharp tips, showing no clear evidence for glandularity, perhaps lost in preservation, placement of medial vein appearing to be symmetric, no accessory veins observed. (Upchurch et al. 1994:42–43)

Illustrating Taxonomic Descriptions

The most important rule on illustrating taxonomic descriptions is this: *Do it!* The Caterpillar Tractor Company's 800-word vocabulary manuals would not have worked nearly so well if they had not included in them illustrated lists of parts and detailed diagrams of all the equipment (Kirkman 1992). Illustration is not an addition to your description. It is an integral part of it. You may not be a great artist (although many biologists are talented artists), but you can learn to do adequate drawings. The great invertebrate zoologist Libbie Hyman was no artist, as she herself admitted in her youth. But she took drawing lessons and she practiced for years, and in midcareer, a visitor described her as drawing rapidly and tirelessly all afternoon (Winston in press). If drawing is impossible for you, you can pay a scientific artist to do your illustrations for you. Graduate students without the means to pay might find an art student who would do the work just for the opportunity to see their work in print. A number of helpful guides to preparing scientific illustrations are currently in print. They are listed in this chapter's *Sources.*

The kind of illustrations you use depends on both the organism involved and the journal in which you publish. All journals can handle line drawings, but the quality of reproduction of half-tones (drawings with gray shading and photographs) varies from journal to journal.

Line drawings are perfectly acceptable for many descriptions. They can be produced by inexpensive low-tech means (ink and paper) or generated using various computer drawing programs. As the Council of Biology Editors' guide, *Illustrating Science,* states, "A good pen and ink drawing is pleasing, informative and reproduces well, even on poor grades of paper" (CBE 1988:35). Good composition is important. The drawing should be organized so that the eye is led to the main area or point. It should not be cluttered with too much detail, and there should be a pleasing balance between detail and white space (CBE 1988). Line drawings are usually made larger than needed (about 1 1/2 to 2 times the desired final size) and then reduced for printing (Zweifel 1961). This lets the artist put in the details more easily. The reduced final size will look sharper and small imperfections will usually be minimized. Figures 11.3, 11.4, and 11.5 are all examples of line drawings created to illustrate species description.

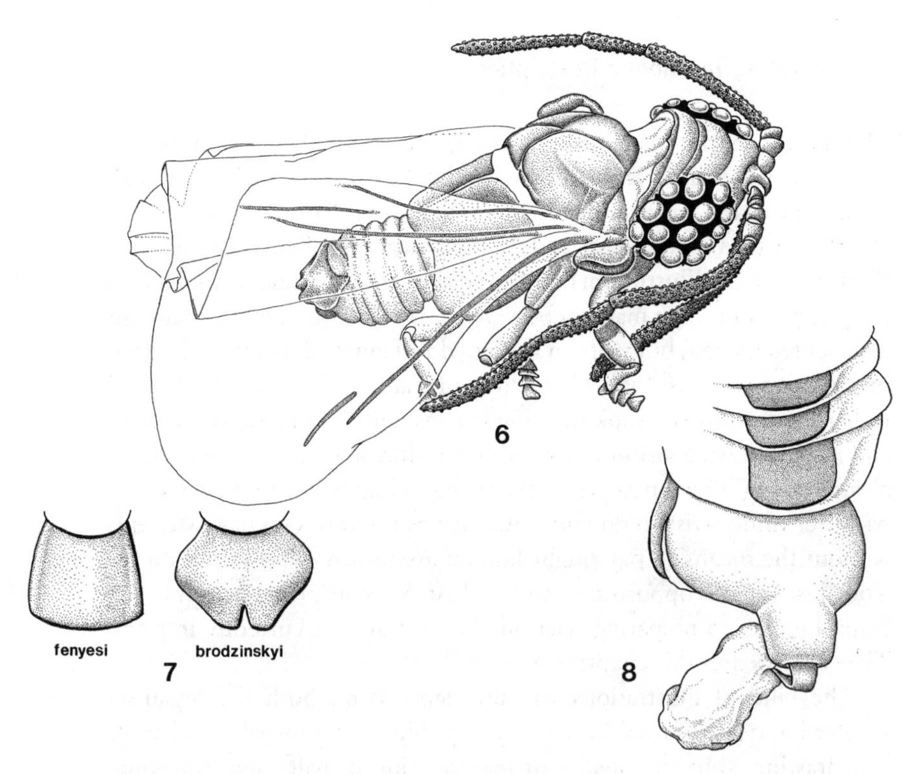

Figure 11.5 Fossil insect, a twisted-winged parasite (Strepisiptera) from 20-m.y.-old Dominican amber. In addition to its aesthetic appeal, a good drawing may actually present more information than a photograph, especially for small subjects, in which photography is limited by narrow depth of field. A drawing can show the entire organism in focus. *–Source: Line drawing with stippled detail by D. Grimaldi, AMNH.*

Technical pens (Kohninor, Rapidograph, Castell) and India ink are recommended. Some new artists' or technical markers use permanent ink. They can be used for drawings, but they may not produce as sharp a line as a regular refillable technical pen.

Drawings are usually made in first in pencil on drafting paper (a heavy high-quality tracing paper), then inked in. The drawing is then mounted on Bristol board, mat board, or poster board. Thick, heavy board protects the drawing from being bent, but higher shipping costs have caused some editors to request that you use the lighter poster board instead. A dry mounting press and tissue, spray adhesive, rubber cement, or a glue stick can be used for mounting the drawing on its backing.

After the drawing is attached to the backing board, it should be covered with a protective overlay of drafting or bond paper, tape-hinged at the top so that it can be folded back when the figure is copied for publication. Label the figure, either well outside the margin of the drawing on the front, or on the back of the mounting board, with the author's name, article title, and figure number. The desired reduction can be marked on the overlay (if it is transparent enough) or in nonphotographic blue pencil in the margin of the illustration.

It is generally recommended that illustrations be submitted at no more than twice the final reproduction size (CBE 1988). It may be necessary to have a large composite plate photographed professionally and the print sent, rather than the very large figure. It is probably always best, when possible, to send a good photographic reproduction of the plate, as it avoids the risk of having the original lost. Editors will request the original if they need it. It is a good idea to try photocopying your illustration at a reduction as close as you can make it to the probable final size before sending it, even if you send the original. This will ensure that labeling and symbols are big enough and that detail is not lost. A stippled drawing may look well shaded at the size you have drawn it, but when you reduce it you may find that closely spaced dots have merged into an ugly blob or that gaps take on undesirable shapes. It is better to learn that before you send the paper in than when you see the proofs. Most mistakes can be corrected with white correction fluid. Some publications ask for camera-ready copy. In this case you must either produce the illustration at the final size or have it photographed and reduced.

Good photographic reproduction is desirable to most scientists, but taxonomists and workers on ultrastructure are probably the pickiest about publishing in journals that keep to a high standard of reproduction of plates. Half-tone (continuous tone) illustrations are reproduced for publication by breaking them into dots using a screen or grid pattern. The number of lines used greatly affects the quality of the final product. Newspaper quality is 85 lines (85 dots/inch or 8500 dots/square inch). Scientific publications normally use 100–150 line screens. A few journals (such as the publication series of the American Museum of Natural History) use 300 line screens. The finer the screen used, the more difficult it is to print and the higher the cost in higher-quality ink, paper, and print control (CBE 1988; Wood 1994).

General advice on illustrating descriptions can be reduced to the following important points:

Illustrate the important distinguishing characters in however many plates or figures it takes, but remember that illustrations, especially photographic illustrations, are expensive to print. Make sure they tell the story, but try to tell it with the minimum number necessary. Related illustrations should be grouped into a figure or plate. They may be organized to show all the important features of a species (e.g. figure 11.4) or to compare the same structure in different taxa (figure 11.6). For micrographs (SEM, TEM) it may be especially important to look for a journal with high-quality reproduction (half-tone screens of at least 150–200 lines, on coated paper stock). Pictures should be cropped to best show important features, as in figure 11.7, a plate illustrating the features of an Australian sponge, *Collospongia auris,* in which the tightly cropped photographs include the holotype specimens, plus SEM and light micrographs of important structural characters.

Size the illustration properly for the page size of the journal. Most scientific journals have a two-column format and many are published in a standard 7- by 10-inch size, with columns running about 2 1/2 to 2 3/4 inches (6–8 cm) in width. Single graphs and figures with a square or vertical format (taller than they are wide) are usually reduced to single column width. Horizontally rectangular figures are more often kept full page width, as are plates scaled to the full page size.

Make sure letters of labels are big enough to see (at least 8 point) in the final reduction. The CBE guide recommends using labels that use both upper- and lowercase because they are easiest to read. Save the all-capitals labels for titles or headings. If illustrations are

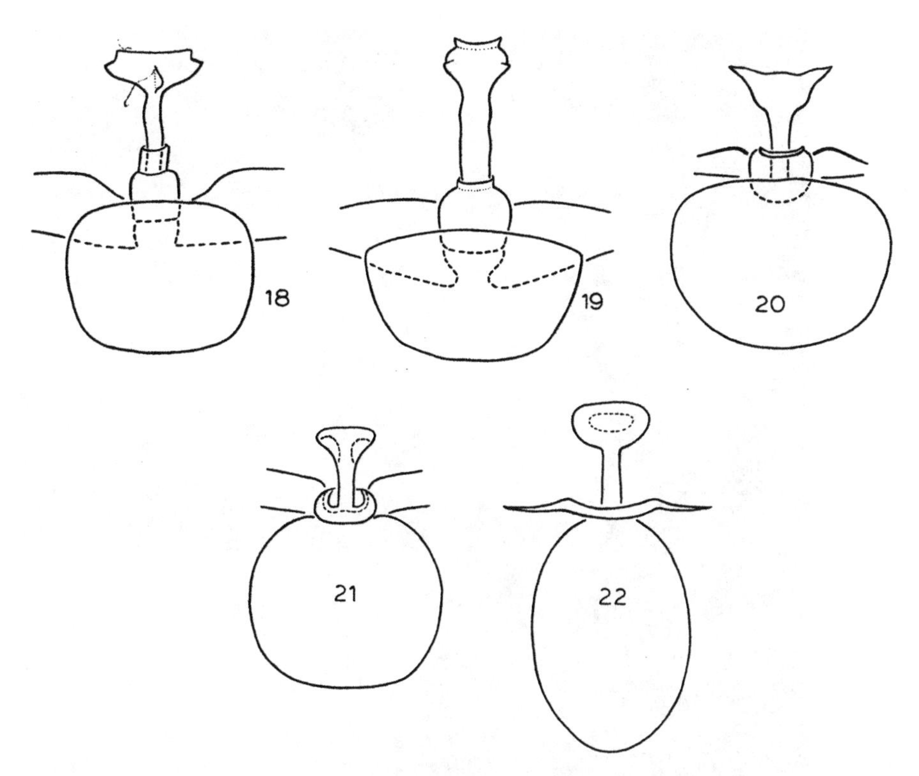

Figure 11.6 Comparative plate showing a taxonomically important character of all the species of a new spider genus, *Unicorn.* **18,** *Unicorn catleyi.* **19,** *U. conao.* **20,** *U. socos.* **21,** *U. chacabuco,* along with **22,** *Xiobarg plaumanni,* a species of a closely related genus for comparison. *–Source: From Platnick and Brescovit (1995), courtesy of the American Museum of Natural History.*

prepared by hand, well-spaced transfer letters, uniform in style and size, do the job well. Nonserif type faces, such as Helvetica, are recommended because they take reduction better. Most graphics and presentation programs also have a selection of sizes and type fonts available. Labels can be applied directly to the photograph or illustration or to an acetate overlay (transparent acetate sheets can be purchased from art supply stores or catalogs).

Don't leave out scales. Make the size of the object illustrated clear by using a line (scale bar) showing its actual size. This may be labeled with transfer letters and numbers or its size may be indicated in the figure legend (e.g., "scale bar = 100 μm"). The legend can show how much the figure has been reduced or magnified (e.g., "× 500," "1/2

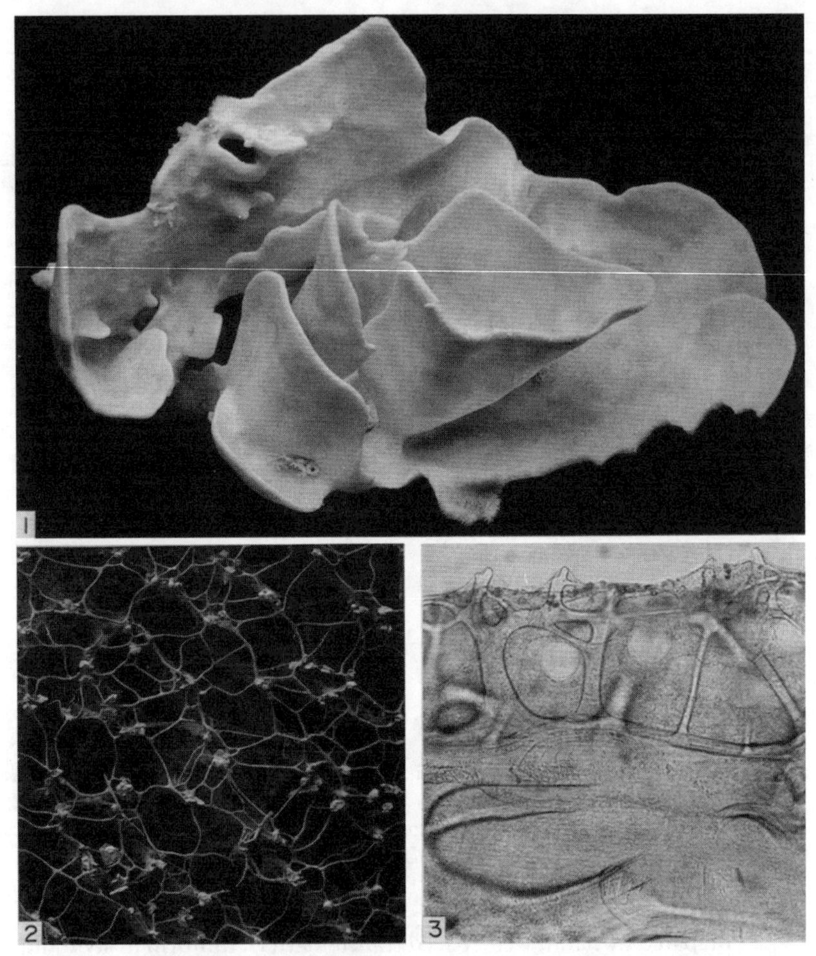

Figure 11.7 *Collospongia auris.* A composite plate illustrating morphology of holotype specimen as well as taxonomically important fiber characters as seen in SEM and light micrographs. **1,** Upper surface of holotype of *Collospongia auris* Australian Museum Z5035. Spirit-preserved specimen 50 percent natural size. **2,** Scanning electron micrograph of the tangential surface network of fine fibers of *Collospongia auris* (× 40). **3,** Light micrograph showing the strongly laminated primary fibers in a central position in the sponge body of *Collospongia auris* and secondary elements extending to the poral surface and intersecting with the fine network in the plane of the surface (× 80). *–Source: Reprinted from Bergquist et al. (1990), p. 351, with permission from Elsevier Science.*

actual size," or "life-size"). Obviously, for this approach you need to know exactly what size the final reproduction on the printed page will be; for this reason most nonprofessional illustrators prefer the first method.

Figure 11.8 Holotype of *Aethopyga linarabonae.* A new sunbird from the Philippines (Kennedy et al. 1997) For larger organisms, a sharp black and white photograph should be a part of the species description. *–Photo: A. P. Sutherland, © Cleveland Museum Center.*

Provide in text or legend the information necessary to link the illustration to the specimen. Remember to allow space for the legend when sizing the illustration, particularly if it is to be a full-page plate.

Make sure the illustrations are as clear and sharp as you can make them. If there is a choice, go for more rather than less contrast. Reprints and journal issues run out. Even if you have done your best to publish a high-quality photograph, later readers will be stuck with a photocopy of your illustration, and a higher-contrast photograph seems to copy best.

Figure legends should not be put on the illustration itself. Some journals specify that the legend for each figure be typed on a separate page. Others want all the figure legends grouped consecutively. Check the instructions for authors. Any symbols other than common ones should be identified in a key on the illustration. Printers may not have the type to reproduce uncommon symbols in the legend.

Finally, even today when every magazine and junk mail advertisement we receive is crammed with color photographs, black and white illustrations are still the standard for scientific publications. Color photographs are prohibitively expensive for publications with the limited circulation (and budgets) of most scientific journals. Very few journals will allow you to have a color plate unless you can pay some or all of its cost. But keep pushing for it. Raise the money if you can. As a field biologist, part of your contribution to systematics is to integrate the living animal or plant into the field of systematics. Color photographs can draw attention to the whole organism. They can put it in context within its environment, and they may make it possible for later ecologists to make identifications in the field without catching or harming the study organism.

Sources

Writing Guides and Style Manuals: General

Cook, C. K. 1986. *Line by Line: How to Improve Your Own Writing.* Boston, Houghton Mifflin.

Kane, T. S. 1988. *The New Oxford Guide to Writing.* New York, Oxford University Press.

van Leunen, M. C. 1992. *A Handbook for Scholars,* rev. ed. New York, Oxford University Press.

Zinsser, W. 1988. *On Writing Well,* 3d ed. New York, Harper & Row.

Writing Guides and Style Manuals: Scientific

Council of Biology Editors, Style Manual Committee. 1983. *CBE Style Manual,* 5th ed. Bethesda, Md., Council of Biology Editors.

Council of Biology Editors, Style Manual Committee. 1994. *Scientific Style and Format: The CBE Style Manual for Authors, Editors, and Publishers,* 6th ed. Cambridge, U.K., Cambridge University Press.

Day, R. A. 1988. *How to Write & Publish a Scientific Paper,* 3d ed. Phoenix, Oryx Press.

Jones, B. 1988. *Style Manual. Scientific Publication Series for the American Museum of Natural History,* 3d ed. New York, American Museum of Natural History.

Kirkman, J. 1992. *Good Style: Writing for Science and Technology.* London, E. & F.N. Spon.

Mancuso, J. C. 1990. *Mastering Technical Writing.* Reading, Mass., Addison-Wesley.

Matthew, J. R., J. M. Bowen, and R. W. Matthews. 1996. *Successful Scientific Writing.* New York, Cambridge University Press.

O'Connor, M. 1991. *Writing Successfully in Science.* London, Chapman & Hall.
Turk, C. and J. Kirkman. 1989. *Effective Writing: Improving Scientific, Technical and Business Communication.* London, E. & F.N. Spon.

Color Charts

GSA Rock Color Chart; with Genuine Munsell Color Chips, rev. text. 1991. Boulder, Colo., Geological Society of America. (For paleontology.)
Kornerup, A. and J. H. Wanscher. 1984. *Methuen Handbook of Colour,* 3d ed., rev. by Don Pevey. New York, Hastings House. (A standard reference.)
Munsell Color Company. 1976. *Munsell Book of Color: Glossy Finish Collection,* 2 vols. Baltimore, Munsell Color.
Munsell Color Company. 1976. *Munsell Book of Color: Matte Finish Collection.* Baltimore, Munsell Color. (With above, a standard reference.)
Munsell Color Charts for Plant Tissues, 2d ed. rev. 1977. Baltimore, Munsell Color. (17 color charts; main use is diagnosing plant diseases.)
Munsell Soil Color Charts, rev. ed. 1994. Baltimore, Munsell Color.
Pantone Color Selector: Designers' Guide. 1996. 2 vols. Carlstadt, N.J., Pantone.
Pantone Color Selector 1000. 1993. Carlstadt, N.J., Pantone.
Pantone Color Survival Kit. 1995. 3 vols. Carlstadt, N.J., Pantone. (Pantone volumes intended mainly for designers and graphic artists.)
Ridgeway, R. 1912. *Color Standard and Color Nomenclature.* Washington, D.C., A. Hoen. (Classic reference, long out of print.)
Royal Horticultural Society. 1995. *RHS Colour Chart,* 3d ed. London, The Royal Horticultural Society. (In print; standard for botanists, but has all colors in four color fans, plastic container.)
Smithe, F. B. 1974. *Naturalist's Color Guide Supplement.* New York, American Museum of Natural History.
Smithe, F. B. 1975. *Naturalist's Color Guide.* New York, American Museum of Natural History.
Smithe, F. B. 1981. *Naturalist's Color Guide Part III.* New York, American Museum of Natural History. (Used mainly by ornithologists.)
Villalobos-Domingues, C. and J. Villalobos. 1947. *Atlas de Colores.* Buenos Aires, El Ateneo. (Another classic color reference; text is in Spanish and English.)

Scientific Illustration

Briscoe, M. H. 1996. *Preparing Scientific Illustrations: A Guide to Better Posters, Presentations and Publications,* 2d ed. New York, Springer-Verlag.
Cook, C. D. K. 1998. A quick method for making accurate botanical illustrations. *Taxon* 47:371-380.
Council of Biology Editors, Scientific Illustration Committee. 1988. *Illustrating Science: Standards for Publication.* Bethesda, Md., Council of Biology Editors.

Dalby, C. and D. H. Dalby. 1980. *Biological Illustration: A Guide to Drawing in Black and White.* Field Studies no. 5. London, Field Studies Council.

Downey, J. C. and J. L. Kelly. 1982. *Biological Illustration: Techniques and Exercises.* Ames, Iowa State University Press. (A course in biological illustration, illustrated by students' work.)

Hodges, E. R. S. et al., eds. 1989. *The Guild Handbook of Scientific Illustration.* New York, Van Nostrand Reinhold.

Holgren, N. H. and B. Angell. 1986. *Botanical Illustration. Preparation for Publication.* New York, New York Botanical Garden.

Kelvin, G. V. 1992. *Illustrating for Science.* New York, Watson-Guptill.

Nield, E. W. 1987. *Drawing and Understanding Fossils.* Oxford, U.K., Pergamon.

West, K. 1983. *How to Draw Plants. The Techniques of Botanical Illustration.* The Herbert Press in Association with the British Museum (Natural History), London.

Wood, P. 1994. *Scientific Illustration.* New York, Van Nostrand Reinhold.

Wunderlich, E. B. 1991. *Botanical Illustration in Watercolor.* New York, Watson-Guptill.

Zweifel, F. W. 1988. *A Handbook of Biological Illustration,* 2d ed. Chicago, University of Chicago Press.

Chapter 12

Taxonomic Discussion Section

The Code refrains from infringing on taxonomic judgment, which must not be made subject to regulation or restraint.

–ICZN (1985:xiii)

Of course speculation *is in order in a Discussion, but it must be reasonable, firmly founded on observation, and subject to test, if it is to get past a responsible editorial board.*

–Woodford (1986:29)

Purpose of the Discussion Section

The discussion section answers the basic question, What do the results mean? It should take the reader back to the introductory statement of the purpose of the paper and answer the question or questions raised according to the results given in the body of the paper. A complete discussion section should cover the theoretical and practical implications of the results, point out exceptions or anomalies and still-existing gaps in knowledge, show how the results agree or disagree with other work, and state the conclusions, giving (succinctly) the evidence for each (Woodford 1986; Day 1988; O'Connor 1991).

In a taxonomic paper the discussion is the section in which you give your evidence and line of reasoning for the taxonomic judgment you have made (e.g., in naming a new species or assigning your specimens to a particular taxon). As the first quotation in this chapter states, your judgment is not subject to regulation by the codes, but it *is* subject to the same code as any scientific discussion: Your decision must follow logically from your results and be upheld by your evidence. The discussion is usually the most enjoyable part of the taxonomic paper to write–the reward for hours of often tedious work. It is also the section that is most likely to get your paper rejected if you do not stick to the guidelines given in the second introductory quotation.

Finally, it is the section that places the heaviest ethical responsibility on the author, not just in terms of general scientific ethics, but because survival or extinction of a species may depend on decisions made about its taxonomy. Mistakes in taxonomy can make the most careful conservation plan useless.

For example, the tuatara, *Sphenodon,* is the only surviving genus of an order of reptiles that flourished during the Triassic period, and thus represents an important proportion of reptilian genetic diversity. Daugherty et al. (1990) have shown the difference taxonomic neglect has made in the survival of species of tuatara in their last remaining habitats, islands off the New Zealand coast. Two still extant species were described in the nineteenth century and a third was described as a subspecies in 1943. But protective legislation put in place in 1895, although well-intentioned and advanced for its time, recognized only one species, *Sphenodon punctatus.* Conservation management since then has been based on the assumption that only one species existed. If there had really been only one species, the extinction of a quarter of its population during the last century would have been, if not desirable, at least a tolerable loss. However, Daugherty et al.'s analysis of allozyme and morphometric data showed the existence of two species. One, *Sphenodon punctatus,* seems secure, with populations still inhabiting 28 islands. But the other species, *S. guntheri,* is now reduced to a population of less than 300 individuals, occurring on just one island. Its survival is questionable. No individual of a subspecies of *S. punctatus* described in 1943, *S. p. reischeki* (which was also found on only one island), has been seen since the late 1970s. It is probably extinct. The authors recommended increased conservation measures on those islands, but admitted it might be too late (Daugherty et al. 1990; May 1990).

Discussion in Descriptions of New Species

The discussion section in a new species description usually follows the description section and comes before the ecology and distribution sections. It can be variously titled (e.g., "Discussion," "Remarks," Taxonomic Discussion," "Taxonomic Implications," "Relationships," "Comments," "Comparisons"), follow untitled, or, more rarely, be mixed in with the description (as a "Results and Discussion" section). Occasionally it follows the entire description as the final section of the paper. This is rare in a description of a single species. It is more common in composite papers in which additional biological or other evidence is being included, in a redescription of a species, or in a revision of a group. Obviously, as with all other sections of your paper, you should consult the journal in which you hope to publish for style before you start to write.

A description of a new species may not need a discussion section at all. In some instances its taxonomic position and the rationale for its publication may be obvious from the diagnosis and description alone. In other cases, as shown in chapter 10, the remarks section may be used in place of a discussion section, supplying the diagnosis of the new species and limited to a brief discussion of those diagnostic characters.

However, you usually do include a separate discussion section to give the reasons for your decision to describe a species as new. Here you present your comparative evidence, including charts, graphs, or diagrams, that show how you distinguished the species from its relatives. Your discussion can be simple or elaborate, depending on the amount of information you have. It may be a single paragraph, perhaps illustrated by a diagram or table (figure 12.1), as in this example from the discussion of a new marine polychaete, a scale worm.

828 PROCEEDINGS OF THE BIOLOGICAL SOCIETY OF WASHINGTON

Table 1.—Comparison of *Harmothoe globosa,* new species and *H. macnabi* Pettibone.

	Harmothoe globosa	*Harmothoe macnabi*
Elytral macrotubercles	balloonlike, near border and on surface (Fig. 1D–F)	extensions on posterior border (Pettibone 1985c, Fig. 6C)
Prostomium	large cephalic peaks; 2 pairs of large eyes (Fig. 1A)	small cephalic peaks; without eyes (Pettibone 1985c, Fig. 6A)
Tentaculophore	with 2 setae (Fig. 1A)	with 6–8 setae (Pettibone 1985c, Fig. 6A, B)
Neuropodial presetal acicular lobe	with well-developed supraacicular process (Fig. 2A)	without supraacicular process (Pettibone 1985c, Fig. 7B)

Figure 12.1 A table can be an effective way to summarize the evidence used to distinguish a new species from others, as was done here for a new species of scaleworm. *–Source: From Pettibone (1990).*

> *Remarks:*–The unusual balloonlike macrotubercles found on the elytra of *Harmothoe globosa* suggest a greater development of the macrotubercles than those found on *H. macnabi* Pettibone (1985c: 749, figs. 6, 7), described from the Galapagos Rift in the Eastern Central Pacific. The two species differ as shown in Table 1. (Pettibone 1990:827–828)

Or it may be pages of text, statistics, and graphics.

You may use the discussion section to discuss the generic position of your new species, giving the evidence on which you have based your decision to assign it to a particular genus, as in this discussion of a new hydroid from the Great Barrier Reef.

> REMARKS.
>
> The lack of prominent marginal cusps is a feature shared with *C. hummelincki* (Leloup, 1935), *Orthopyxis integra* (Macgillivray, 1842), and *O. crenata* (Hartlaub, 1901b). In comparison with the present species, these have much less annulation of the pedicel, and the gonothecae are markedly different. Furthermore, species of *Orthopyxis* L. Agassiz, 1862, are characterized by presence of a sub-hyrothecal spherule and absence of a diaphragm (Cornelius, 1982). Using these criteria, therefore, despite the superficial similarity, the present species should not be referred to *Orthopyxis.* The gonothecae in our material do not clearly show the formation of medusae but the wide aperture suggests that free medusae are released. (Gibbons and Ryland 1989:399–400)

You may also want to compare the new species with relatives within the same genus, summarizing again its unique and diagnostic features, as done by Goodbody and Cole (1987) in their description of a new ascidian from Belize.

> The species seems most closely related to *Perophora modificata,* recently described from Australia by Kott (1985). Both species have large zooids arising from a tangled mat of stolons and are similar in the orientation of the zooid on the stalk, large number of long stigmata and large number of testis lobes. *P. modificata* differs from *P. regina* in having a thickened test, longitudinal rather than oblique muscles, and in the possession of a central vascularized extension of the body with paired stolonic vessels at the posterior end. The trifid nature of the branchial papillae in *P. regina* appears to be unique among the members of the genus. *P. regina* differs

> from the other western Atlantic species, *P. formosana* and *P. viridis,* in its size, large number of testis lobes and in the form and extent of its mantle musculature. (Goodbody and Cole 1987:252)

You may include a key to help in others distinguish your new species from similar species, as was done in this paper describing a new Brazilian bee fly.

> Discussion.–The color pattern of the wing easily distinguishes *H. diminuta* from other species in the genus. In Painter & Painter's (1968) key, this species is near *H. surinamensis,* as in both, the line between hyaline and dark parts at tip of wing is straight. The following key separates these two species.
> Triangular hyaline area in the end of second basal cell; a broad hyaline area in center of wing separated by a band of black bordering the third posterior cell, this inferior part reaches margin of wing (Fig. 3) . . . .
> . . . . *H. surinamensis* Rondani
> Round hyaline area at second basal cell; narrow hyaline area in center of wing, separated by a large band from a small hyaline area in third posterior cell, which does not reach margin of wing (Fig. 1) . . . .
> . . . . *H. diminuta.* (Souto Couri and Einicker Lamas 1994:121)

Or a cladogram that places the new species phylogenetically, as Norman Platnick did for the species in a South American spider genus (figure 12.2).

Of course, many new species discovered by ecologists are actually not new in the sense of never having been seen by human eyes, but instead are rare or even common species that have simply been overlooked or misidentified for years. In such cases, you may want to use the discussion section to point out how others failed to distinguish the species in the past, as was done in this discussion of a new fungiid coral, *Ctenactis albitentaculata.*

> Remarks.–Specimens have been collected and photographed at several locations (see the literature mentioned above), but have not been recognized as belonging to a still undescribed species. However, the consistently white tentacles, which make the species easily recognizable in the field, are not the only character distinguishing it from other species. . . . (Hoeksema 1989:158)

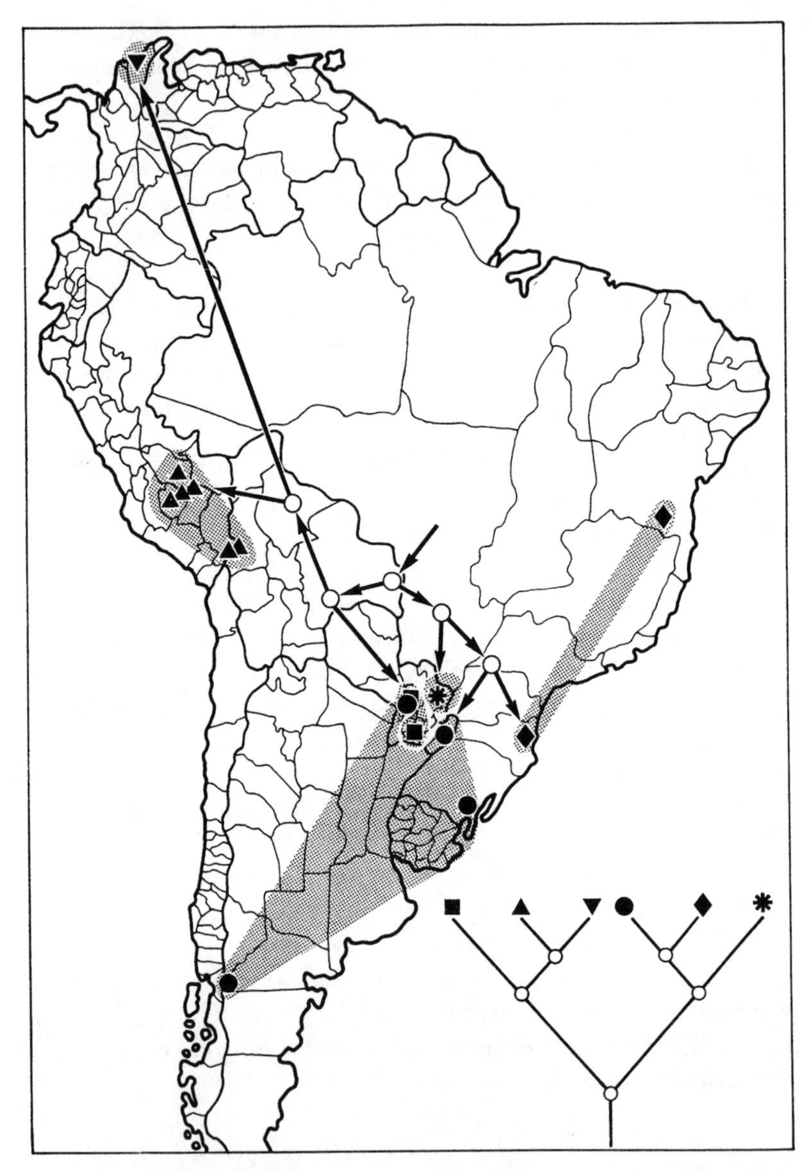

Figure 12.2 A cladogram placing your new species in phylogenetic context might be part of your discussion section. Cladogram of the *keyserlingi* group of *Trachelopachys* (lower right), superimposed on map of South America showing distributions of *T. keyserlingi* (circles), *T. gracilis* (diamonds), *T. aemulatus* (asterisks), *T. magdalena* (inverted triangles), *T. bicolor* (upright triangles), and *T. ignacio* (squares). *–Source: From Platnick (1975), courtesy of the American Museum of Natural History.*

If your new species is one that has been misidentified in the past, your taxonomic discussion may even serve as the introduction to the paper, as in the following example:

> Until recently, several species of geryonid crabs from different localities had been identified with either *Geryon quinquedens* Smith, 1879 or with *Geryon affinis* Milne Edwards & Bouvier, 1894; both of these species were transferred by Manning & Holthius (1989) to the genus *Chaceon.* Of the 24 species of *Chaceon* now recognized, 14 at one time in the past had been identified with either *C. quinquedens* or with *C. affinis,* and six species had been identified with both *C. quinquedens* and *C. affinis* (Manning and Holthius 1988, 1989). In their revision of the geryonids, Manning & Holthius (1989: 76) pointed out that the status of a species identified with *Geryon affinis* by Alcock (1899) needed to be determined. That species, which proved to be new, is described below. (Ghosh and Manning 1993:714)

Evidence to Include

In your comparative discussion you may draw upon any sources of information available to you. The following examples show some categories of evidence that are often used, but remember that taxonomic characters can include any useful aspect of an organism's biology or chemistry.

One of the first lines of evidence is the published literature, the information you found in publications by other authors during your library search. In this way you can incorporate as evidence information obtained by other workers and used in their publications. Although it is always best to check the actual specimens for yourself, sometimes you cannot do so. In that case, you may want to use other people's observations or measurements on museum specimens to uphold your argument (provided, of course, that you believe they are reliable). For example, in this description of an extinct coua, a bird of the cuckoo family endemic to Madagascar, old published measurements on a similar species were used for comparison.

> *Coua delalandei,* a species that has gone extinct in the past 150 years, has the longest tarsometatarsus of any known recent *Coua.* It is represented by less than 15 skin specimens in museums, and no skeletal material

> is available. The tarsometatarsus length of this species, as measured from museum skins, is 70 mm (Milne-Edwards & Grandidier 1895), considerably smaller than the tarsometatarsus measurement of *primavea* or *berthae*. (Goodman and Ravoavy 1993:29)

You can also discuss identifications and taxonomic assignments made by others to point out how a previously described species differs from your new one.

> *Distaplia yezoensis* Tokioka, 1951a, from Japan, has both colony and zooids (including the reticular pattern in the internal stomach lining) that resemble those of the present species. The Japanese species has fewer stigmata (12 to 13) and it has conspicuous parastigmatic vessels. It appears to be a separate species. (Kott 1990:125)

In the case of a species that has a long history in the scientific literature but has been misidentified or unidentified during that time (e.g., studied by biologists and even published on as *Xxxxus* sp., sp. A), it is particularly important to point out papers that have been based on the newly described species before it was recognized and described as new. For example, the marine gastrotrich *Neodasys ciritus* is found from Massachusetts to Florida, but it had not been given a scientific name, even though a considerable amount of work had been done on its biology. Its describer (Evans 1992) discusses this work and links it to his new species.

> *Remarks.–Neodasys ciritus,* the "red" gastrotrich, has a long history in the gastrotrich literature and has been described as *Neodasys* sp. by several workers. . . . Hummon (1969) gives a detailed description of adult and subadult specimens found in Massachusetts beaches, describing the red-pigmented cells as lying in two longitudinal bands adjacent to the digestive tract. He also describes the reproductive system and sperm morphology. Ruppert (1977) includes drawings and photomicrographs of both whole animals and of the reproductive system, specifically the testes and spermatozoa. He describes the red-pigmented cells as "Y-cells," compares *Neodasys* sp. and *N. uchidai* Remane, 1961 with respect to their general morphology, and provides information on the ecological and geographic distributions of both species. Kraus et al. (1981) were the first to show that the red color in *Neodasys* sp. is due to the presence of hemoglobin in the Y-cells. Ruppert and Travis (1983) give the most complete description

of *Neodasys* sp. to date and include a detailed cytological description of the hemoglobin-containing cells. . . . (Evans 1992:322–323)

Other important evidence consists of the results of your own background museum research, including the examination of types or other specimens, as Kott did in her discussion of a new ascidian.

REMARKS. The zooids of this species closely resemble those of *Clavelina viola* Tokioka and Nishikawa, 1976 from Sagami Bay, Japan. The holotype of the Japanese species is a large colony of 164 zooids, and this constitutes its major distinction from the present species which never forms large colonies. (Kott 1990:57)

You may be able to gain new information even from specimens someone has already studied, as Bamber did in presenting evidence for a new species of tanaid crustacean.

Examination of 76 paratypes of *K. bahamensis,* kindly loaned by the National Museum of Natural History, Washington (USNM 181901) revealed that, despite the description in Sieg (1982), nearly half of these specimens were male with genital cones. This allowed the examination of sexual dimorphism and variability in this species. (Bamber 1993:128)

You should also include results from your own study of material you obtained in some other way than from your own collections (e.g., material obtained from another's personal collection), as in this description by Fugate of a new species of fairy shrimp from California.

Branchinecta sandiegonensis has been reported as *Branchinecta lindahli* Packard, 1883, a common species throughout western North America, by Evert & Balko (1987) in a study of the vernal pools on Kearney Mesa, San Diego County, California (D. Belk, pers. comm.; vouchers in personal collection of D. Belk). . . . (Fugate 1993:303)

Of course, you should present evidence from your own collections and studies (after all, they were very likely what alerted to you to the possibility of a new species in the first place).

Their description of this species, and that of Ruppert (1977), are in general agreement with my observations of *N. ciritus.* However, no previous workers have observed the tuft of long ventral cilia, probably due to the fact that these cilia are only visible in a lateral view of the animal (seldom obtained with a glass slide and coverslip preparation). (Evans 1992:323)

Naturally, the discussion section is the place for comparisons of any relevant taxonomic characters. These may include evidence from gross morphology, as in this paper on a new species of possum shrimp.

> *Comparisons.–* Nearly 40 species of *Boreomysis* are recognized currently, but it is uncertain how many of them will prove eventually to be valid. Some are based only on females or immature males, hence the structure of the mature male exopod of pleopod 3, a character of high taxonomic value, is unknown. The condition in *B. oparva,* the 8 distal segments each with 3 simple setae, is at present unique; other species in which this pleopod has been described have two such setae. This feature, combined with the upturned rostrum, the large eyes lacking papillae, and the 2 spines on both the endopod and exopod of the uropod, readily distinguish *B. oparva* from similar species. (Saltzman and Bowman 1993:330)

They may also include ultrastructural evidence. In the last 30 years ultrastructural studies using scanning and transmission electron microscopy (SEM and TEM) have led to the discovery or corroboration of many new species and changed our conceptions of many higher-level groups. Of the two, SEM is more commonly used in taxonomic work, as in this excerpt from a discussion of the evidence for creating a new genus for some members of a South American plant family.

> . . . the firmness of attachment of the crests of the lophate pollen indicates baculate, non-rhizomatous pollen like that of the latter two genera, a feature confirmed by SEM observation (Figure 21). (Robinson 1994:561)

But TEM is used also, as in this description of *Minchinia teredinis,* a new parasite of *Teredo* shipworms.

> The spore wall consisted of two electron-dense layers, each 45 to 70 nm thick, sandwiching an electron-light layer about 20 nm thick (Fig. 8). The wall was devoid of periodic striations such as those reported for *H. nelsoni* [23] and of *H.* sp. (parasite of the crab *Panopeus herbstii*) [21]. (Hillman et al. 1990:367)

Naturally, genetic evidence is of primary importance. This can include chromosomal differences, as in this excerpt from a paper on a new red alga from the Pacific Northwest. The new species had previously been confused with *Poryphyra perforata,* a well-known species from the area, but chromosomal evidence, electrophoresis, morphology, seasonality, and habitat helped to distinguish the two.

Discussion

> A chromosome number of $n = 2$ distinguishes *Porphyra fallax* from most other species of *Porphyra.* Only *Porphyra schizophylla* HOLLENBERG, among the species that have been studied, has been consistently reported to have a haploid chromosome count of 2, but the large size of its chromosomes clearly distinguishes it from its congeners that have been studied to date. . . . (Lindstrom and Cole 1990:374)

Enzymatic differences (revealed by protein electrophoresis, for example) have provided much evidence useful in distinguishing species, particularly in cases of sibling species that are difficult to separate by morphological characters, as in this study of two species of commercially important cockles from the coast of Portugal.

> The electrophoretic studies of the present work provide additional evidence that *Cardium edule* and *C. glaucum* are two separate and valid species. Indeed, intensive sampling and experimentation on Portuguese *Cardium* spp. has confirmed the universal natural of the MDH electrophoretic pattern first established by Brock in 1978. In addition, interspecific differences in hepatopancreatic LAP and esterase electrophoretic patterns were also clearly demonstrated. . . . (Machado and Costa 1994:541)

Electrophoresis is still the cheapest molecular method for distinguishing species, but the results of DNA analyses are being used with increasing frequency, as in this discussion of a new species of acorn worm.

> MtDNA analyses also revealed substantial divergence (Fig. 4). No common DNA fragments were found in comparisons of restriction endonuclease digests of the mtDNA of the two taxa based on *Hind* III, *Xba* I, *Nde* I, *Sty* I, *Stu* I, *Dra* I or *Ava* I. In contrast, mtDNA haplotypes for populations of each taxon were homogeneous. For example, an *Xba* I

digest yielded the following approximate fragment molecular weights: *S. bromophenolosus*–8750, 1075, 925, and 625. *S. kowalevskii*–5000, 3225, 1875, and 1000. (King et al. 1994:389)

Information on chemical composition may also provide key taxonomic characters, as in the following study, which distinguishes a new species of liverwort.

The phytochemical and morphological data clearly indicate the existence of more than one taxon in *Monoclea.* The differences in the sporophyte, the antheridial receptacles and the chemical constitution between plants from New Zealand and the New World fully justify the recognition of two different species, *M. forsteri* HOOK. in New Zealand and *M. gottschei* LINDB. in the New World. The synonymizing of these two species by CAMPBELL (1987) and others is not supported by our data. (Gradstein et al. 1992:131)

Consider this comparative discussion of the taxonomic placement of a new species of sponge.

The isolation of scalarane sesterterpenes and norsesterterpenes from *Collospongia auris* is consistent with the morphology of the sponge. *Collospongia* in terms of terpene chemistry belongs to a group of genera within the family Thorectidae that includes *Cacospongia* and *Hyrtios,* a group which exhibits chemical affinity with another group including *Phyllospongia, Cartieriospongia* and *Strepischordaia,* presently classified within the family Spongiidae. *Collospongia* in general morphology and in terpene composition has affinities with both groups. (Bergquist et al. 1990:355–356)

Finally, biological, ecological, or life history information may be vital to your argument, as in the following examples taken from descriptions of a new species of ascidian.

The functional purpose of the orientation of the zooids in the colony in both *P. regina* and *P. modificata* is probably related to crowding and the necessity of ensuring that the atrial aperture can discharge its contents well clear of the colony. As a consequence of the angled orientation the neural complex is fully exposed to the exterior and the development of pigment in this region is probably protective. *P. regina* is a shallow water species and in certain environs is fully exposed to bright illumination. The pigment

> is most well developed in colonies living in such situations and is less well developed in those living in shaded places. Goodbody (unpubl. obs.) has observed a similar development of pigment around the neural complex of *Corella minuta* if it is removed from its normal shaded environment to a brightly illuminated position. (Goodbody and Cole 1987:252)

And a new species of scale worm:

> *Taxonomical Remarks.*–The genus *Harmothoe* now contains in excess of 150 species (Hanley, pers. comm.). Many of these species are poorly described and illustrated and their distributions are poorly known. As a consequence, it is difficult to adequately define the basis for new species in this genus. However, in this case, the unusual habit of commensalism with bivalve molluscs, the lack of evidence of free-living individuals, and the lack of ornamentation on the elytra (unusual in the genus, but often typical of commensals) provide strong grounds for the erection of a new species. (Rozbaczylo and Cañete 1993:668–669)

Again, such differences–the ones you first noticed during your field work–may have initially triggered your suspicions that you were dealing with something new. It cannot be stressed too strongly that this is where biologists make a real contribution to systematics: by incorporating as much biology as possible into the taxonomic process.

The discussion section is also an appropriate place to point out anomalies or information still to be discovered.

> Three male specimens were found to have both male and female gonopores. One male (SL 5.2 mm, UMML 32:1645) had female gonopores on the right and left side, and two males (LS 5.1 mm, 5.6 mm, UMML 32:2216) had only one female gonopore on either the left or right side. McLaughlin & Lemaitre (1993) have reported specimens with both male and female gonopores in the tube-dwelling hermit crab *Paguritta kroppi* McLaughlin & Lemaitre. As in *P. kroppi,* it is unclear whether the condition observed in the specimens of *C. verrilli* represents an aberration or a reproductive adaptation, such as protandry or protogyny, to a restricted mode of life. (Campos and Lemaitre 1994:147)

In addition, if you are a paleontologist, the discussion section is the place to add necessary evidence, stratigraphic or otherwise, to deal with two additional problems. One is the problem of deciding whether the specimens in your sample were likely to have belonged to the same population (that they probably all lived and died within a short time of each other). The other is the problem of deciding whether different fragmentary parts, even if chronologically identical, could have belonged to the same species.

Discussion

CONSPECIFICITY: Are the three specimens described above representatives of the same biospecies? For the purposes of this discussion, we exclude the second paratype, which is too fragmentary and distorted to provide much information on the question.

All three specimens came from the same small subsample of black shale (#411–19), and therefore are almost certainly exact contemporaries, because the subsample represents (at most) only part of a single depositional event. Thus we may exclude the concepts of chronospecies or paleospecies. Modern soil and litter communities, however, may support several syntopic species of pseudoscorpions. The two more complete specimens are from different instars and there is only some overlap between what is preserved on one and what is preserved on the other.

Arguing for conspecificity on the basis of the information available, we may adduce the following. The pedipalp chelae closely correspond in general shape, and the dentition is identical. The bothria of the fingers are similarly placed (a basal pair and a single distal bothrium on the fixed finger, a distal bothrium on the movable finger). The serrulae exterior and interior are similar, respectively, in both specimens, and so is the denticulation of the cheliceral fingers.

Against this we can only draw attention to the fact that no galeae can be observed on the protonymphal chelicerae. However, the galeae, a delicate structure, may have been lost in the fossilization process on this tiny, poorly sclerotized specimen. There is also the fact that some instars of living pseudoscorpions have galeae, and others, from the same species, do not.

The preponderance of evidence available leads us to the conclusion (obvious from our description) that both more complete fossils are different instars of the same biospecies. (Schawaller et al. 1991:7–8)

Composite Papers

As pointed out in chapter 6, there is also a category of publications intermediate between a straight taxonomic description and a research paper on the biology or ecology of a species. In such a composite paper the biological research on the species leads to a conclusion that affects the taxonomy of one or more species, and much of the discussion concerns the taxonomic implications of that work. Some examples:

Edwards, C. 1983. The hydroids and medusae *Sarsia piriforma* sp. nov. and *Sarsia striata* sp. nov. from the west coast of Scotland, with observations on other species. *Journal of the Marine Biological Association of the United Kingdom* 63: 49–60.

Goodbody, I. and L. Cole. 1987. A new species of *Perophora* (Ascidiacea) from the western Atlantic, including observations on muscle action in related species. *Bulletin of Marine Science* 40: 246–254.

Ohtsuka, S. and C. Mitzumi. 1990. A new asymmetrical near-bottom calanoid copepod, *Paramisophria platysoma,* with observations of its integumental organs, behavior and in-situ feeding habit. *Bulletin of the Plankton Society of Japan* 36: 87–101.

Riemann-Zürneck, R. and V. A. Gallardo. 1990. A new species of sea anemone (*Saccactis coliumensis* n. sp.) living under hypoxic conditions on the central Chilean shelf. *Helgoländer Meersuntersuchungen* 44: 445–457.

Safriel, U. and M. G. Hadfield. 1988. Sibling speciation by life history divergence in *Dendropoma* (Gastropoda; Vermetidae). *Biological Journal of the Linnaean Society* 35: 1–13.

Spoon, D. M., C. J. Hogan, and G. B. Chapman. 1995. Ultrastructure of a primitive, multinucleate, marine, cyanobacteriophagous ameba (*Euhyperamoeba biospherica,* n. sp.) and its possible significance in the evolution of eukaryotes. *Invertebrate Biology* 114: 189–201.

However, when you publish this kind of paper, it is especially important to carry it through and not leave the undescribed species you studied as "*Xxxxus* sp., or *Xxxxus* sp. A." The only justification for doing so would be in the case in which you have another paper, a straight species description, in press, and you are concerned that the biological paper might be published before the taxonomic description. If the biological paper comes out first it can cause nomenclatural headaches for you and others. If the new name is used, but without description and illustration, the new name becomes what taxonomists call a *nomen nudum.* Such a name is not considered validly published, and although you could use it in describing the species in a later publication, someone else could, meanwhile, use it for another organism. On the other hand, if you do publish an adequate description in your biological paper, then the date that paper is published becomes the date of publication of the name; *that* description becomes its official first description, making your second paper a redundant effort. If you happen to have two papers like this scheduled to be published close together, and you fear that the popular or biological one might be published first, you can get around the problem by clearly stating in your biological paper that the official taxonomic description will follow and that the information being published now is not intended to be the description. But you can also choose to avoid the problem by leaving the new species name out of the biological paper (i.e., leaving it as *Xxxxus* sp.) and referring back to that paper in your taxonomic paper, making it clear that it is about the biology of your new species. Such bad timing of publications doesn't happen often, but it (not laziness) is the only good reason for not naming and describing a species to which you have devoted your scientific efforts for months or years.

Many biologists think they can't publish such a hybrid paper, one that includes both a species description and biological research on that species, but there is no rule against it. In fact, it is done all the time. Indeed, it is far better to include the description of your new species along with the rest of your research on the species in a publication in a reputable journal than to let the organism you've studied for years languish as "species X" until someone (who?) gets around describing it.

The Discussion Section in Other Species Descriptions

Taxonomic discussion sections are also useful when redescribing species, to add any new information that you or others have come up with since the species was originally described or last revised, as was done by Ted Bayer in the following example. In his redescription of an Antarctic gorgonian he used SEM to more accurately describe its morphology.

> The original description of *A. splendens* is accurate except in regard to the form of the polyp sclerites: "the striking peculiarity is that the main sclerites of the vertical rows have a distinct ascus-like or chalice-like form. That is to say, the basal portion is the substantial very warty support of a delicate cup whose cavity is open to the exterior! The delicate edges of the cups are weakly notched, sometimes almost entire. The cup of the sclerite is broader than the substantial knobbed support, so that the appearance

> is somewhat like a short-stalked chalice or fruit-basket." The authors go on to say that "We must emphasise the point that the specimen is on the whole like a *Caligorgia,* but its hollow ascus-like sclerites are very far from the ctenoid-scale type" (Thomason & Rennet 1931: 20–21).
>
> As is immediately apparent from scanning electron micrographs (Figs. 3–5), the "delicate" cups are an optical illusion. As the calcite of which the sclerites are composed is almost glassy clear, only the edges of the smooth outer portion of the body sclerites are visible as a bright line under the microscope; the fact that it is solid is not easily detected in a bulk preparation of sclerites. (Bayer 1990:775)

Taxonomic Ethics

Some of the important points of taxonomic ethics are covered here because it is in the discussion sections of taxonomic papers that bitter personal battles have been fought and where bad temper and irrationality occasionally still surface today. The International Code of Zoological Nomenclature (1985) includes a Code of Ethics (Appendix A), which gives excellent guidelines for polite taxonomic behavior that apply just as well to botanical taxonomy. As Rule 6 of this code states, "Intemperate language should not be used in the discussion of zoological nomenclature, which should be debated in a courteous and friendly manner. Difficult problems are most readily and quickly solved by respecting the rules of courtesy in discussing the views of others."

Here are a few negative examples from the older (and not so old) myriapod literature:

> I was unable even after intensive study of the Chamberlinian work, to identify a single specimen of the Congo material with the new descriptions of Chamberlin, with exception of *Habrodesmus latilobus,* which is identical with the much earlier described, *Habrodesmus cagnii* Silv. In part these results go back to the fact that Chamberlin never gave comparative observations in his brief descriptions, so that it is difficult or even impossible to relate his species to those already known. (Attems 1929:253; translated by R. Hoffman)
>
> [Following a discussion of the anatomy and homology of genital segments in geophilomorph centipeds, Attems concludes]: "Verhoeff, then, by an arbitrary and illogical change, gave to the praegenital segment the name of 'genital segment' and to the genital segment (Heymons) the name of

'postgenital segment,' but why, heaven alone knows." (Attems 1928:111; translated by R. Hoffman)
"About the confines of this family [Miridae], as represented not only in this country but in the world at large, he has attempted to construct a hog-tight, bull-proof fence and woe be unto any person who attempts to root his way under." (Blatchley 1928:16)
"It will be shown that the solenite is present in perfectly normal configuration, and that the genus is referable in every respect to the Chelodesmidae. Verhoeff simply damaged the gonopods in making his preparation, and was then, because of his ignorance of chelodesmoid morphology, unable to realize that the solenite had not once been present. Thus the genus *Telonychopus* and family Telonychopidae were based upon a gross and palpable error. (Hoffman 1965:244)

Reading these comments on the mistakes of early workers can make you smile, but when derogatory remarks are applied to living scientists, those being disparaged are not likely to be amused.

Preferably, the discussion section is where you show off both your critical ability and your good taste. Presenting the reasons for your decision does not necessitate a personal attack on previous workers. Contrast these two hypothetical examples:

1. "The specimens collected by Dr. Smith from the Beaufort Sea (Smith 1993) and identified by him as *Xxxxus darwini* are similar in most respects to my own material of *Xxxxus darwini* from Lizard Island, Australia (the type locality). However, reexamination of the septate scales of one of the Alaskan specimens by SEM (specimen kindly provided by Dr. Smith) shows that they possess rows of fimbriate teeth (Fig. 1A). Micrographs of scales from topotypic material of *Xxxxus darwini* (Fig. 1B) show clearly that the scales of this species have no indentations, but are gently rounded in form. For that reason I consider the Arctic material to represent a new species, which I have named in honor of Dr. Smith."
2. "Dr. Smith, who describes his organisms using only a magnifying glass, identified some specimens collected from the Beaufort Sea as *Xxxxus darwini,* a species originally described by Jones (1926a) from the Great Barrier Reef of Australia. In view of the barriers presented to a nonplanktonic marine organism by current patterns, deep ocean trenches, and temperature differences between these two regions, I strongly doubted the validity of his identification. Because Dr. Smith is notorious for not depositing his specimens in a museum, I contacted a colleague at the

> University of Alaska, who was able to obtain a specimen for me, and I compared material of both species using SEM (Fig. 1A and B). It is hard to see how even a fool with a magnifying glass could have failed to distinguished rounded scales from serrated! I have therefore redescribed the Arctic material as *Xxxxus smithistupidus.*"

The first example may not make you a friend–taxonomists don't like being wrong any more than anyone else does–but at least it won't make you an enemy. The second example might keep your paper from being published at all. At best, your personal comments will be heavily edited or deleted by the editor, and you will probably be asked to change the name to something less offensive. Although the second hypothetical example is an extreme, you do sometimes see in discussion sections, "It is hard to see how . . . ," or "It is a mystery why . . . ," and other statements that can definitely create ill will when aimed at authors still alive, and could easily be phrased less personally without losing their point. This is not to say that you have to avoid controversy at all costs. As Blackwelder (1967:313) said, "Although personal attacks are never justified in taxonomic papers, emotion is not necessarily bad and may be worth communicating. Controversy, furthermore, is the source of much clarification of ideas; the admonition should be rather not to take offense at argument or criticism, but to study it for its possible contribution."

Chapter 13

The Ecology Section

Until you know what grows and lives in a particular place, and recognize its position in the biosphere, you can neither exploit nor conserve those biological resources properly.

–Lichen systematist D. Galloway,
quoted in Pain (1988:49)

Don't trust your memory; it will trip you up; what is clear now will grow obscure; what is found will be lost. Write down everything while it is fresh in your mind; write it out in full. Time so spent will be time saved in the end, when you offer your researches to the discriminating public.

–Coues (1874:44–45)

Ecology in Species Descriptions

Some ecological variation reflects subpopulational, subspecies, or individual differences (see chapter 17), but many aspects of an organism's ecology and behavior, habitat or host, feeding, reproductive patterns, seasonal patterns, mortality patterns, and so on are species-specific. It will be no news to an ecologist reading this book that the ecology of an organism can be important taxonomically because you may have become aware that you might be dealing with a new species when some characteristic of the ecology of your study organism did not fit any of the patterns described for known species. However, you may not have realized the importance of including ecological information in your species description.

In fact, the use of ecological information in taxonomic descriptions goes back to folk taxonomy. Information on habitat preference, seasonality, and other factors has always been vital to the hunter or herbalist. In his *Methodus* (1736) Linnaeus included his own set of rules for describing plants and animals, this section on attributes, which could serve just as well today:

> V. ATTRIBUTES
>
> 22. Include what is known about the *season* of birth, growth and maturity, with mode of breeding and of birth or hatching, old age, and death.
> 23. State the *locality*, giving the geographic region and political province.
> 24. Give the latitude and longitude.
> 25. Describe the climate and soil.
> 26. Give an account of the diet, habits, and temperament.
> 27. Describe the anatomy of the body, particularly any remarkable features, together with a microscopic examination. (Linnaeus 1736, Trans. by Schmidt 1952:373)

One of the most important reasons to include ecological information in species descriptions is to enable future workers (or you) to find the organism again. Even if the type locality is turned into a shopping mall, good habitat data may help locate the other populations of the species somewhere else in the area. Workers wanting to find additional specimens to use in systematic studies or biological research will also greatly appreciate any information on months of peak abundance, reproductive season, and other features.

Demographic and ecological data are also critical to understanding conservation needs and documenting the impact of environmental change. For example, Barry et al. (1995) recently showed the effects of 61 years of ocean warming on the invertebrate fauna of the rocky intertidal community in Monterey Bay, California. They found large changes in community structure, with abundances of southern species increasing and northern species either decreasing or moving lower in the intertidal zone, suggesting a northward shift of species ranges, a conclusion that agreed with predictions of the changes to be expected with climate warming. They were able to reach this sweeping conclusion thanks to painstaking taxonomic and ecological work: a 1931–1933 transect study of abundances and identities of all organisms found along an intertidal transects (Hewatt 1937) and their own restudy of the same transect.

Including habitat and ecological data in taxonomic descriptions can give conservation biologists the information they need to locate protected sites for long-term species management and protection. For example, in their description of three Mexican goodeid fish species, Smith and Miller concluded with a discussion of their conservation status:

> . . . *Characodon lateralis* has seriously declined in abundance and distribution during the last 25 years; it is listed as a threatened species by Deacon et al. (1979). . . . In 1982 and 1983, a survey of the habitats of *Cyprinodon meeki* (which is frequently an associate of *Characodon lateralis*) revealed the goodeid at only seven springs or spring-fed habitats, except for one sample from an upper tributary to Río Canatlán, about 17 km north of Canatlán. Factors responsible for the depletion of *C. lateralis* apparently include pollution, habitat destruction, and competition with exotic fishes. Although we know of no immediate threat to *C. audax,* its very restricted range is cause for concern. (Smith and Miller 1986:12)

Information on distributions and habitat preference of native species is invaluable when an exotic species is introduced into an area. For example, the effects of the introduction of the Asian clam *Potamocorbula amurensis* to San Francisco Bay could be studied in detail Carlton et al. (1990) because previous work (Nichols and Pamatmat 1988) had documented the benthic and intertidal communities of the bay.

For plants, data on life form, ecological parameters, reproductive ecology, and climatic needs are also essential in agriculture, horticulture, and land reclamation (Morin et al. 1989) as well as in conservation biology, ecology, and physiology.

Taxonomically, you must be sure you have covered all the different life cycle stages and phenotypes found in the population you have studied. In an extreme case, it may be difficult or impossible to find all the *phena,* meaning the different sexes, life cycle stages, larvae, instars, juveniles, seasonal variations in phenotype, and individual variants (Mayr 1969) that occur in the organism you are attempting to describe. Sexually dimorphic taxa, when only one sex has been identified (e.g., some spiders) can be particularly difficult, and have led to embarrassing mistakes (e.g., males and females being described as different species).

Sometimes it takes careful ecological study to determine what the different phena represent biologically. The example of the giant kelp fish shows how complicated this can be even when only two such differences (color and sexual dimorphism) are involved. *Heterostichus rostratus* occur in three colors–red, brown, and green–which usually match the color of their algal habitat. Juveniles are green or brown and are not sexually dimorphic in color. Adult males are usually brown, rarely olive green. Adult females can be found in all three colors. However, brown adult males can be distinguished from brown adult females by sexually dimorphic melanin patterns. The different patterns are involved in both cryptic coloration and intraspecific communication, with adult males and females occupying plant habitats different in depth and color background, but with females leaving their cryptic habitats during the spawning season to visit male territories (Stepien 1988). Similar situations occur in other blennioid fish (Stepien 1987).

Finding and matching up all the life history stages in parasites or other plants and animals with complex life histories can be a difficult and time-consuming process. Yet the larval or juvenile states may contain taxonomically useful, even essential information. Marine biologists who study the taxonomy of organisms with long-lived pelagic larvae spend years completing life history studies, which may have to be carried out in several phases. For example, most sipunculan larvae undergo a double transformation, developing first into a nonfeeding trochophore larva, then metamorphosing into a planktotrophic pelagosphera larva that can remain in the plankton for months before finally settling and metamorphosing into a juvenile worm (Rice 1978). Mary Rice (1981) raised sipunculan larvae collected from the open ocean waters of the Florida Current in her laboratory in Florida, and on the basis of juvenile morphology tentatively identified them as *Siphonosoma cumanense,* a Caribbean species. The next step was to collect known adults of *S. cumanense* from Puerto Rico, induce them to spawn in the laboratory, and raise the eggs and larvae to juvenile worms. This confirmed that the ocean-collected larvae did indeed belong to that species (Rice 1988). In addition to their practical role in allowing identification of an unknown planktonic larva, studies of larval and juvenile stages turned out to provide taxonomically useful information. In this case the development of the species was found to be similar to that of members of other families, and distinct from the developmental pattern of *Sipunculus nudus,* another species from the family in which *Siphonosoma cumanense* had been placed (Rice 1988).

Complex life cycles also occur in insects. Larvae of cleptoparasitic bees have been found to have features useful for phylogenetic analysis (Rozen 1989). Other aspects of their biology (nest site requirements, nest architecture, cocoon construction, and nest parasitism) may also show significant interspecific differences (Rozen and Buchmann 1990).

Analysis of Ecological Variation

Ecologists can often follow populations long enough and study them in enough depth biologically to figure out whether the variations they have observed represent good species or subspecies of a geographically variable species or species complex. Taxonomists working on collected material alone cannot observe the processes of reproductive isolation or species formation in action. Ecologists sometimes can, as in the studies of Darwin's finches by Grant and Grant (1993), or of stickleback fish by Schluter (1994).

Analysis of ecological variation involves many of the methods mentioned in the section on morphological variation (chapter 11). Those methods are beyond the scope of this book, but are found in textbooks on ecology, comparative biology, and statistics, such as those given in this chapter's *Sources.*

Field Records: Getting the Most from Field Work

Whether you have any ecology to include in your description depends not only on how long you studied that particular organism or community, but on what kind of field records you kept. However, instruction on how to keep field records is difficult to find. Whereas there are numerous books on experimental methodology, design of experiments, and statistical analysis, there is little written about observational methods and record keeping. Almost any kind of record is better than none, but following certain guidelines will help you maximize the value of your observations.

The naturalist's journal, a field record of the naturalist's daily activities, was a feature of natural history from the 1700s to the early 1900s. It was a tradition followed by most of the great explorers and naturalists, from Linnaeus's *Journey to Lapland* to Darwin's *Voyage of the Beagle* (Herman 1980). Thoreau's *Walden* was a literary version of this tradition, combining natural history and social commentary. That literary form is still vibrant today, with practitioners such as Dillard, Evans, Hubbard, Quammen, Lopez, Wilson, and Zwinger (Lueders 1989).

With present-day biology divided into many specialized subfields, and with experimentation, rather than observation, paramount, not many biologists keep field journals. But they can still be extremely important to systematists, ecologists, and environmental biologists. As a graduate student, I was fortunate in having access to the research notebooks of one of my predecessors. I studied them, and learned, and slowly began to work out my own system. Other colleagues tell me that they learned to keep field records from their graduate teachers and advisors, by going in the field with them and duplicating what they saw done, either informally or as part of a field biology or geology course.

Some vertebrate zoologists and ornithologists use a system established by Joseph Grinnell and taught at the University of California, Berkeley for many years. It is explained in detail in *The Naturalist's Field Journal* by Steven G. Herman (1980). This system consists of three sets of records: a field *journal, species accounts,* and a *catalog,* all supported by field notebooks. Details are recorded in the field notebook during the fieldwork, then rerecorded into the species accounts or catalog. They are compiled on a yearly basis, whereas the number of field notebooks varies with time spent in the field. His style recommendations are useful to anyone planning to keep such records. My own additions are in brackets.

All pages should be numbered and dated.

Write text on one side of paper only (save other side for drawings and maps).

Write neatly in India ink [or some other indelible medium in a color that will photocopy well].

Record the locality in every entry.

Use abbreviations as little as possible, but use numerals [1, 2, 3] instead of written-out numbers.

Write date as follows: 3 March 1995, and times following international convention, e.g. 1500 for 3 P.M.

Journal entries are the core of the field records. They should always be written as soon as possible after the event. They should include this basic information: date, locality, route (if applicable), weather, habitats, including topography, vegetation, species lists, and commentary (Herman 1980). According to Herman, *species accounts* were developed for observations on birds and were originally based on collected specimens in hand. Sight records were not scientifically acceptable in ornithology until the 1950s. Now, if properly documented, they are acceptable, even invaluable, in describing species' ecology and distributions, particularly of threatened or endangered species of animals or plants, and are used by biologists working for Heritage programs and other conservation organizations.

In the case of a new species, any information you have recorded is going to be useful, but obviously species accounts are of greatest value if they can be carried out at regular intervals over a period of months or years. The full value of species accounts lies in the depth of knowledge that can be built up over time. As Imes says in *The Practical Botanist* (1990:43), "Think of each one as a jigsaw puzzle with the pieces askew; with each rendezvous you can put a few more pieces in place. Gradually, over weeks, months, or even years, a picture begins to emerge." The observations you choose to record depend on the kind of organism, you are studying. For birds, observations on nesting, abundance, inter- and intraspecific behavior, and distribution are typical. For plants, distribution information and *phenology* (the annual cycle of growth, leaf development, flowering, fruiting, leaf fall, and dormancy) are important.

The *catalog* section provides a record of specimens collected to document the study. Lots (groups of specimens from a single locality or sample) or individual specimens are labeled with at least the date collected, collection locality, collector's name, and collector's field number. The same information is recorded in the catalog and cross-referenced into journal and species accounts. Additional information may be included; for mammals, standard measurements, weight, and sex are usually taken from fresh specimens (Herman 1980).

Many modifications of this system are possible, and you will work out one that fits your needs. For example, as a marine biologist, I have to deal with the fact that my notes may be subjected to salt water. I take only a small field notebook out to the shore or on a boat, and use a dive slate to record essential information when working underwater. I transcribe the information into my journal when I return to the lab. Working in a seawater wet lab means that even the journal notebook can get wet at times. I like to use French-made quadrille-ruled bound notebooks for my journals because their blue lines don't run and their paper doesn't fall apart if spattered with water. I keep one notebook divided into several sections, and in it I include a catalog of photographs taken and video footage, as well as specimens collected, and my field journal and species lists. I use separate specialized forms for species accounts, with spaces for measurements and behavioral data. These may vary according to a particular study's objectives, and I keep them in looseleaf notebooks with plastic covers (they can get wet, too).

If you are carrying out ecological or behavioral observations in the field you may be using tapes (audio or video). Some people keep species information on notecards rather than in notebooks, and today field records are often transferred into computer files. You may even be taking a laptop computer into the field with you to record your observations on site. Use whatever works for you. My only advice, based on my own experience, is that I've never been sorry for taking too many notes–only for the notes I didn't take. Almost every note I've taken was incorporated into a paper sooner or later.

Ecological Information from Museum Specimens

At the opposite extreme from getting the most from your time in the field comes the challenge of obtaining ecological and life history information from museum specimens. As Bock (1985) pointed out, museum collections have been essential to the study of adaptations of living and fossil organisms ever since the first museums began. As they pondered the bizarre creatures in their cabinets of "curiosities," early naturalists were making their first attempts at accounting for the uses and possible advantages of the peculiar structures the organisms possessed. Researchers still use museum specimens to make inferences about structures and their possible functions and to come up with ideas that can be tested in the field.

Studying the specimens in museum collections can also provide good information. For behavior and sociobiology, size differences in sexual dimorphism, not just of the entire organism but even of certain body parts (e.g., canine teeth in primates), can have implications for social structure and mating systems. Caste specializations (e.g., in colonial insects and invertebrates), which may be physical as well as behavioral, can be often detected in museum specimens. Structural specializations, features such as horns, crests, manes, and patterns of pelt or plumage may also be important socially. Growth and ecogeographic variations (e.g., of egg size or clutch size) and changes with age (development of male plumage in birds, increasing UV reflectance in butterflies) are all morphological aspects of geographic variation. Sometime museum specimens can help answer questions about an organism's ecology. At other times they raise questions, enabling you to develop hypotheses, which you can then test by further field or laboratory research (Miller 1985).

Important life history information can also be gathered from museum specimens. The kinds and amount of data available depend on the group being studied. For example, for mammals, teeth and skeletons, as well as ridges on horns, can be used to estimate age. Preserved whole specimens can be used to obtain fecundity information (from placental scars) as well as information on diet and external and internal parasites. For birds, egg collections can be used to trace historical trends in fecundity and hatching success; breeding seasons and lines of migration can also be determined using museum specimens. Fish collections can also yield information on age structure, fecundity, and seasonal timing of reproduction. Large collections of several thousand individuals in organisms such as salamanders have even been used to develop life tables and survivorship curves for a species, or for populations of a species occurring in different habitats (Allen and Cannings 1985).

The Ecology Section

The ecology section usually follows the description or discussion sections and comes before the section on distribution, but its position is variable and the two sections may also be combined. Although I have called this chapter "The Ecology Section," the section's actual title in a species description may vary according to the journal or the type of information the author has to include. It may be titled "Biology," "Habitat," "Habitat and Habits," or "Occurrence and Ecology," or simply included without a title as part of the remarks or discussion. Exactly what points you'll want to cover depend on your study organism. For animals, habitat, microhabitat, substrate, and host, are next to location in importance for anyone who wants to find that animal again. For plants, it is important to consider soil-related (edaphic) parameters, geology, and the plant community with which the species is associated. For example, soil type may determine growth and size of a plant at sexual maturity. One plant may be tall, its neighbor a few yards away still tiny when it flowers. To determine the meaning of the variation, it may be necessary to discover whether there are gradations in varying features and, if so, whether these have environmental causes or are caused by hybridization. With plants also, problems can be caused by having male and female plants that look different, or differences in leaf structure between juvenile and mature trees, as in eucalypts (Pain 1988).

If a new species was collected as part of a broader faunal or floral survey, the discussion of ecology in your new species description may be minimal, covering in a sentence or two anything you were able to learn of the species's ecology as you collected it, as in this description of a new Panamanian lizard.

> FIELD NOTE: The holotype was active during a sunny afternoon, in ground litter on a forested slope. (Myers and Donnelly 1991:5)

If it was a study in which you were able to return to a site several times, you may have a little more information to include. For example, Eckart Håkansson and I found several new species of sand-grain encrusting bryozoans while carrying out a study of the population biology of free-living cupuladriid bryozoans. Although we had not specifically targeted the sand-grain dwellers, we did at least have knowledge of when they occurred in the eight seasonal censuses we made over the 2 years of the population study.

> DISCUSSION: Sexually mature colonies on sand grain substrata are very small, often consisting of the ancestrula, one or two autozooids, and a single ovicelled zooid. . . .
> OCCURRENCE: Found in collections from each census. Living colonies were found in August 1984 and January 1985. (Winston and Håkansson 1986:11–12)

In cases in which ecological as well as morphological variation must be assessed in order to make a species assignment, the ecological section may be long and complex, even comprising most of the paper. Titles such as the following fall under the composite category mentioned previously. They are ecological papers with taxonomic implications.

Corriero, G., A. Balduzzi, and M. Sarà. 1989. Ecological differences in the distribution of two *Tethya* (Porifera, Demospongiae) species coexisting in a Mediterranean coastal lagoon. *Proceedings of the Zoological Station Napoli: Marine Ecology* 10: 303–315.

Cumberlidge, N. and R. Sachs. 1991. Ecology, distribution, and growth in *Globonautes macropus* (Rathbun, 1898), a treeliving freshwater crab from the rain forests of Liberia (Gecarcinucoidea, Gecarcinucidae). *Crustaceana* 61: 55–68.

Emig, C. C. 1970. Remarks on the systematics of Phoronidea. IV. Notes on ecology, morphology and taxonomy of *Phoronis mülleri. Marine Biology* 5: 62–67.

Kaehler, S. and R. G. Hughes. 1992. The distributions and growth patterns of three epiphytic hydroids on the Caribbean seagrass *Thalassia testudinum. Bulletin of Marine Science* 51: 329–336.
Ma, H. H. T., D. Dudgeon, and P. K. S. Lam. 1991. Seasonal changes in populations of three sympatric isopods in a Hong Kong forest. *Journal of Zoology, London* 224: 347–365.

In a simple new species description, most ecological sections fall between these two extremes. They are usually a paragraph or two in length. They incorporate whatever ecological information you have been able to gather on the new species, discuss its implications, and (possibly) suggest future work needed, as in the following examples.

Information on the ecology of a new species of fish, *Catostomus cahita,* was given under a *Habitat* heading:

> HABITAT AND ASSOCIATED SPECIES: The Cahita sucker inhabits streams of variable size and configuration. It often is syntopic with *Catostomus bernardini,* especially in larger systems, and ranges from high-elevation habitats in Madrean Montane Forest to lower elevations in Sinaloan Thornscrub (Brown, 1982). It was recorded in greatest abundance, however, in small streams at high and intermediate elevations. It avoids turbulence and characteristically lives in pools, but sometimes was taken from deeper runs. Associated fishes ranged from Yaqui trout at high elevations, through most cyprinids known from the region, to an undescribed pupfish (*Cyprinodon* sp.) in the Río Papigóchic (table 4). (Siebert and Minckley 1986:12)

Ecological information on a new species of crayfish was given under *Remarks:*

> *Remarks.*–All of the specimens at hand were obtained on warm, rainy nights when they were either at the mouth of a burrow or wandering about the grassy area. Perhaps as many as 100 individuals, including mostly young having total lengths of approximately 25–40 mm, were seen in an area of about 1000 square feet (90 m^2). All eight of the second form males had clean exoskeletons, suggesting a recent molt, that contrasted conspicuously with the crusty ones of the three first form males. (Hobbes and Whiteman 1987:410)

Ecological and biological information on a new species of brittle star was summarized under *Biology:*

BIOLOGY. The species lives on soft-bottom habitats, in sediments with seagrass, amidst rubble, shell, stones, coral, and coralline algae and under sponges. Individuals are cryptic and can conceal themselves beneath a thin layer of sediment. When freshly collected, individuals usually are coated with fine grains of sediment, which appear to adhere to mucus on the disk and arms.

Gonads are present on some individuals as small as 1.0 mm in disk diameter. Several individuals of moderate size were found to have up to 8 ovaries per interradius, up to 76 oocytes per ovary, the oocytes with a mean diameter of 0.15 mm. (Hendler 1995:8)

Information on position of a new species of parasitic copepod on its host fish was found in the discussion section:

DISCUSSION

In each of the host fish, only one specimen of *D. lycenchelus* was found. However, two specimens, an adult female and a much smaller juvenile female, were found in the same nostril of a third host. All of the parasites found were recovered from the right nasal cavity. Infection of the nostril by *D. lycenchelus* caused a large swelling under the skin surrounding the parasite (Fig. 22a, b). This probably seriously impaired nostril function. When dissected from the host each specimen was oriented with the tip of the cephalothorax directed toward the posterior part of the nasal cavity with the second antennae firmly embedded to their bases in the epithelium. The parasite utilizes the anterior nostril as a duct for release of eggs (Fig. 22 b). . . . (Hogans and Sulak 1992:306)

If you are describing a species that has been misidentified or was described previously, you may include ecological information from the literature as well as from your own observations, as in these examples:

ECOLOGY: As noted under Distribution above, *Mesopristes elongatus* is evidently limited to freshwater localities. Petit (1937: 27–29) in his discussion of this species [identified as *Therapon (Datnia) argenteus*] noted that it was limited to fresh water on Madagascar. . . . (Vari 1992:5)

OCCURRENCE AND ECOLOGY: When living, colonies from the Indian River are ivory to beige in color. . . . One of the most abundant bryozoans recruited on panels in the river in 1974 and 1975 (Mook,

1976). Eurythermal, but reproduction heaviest in fall (October to January) though colonies settled in bryozoan traps through June. (Winston 1982:142)

Sources

Ecological and Evolutionary Studies

Brooks, D. R. and D. A. McLennan. 1991. *Phylogeny, Ecology, and Behavior.* Chicago, University of Chicago Press.

Dirso, R. and J. Sarukhán. 1984. *Perspectives on Plant Population Ecology.* Sunderland, Mass., Sinauer.

Eggleton, P. and R. I. Vane-Wright. 1994. *Phylogenetics and Ecology.* London, Academic Press.

Gaston, K. J., ed. 1996. *Biodiversity: A Biology of Numbers and Difference.* Oxford, U.K., Blackwell.

Grant, V. 1981. *Plant Speciation,* 2d ed. New York, Columbia University Press.

Grant, V. 1991. *The Evolutionary Process,* 2d ed. New York, Columbia University Press.

Harvey, P. H. 1996. Phylogenies for ecologists. *Journal of Animal Ecology* 65: 255–263.

Stearns, S. C. 1992. *The Evolution of Life Histories.* Oxford, U.K., Oxford University Press.

Tilman, D. 1988. *Plant Strategies and the Dynamics and Structure of Plant Communities.* Princeton, N.J., Princeton University Press.

Vane-Wright, R. I., ed. 1994. *Systematics and Conservation Evaluation.* Oxford, U.K., Clarendon.

Methods of Analysis and Field Techniques

Ecological Census Techniques: A Handbook. Cambridge, U.K., Cambridge University Press.

Feller, G. H. 1995. *A Standardized Protocol for Surveying Aquatic Amphibians.* Davis, Calif., National Biological Service. Cooperative Park Studies Unit, University of California, Division of Environmental Studies.

Fowler, J. 1998. *Practical Statistics for Field Biology,* 2d ed. New York, Wiley.

Fry, J. C., ed. 1993. *Biological Data Analysis: A Practical Approach.* Oxford, U.K., IRL Press at Oxford University Press.

Grub, T. C. 1986. *Beyond Birding: Field Projects for the Inquisitive Birder.* Pacific Grove, Calif., Boxwood Press.

Haccou, P. and E. Meelis. 1994. *Statistical Analysis of Behavioural Data.* Oxford, U.K., Oxford University Press.

Handbook of Field Methods for Monitoring Landbirds. 1993. Albany, Calif., Pacific Southwest Research Station.

Harper, J. L. 1977. *Population Biology of Plants.* New York, Academic Press.

Hayek, L.-A. and M. A. Buzas. 1997. *Surveying Natural Populations.* New York, Columbia University Press.

Higgins, R. P. and H. Thiel, eds. 1988. *Introduction to the Study of Meiofauna.* Washington, D.C., Smithsonian Institution Press.

Holme, N. A. and A. D. McIntyre, eds. 1984. *Methods for the Study of Marine Benthos.* London, Blackwell.

Magurran, A. E. 1988. *Ecological Diversity and Its Measurement.* Princeton, N.J., Princeton University Press.

Measuring and Monitoring Biological Diversity. Standard Methods for Mammals. 1996. Washington, D.C., Smithsonian Institution Press.

Pettingill, O. S. 1984. *Ornithology in Laboratory and Field,* 4th ed. London, Academic Press.

Smith, R. L. 1990. *Ecology and Field Biology,* 4th ed. New York, HarperCollins.

Socha, L. O. 1987. *A Birdwatcher's Handbook: Field Ornithology for Backyard Naturalists.* New York, Dodd, Mead.

Stoddart, D. R. and R. E. Johannes, eds. 1978. *Coral Reefs: Research Methods.* Paris, UNESCO.

Stuessy, T. F. 1990. *Plant Taxonomy: The Systematic Evaluation of Comparative Data.* New York: Columbia University Press.

Keeping Field Records

Herman, S. G. 1980. *The Naturalist's Field Journal: A Manual of Instruction Based on a System Established by Joseph Grinnell.* Vermillion, S.D., Buteo Books.

Literary Natural History: Using Journals

Nabhan, G. P. and A. Zwinger. 1989. Field notes and the literary process, pp. 69–90 in E. Lueders, ed. *Writing Natural History: Dialogues with Authors.* Salt Lake City, University of Utah Press.

Some nature writers depend largely on library research for their essays. Other works clearly show the influence of notes or journals recording field time by their authors. Those listed below are just a few samples, some famous, others personal favorites.

Bonta, M. M., ed. 1995. *American Women Afield.* College Station, Texas A&M University Press.

Dillard, A. 1988. *Pilgrim at Tinker Creek.* New York, Harper Perennial.

Gantz, C. O. 1971. *A Naturalist in Southern Florida.* Miami, University of Miami Press.

Holden, E. 1991. *The Nature Notes of an Edwardian Lady.* New York, Penguin.

Hubbell, S. 1986. *A Country Year: Living the Questions.* New York, Random House.

Leopold, A. 1972. *Round River: From the Journals of Aldo Leopold.* London, Oxford University Press.

Ricketts, E. F. 1978. *The Outer Shores,* 2 vols., ed. by J. W. Hedgpeth. Eureka, Calif., Mad River Press.

Steinbeck, J. 1969. *The Log from the Sea of Cortez: The Narrative Portion of the Book, Sea of Cortez, by John Steinbeck and E. F. Ricketts, 1941.* New York, Viking Press.

Thoreau, H. D. 1995. *Selections from the Journals.* New York, Dover.

White, G. 1986. *The Journals of Gilbert White,* 2 vols. London, Century.

Chapter 14

Occurrence and Distribution

> *The label is so important that some taxonomists joke that to other taxonomists, it is more important than the specimen. Many kinds of information are desirable, but by far the most important single piece of information is the exact locality of a collection.*
> —MAYR AND ASHLOCK (1991:330)

> *There is a popular superstition to the effect that museum entomologists devote their time to the fabrication of mommets. These are small, artificially constructed images into which pins are stuck, mainly with a view to making life less tolerable for other entomologists. The museum entomologist calls them species, but they bear little resemblance to the species actually encountered in nature. Such a view is exaggerated, but it may contain a grain of truth. Most of us could probably do more to give our species reality by giving thought to the distributional data available to us than in fact we do.* —MATTINGLY (1962:17)

Distributional Information in Species Descriptions

A species description usually includes information on the distribution of the organism, covering its *occurrence,* the sites at which it was found in your study, as well as the localities in which the organism has been reported by others, if applicable, and its *range,* the general geographic area in which populations of the species occur. For a new species this section may be simple. Often it is only the type locality because that is the only place the organism has so far been found. In a monograph or revisionary study, however, the distribution may be a long section with lists of sites and maps or diagrams of ranges.

Parameters of Species Distributions

There are two important points to remember with regard to distributional information. The first is to describe localities as accurately as possible so that they can found again. Documentation of the locality begins with accurate and complete specimen labels and field records, backed up with maps, photographs, and other data as required.

Terrestrial and Freshwater Habitats

These environments, especially in heavily populated regions, are subject to changes caused by human activity. Road numbering systems and routes of the roads themselves can be altered, bridges can be swept away by floods and not replaced at the same spot, housing developments and shopping malls can overrun what used to be fields, swamps, and forests, and records from 50 or even 5 years ago suddenly become useless. Yet, as the most likely to disappear, these habitats are some of the most important to study now, before the organisms inhabiting them become locally or entirely extinct. As Mayr and Ashlock (1991:328) put it, "*At the present time this task is far more urgent than is collecting in remote uninhabited areas.*" Freshwater habitats have been given priority in the programs of the National Biological Service, in recognition of this fact. Recent statistics show that the United States has lost of 55 percent of its wetlands since European colonization. Some states (California and Ohio) have lost 90 percent or more (Anderson 1996).

To locate your collecting site, you will need to use a map of the area of study to get highway route numbers and names of cities, towns, and counties. Many states are covered by large paperback atlases, with detailed maps of each section of the state, and these can be very helpful, as can county and city maps. Atlases and county maps are can often be purchased at convenience stores, discount stores, and even supermarkets in the study area. Maps of cities and counties also may be available from realtors' offices and chambers of commerce.

You should record locality information from most detailed to most general, from the specific site, (i.e., from the particular mountain, river, reef, etc., to the town, county, topographic map coordinates, latitude and longitude, state, and country in which the site is located, e.g.,

> Buffalo Mountain, 3500 ft., 3 miles south of Willis, Floyd County, Virginia (Pereira and Hoffman 1993:30)
> n. side of n.w. passage, Puerto Galera, Mindoro, Philippine Islands, 10 m depth. . . . (Gosliner 1995:31)
> Hillside above State Highway 79 just south of the bridge over Buffalo Creek, NW1/4 sec. 28, T54N, R1W, Pike County , Missouri. (Hoare and Mapes 1995:122)

Record mileages to the site along roads and highways, especially when this can't easily be figured from a map. It's easy enough to look up the distance from Atlanta to Mobile, if you need it later, but you should record distances along secondary roads, distance to a base camp, or a particular patch of plants or a bird's nesting spot, which won't be found on any map and cannot be easily calculated later. Mileages and direction relative to some prominent landmark are useful in describing a site such as a particular farm or hillside. Town post offices are often used as centering landmarks for such directions (Blackwelder 1967; Herman 1980; Mayr and Ashlock 1991). Localities where you stay and work for a longer period should be described in greater detail, including altitude, township, range and section, latitude, and longitude. This is where topographic maps are the best resource, as all this information can be obtained from them. The U.S. Geological Survey publishes catalogs of state maps. The USGS Quadrangle 7.5 minute 1:24000 maps are recommended for ease of use. They can be ordered from the USGS Publication Office (see this chapter's *Sources*) or obtained from the USGS Map Store in Reston, Virginia. Topographic maps for areas near or in national parks can sometimes be purchased at the National Park Service Headquarters for that particular park, and they are also carried by some outdoor sports outfitters and travel book stores. In addition, mapping software is available for personal computers. If you use a laptop computer in the field, you may be able to plot your localities while on site.

One problem that arises with the use of GIS mapping systems for recording locality data (and it applies to paleontological and marine studies as well) is that they use a different notation system for recording location. Topographic maps and hydrographic charts, as well as conventional printed atlases and gazetteers, give latitudes and longitudes in degrees, minutes, and seconds. GIS systems use decimal degrees for location information and conversions must be made via the program software or by the user before the data are entered.

Paleontology

Paleontologists also use topographic maps. They recommend plotting locality information (as well as field numbers of specimens) on a topographical map, as well as taking photographs of the site (with roll, frame number, and description being recorded in field notebook). For investigators who can spend a few hundred dollars for field equipment, a global positioning system (GPS) is the best way to go. These portable units use satellite data to plot position to about 100 meters. Some models can even download their information right into your field computer (Leiggi et al. 1994).

Marine Habitats

For coastal intertidal collecting, hydrographic charts or topographic maps can be used. For offshore sites, locations should be plotted on hydrographic charts. Charts of a particular area can be obtained at marinas and marine supply stores in that area. In marine environments, Loran-C (a marine radio navigation system) or GPS makes it possible to return to the same spot, even if there are no landmarks to go by. When trawls are taken from a ship, the starting time, latitude and longitude, beginning depth, time on the bottom, and depth, time, latitude, and longitude when trawl is brought off the bottom are recorded. For marine sites, depth is the most important factor to record after location, then water temperature, salinity, bottom type (mud, sand, hard ground, reef, etc.), and exposure to wave action.

An organism's range can be assessed as the area that includes all the localities where it has been found. Needless to say, local populations of the species do not occur everywhere in that area, but only in habitats that are suitable, and probably not in all of those (Wiley 1981). If you have sufficient data, your information should be presented as a distribution map. Such maps are important tools for managing and monitoring a species over time (Loeng 1994).

A large amount of distributional data on a species is best kept on a computerized database, but even the most sophisticated database is only as good as the data that went into it. When you put together a map or list a species's range, you should base it on the specimens you studied, plus information from reliable reports by others. If a record from the literature seems unlikely, and you can't get the material to confirm or eliminate it, leave it out of the distribution, but bring it up in your discussion of the species range (Hawksworth 1974). Records that seem likely but cannot be proven from your evidence can also be discussed here. For example:

> Distribution: Aside from the type locality this species is known only from a collection site identified by Chamberlin (1949: 15) only by the coordinates 68.20N and 151.30W. This locality is on the north side of Anaktuvuk Pass in the Brooks Range, and if the data and identification are both correct, would suggest a range over much of eastern Alaska and southwestern Yukon Territory as the two localities are 750 miles apart. (Pereira and Hoffman 1993:49)

How do you know when your distribution is complete? According to Hawksworth (1974:81) "when additional records almost entirely fall within the already established range and add little to the general pattern." He gives as his example a case in which you have only records of a species from Europe, but then get 20 from South America; that is not the time to stop. Of course, sometimes you might have to stop because of time deadlines, but you would do so knowing that stopping at that point would probably mean a another project later to resolve the problem.

If you are carrying out a faunal or floral study, it is important to record the area covered as well. This is essential information for those attempting to assess such studies from a biodiversity perspective. For instance, if your list of species is larger than that found by a study in another region, it might be just because your study covered a much larger area. Knowing both species richness and area allows you to compare different studies and make decisions about what the data mean. If an area has been thoroughly studied and richness is low, it might truly be an area of low diversity. On the hand, if only one study has been carried out, low species richness might merely indicate a need for additional research (Palmer et al. 1995).

The Distribution Section

Information on a species's occurrence and range may be found in one section, or may be scattered under several headings within the description: "Occurrence," "Stratigraphic Range," "Type Locality," "Material," "Material Examined and Distribution," "Habitat," and so on.

It often turns out that the new species has been collected in only one place. In that case the locality information *is* the type information, as in this example for a new milliped.

> TYPES: Male holotype (ZMUM) from upper reaches of Chashevityi Stream, Mt. Lysaya, Lazovsky State Reserve, Maritime Province, USSR Far East, collected July 16, 1979, by T. S. Vshivkova. (Shear 1990:30)

The distribution section merely refers the reader back to the type section.

> DISTRIBUTION: Known only from the type locality. (Shear 1990:30)

It can also be handled by giving a general distribution in the distribution section:

> DISTRIBUTION: Philippine Islands: Leyte and Negros. (Schuh 1989:15)

and more detailed information in the type section:

> HOLOTYPE: Male, PHILIPPINE ISLANDS: Leyte: Dagami, 14 Mi. SW of Tacloban, July 21, 1961, P. I. Nat. Mus. & AMNH Expedition; deposited in AMNH. . . .
>
> PARATYPE: ♀, PHILIPPINE ISLANDS: Negros Isl: 1300 ft, Camp Lookout, Dumagutete, Feb. 15–April 15, 1961, T. Schneirla and A. Reyes (AMNH). (Schuh 1989:14–15)

Or in an economically worded combination of locality, type information, and distribution, as in this example for a new coralline alga from Gough Island:

> Isolda Rock, very abundant in lower-littoral rock pools, 6 Mar. 1956, *G.I.S.S. 1547* (holotype).
>
> Not known elsewhere. (Chamberlain 1965:213)

Or this example, for a new gobiid fish from Turkey, *Knipowitschia mermere:*

> Geographical distribution (Fig. 5): *K. mermere* is only known from the western Anatolia Lake Marmara (= Mermere), which lies about 100 km inland, cut off from the River Gediz by embankments. (Ahnelt 1995:162)

If the distribution of the new species is more widespread, the distribution section lists the additional localities in which it has been found, as in this description of a new milliped from the United States.

> Distribution: Southcentral coastal plain of Georgia (Fig. 22). In addition to the type locality, the species is known from two samples, one from the Bar M Ranch, near Boston, Thomas County, April 1968, W. Sedgwick (1 male), the other from two miles east of Perry, Houston County, 1 February 1976, J. A. Payne (1 male, 1 female), both VMNH. *Georgiulus paynei* will almost certainly be discovered in northern Florida as well. (Hoffman 1992:12)

Or, this distribution from a description of a new Russian milliped:

> DISTRIBUTION: USSR: Far East, Maritime Province, from rotten wood, Ussuri State Reserve, May 22, 1972, G. F. Kurcheva, 1 ♂ (paratype, SMF); *Fraxinus* litter in valley, November 2, 1978, E. V. Mikhaljova, 11 ♂, Vladivostok, Popova Island, forest litter, October 9, 1979, E. V. Mikhaljova, 1 ♂ (paratype, ZMUC). (Shear 1990:29–30)

When a new species has been recorded but misidentified in the past, it is important to make that clear, as done here for *Thuridilla hoffae,* a new sacoglossan mollusk from New Guinea.

> DISTRIBUTION.–Enewetak, Marshall Islands (Johnson and Boucher, 1983, as *Elysia livida,* S. Johnson, pers. comm.), Guam (Carlson and Hoff, 1978, as *E. livida,* in part), Okinawa (present study), Papua New Guinea (present study). (Gosliner 1995:27)

Height above sea level and water temperature are important habitat descriptors for freshwater fish, as shown in this description of a new species of freshwater fish from Mexico, the Highland Swordtail *Xiphophorus malinche.*

> DISTRIBUTION. Río Claro at 650, Río Moctezuma drainage, Río Pánuco basin, HID; Río Calnalí at 1000 m and Río Conzintla at 1000 and 1140 m (water temp. 15°C), Río Atlapexco drainage, HID; and Arroyo Soyatla, HID, at 1270 m (water temp. 15°C), Río Calabozo drainage, Río Pánuco basin. (Rauchenberger et al. 1990:22)

When an unusual environment may define an organism's distribution, distribution may be given under habitat, as for this new mealy bug.

> Habitat: *Puto kosztarabi* is associated with the grass *Danthonia spicata* on Buffalo Mountain (elevation 3,972 ft.), Floyd County, Virginia. Both mealybug and grass are found near the summit's higher elevations in unique prairie-like glades that contain magnesium-rich soils. (Miller and Miller 1993:10)

In this description of a new harpacticoid copepod from Sri Lanka, locality information is given under *Material.*

> MATERIAL: Four males and two females from the sandy beach looking on Ikkaduwa village (South-western coast of Sri-Lanka), collected on 11/2/1983 (V. Cottarelli leg.) by Karaman-Chappuis method. (Cottarelli and Puccetti 1988:129)

But additional ecological information (that might help future collectors) is given in another section:

> REMARKS ABOUT SAMPLE SITES. The Hikkaduwa coast shows large beaches of medium size sand, that extend for some kilometres. The new species is present only in a set of samples collected 2 meters from the sea-shore at a depth of 20–30 cm; there interstitial water is salty, and the Fauna is qualitatively and quantitatively considerable. . . . (Cottarelli and Puccetti 1988:135)

Even when distributional information on a new species is limited, the distribution section may be the place to speculate on its probable range.

> *Spiculidendron corallicolum* occurs singly or in clusters in semishaded clear-water reef locations, 20–30 m and below. . . . Distribution of the species is presumed to be Caribbean-wide; so far it has been found in the Bahamas (San Salvador), Turks & Caicos Islands, Leeward Islands (Dominica), Colombia, Honduras (Roatán), Belize (Carrie Bow Cay, Lighthouse Reef), and Cayman Islands. (Rützler and Richardson 1996:149)

Vague locality description is practically useless, as anyone knows if they have searched through old species descriptions or examined labels of specimens collected long ago, to find as their only locality data, "St. Peter's Bank," "Utah," or even worse, "China??" But there are two circumstances under which locality descriptions, in contrast to what is usually recommended, may be left vague or general. One of them is biological. In the case of animals that fly, swim, or otherwise roam a large territory, the collecting locality is really just a small part of their local distribution, and for some purposes, a general location is sufficient:

> While many of the localities where Voros collected are specified in considerable detail, I have only given the larger localities, which can be found on maps. For volant mammals, such as bats, which may forage several miles form the roost, this seems adequate. (Koopman 1989:1–2)

The other reason for keeping locality information general is for the protection of the species being described. This applies mainly to the groups of rare or endangered organisms that are seen as desirable acquisitions by collectors, such as mollusks, birds, lizards, turtles, orchids, or other rare wildflowers, species that may be known from only one or two sites. In such cases the exact locality information may be kept with museum records, but would made available only to those with a legitimate need for it. What would be put into the online collection record, or even the scientific publication, would be a less specific location, as in this description of an endangered plant, *Astralagus neglectus:*

> Cooper's milkvetch is extremely rare in Virginia. It is known only from a single locality in Allegheny County where it grows on a moderately steep, forested slope. (Wieboldt, in *Virginia's Endangered Species* 1991:123)

The distribution section may also be the appropriate place to discuss impacts on a species distribution, as in this example:

> DISTRIBUTION: (Figure 17, area 2): *Rheocles alaotrensis* is the most widespread of *Rheocles* species. Kiener (1961) and Maugé (1986) recorded the species as occurring in the fresh waters of central eastern Madagascar including Lake Alaotra. . . . The demise to extinction of the Alaotra populations has been well documented by Moreau (1979), and Reinthal and Stiassny (in rev.) discuss the further attrition of riverine populations due to the combined onslaught of deforestation and the introduction of exotic species. It seems probable that today *R. alaotrensis* is restricted to the

> protected waters of forest reserves, principally the small reserve of Perinet in the district of Moramanga. (Stiassny 1990:26)

For a newly described fossil, distributional information may also be included under two headings, if the species is known only from the type locality.

> Type horizon and locality: Exposure of the Gene Autry Formation in a series of east–west gullies, on the east side of an unnamed tributary of Sycamore Creek on the Daube Ranch, NW1/4, NW1/4, SW1/4 sec. 2, T4S, R4E, Johnson County, southern Oklahoma, Ravia 71/2' quadrangle. . . .

and

> **Occurrence.**–Known only from the type locality. (Hoare and Mapes 1995:124, 126)

If a fossil species occurs more widely, material referred to that species may be listed under *Locality*:

> LOCALITY: Younger World (IGM 3670, 3671) and Too Much Hot (IGM 3672). Las Tetas de Cabra Formation, 40 km south of Punta Prieta region, Baja California Norte, Mexico. Wasatchian (early Eocene). (Novacek et al. 1991:56)

The distribution of a fossil species is often listed under stratigraphic occurrence, as in this example of a Devonian brachiopod.

> STRATIGRAPHIC OCCURRENCE: The shells were collected from two localities, both within the Seneca Member, Onondaga Limestone.
>
> The first (AMNH Loc. 3129A) is in the Jamesville Quarry #3 pit which is the most complete section of the Onondaga in the central part of New York. Collecting in this quarry is difficult since it is very active. Not only is it hard to find weathered bedding surfaces but the rocks are being continually blasted and removed, making it difficult to locate and recover fossil material. The Seneca Member here is 14 ft thick, with the Tioga Bentonite at the base. Ten feet above the Tioga is found the "Pink *Chonetes*" Zone which consists of a fine-grained limestone with a fresh

> dark gray surface. There are wavy contacts between bedding planes and stylolites present. . . . (Racheboeuf and Feldman 1990:8)

When you need considerable discussion of the age of a fossil, it may be included in the discussion section, as in this example from a description of *Pakkokuhyus lahirii,* a new artiodactyl from Burma. Locality information is presented under type locality.

> *Type Locality.*–"K21"-315, three furlongs [approximately 0.6 km] N. W. [north-west] of Thanudaw village, Myaing Township, Pakkoku District, Burma." "K21" is a section designation on Geological Survey of India maps of the region. (Holroyd and Ciochon 1995:178)

The age of the new species is covered in the discussion.

> The Pondaung mammalian fauna occurs within a thick succession of marine and continental rocks that were deposited throughout the Eocene and are now exposed in the Chindwin–Irrawaddy Basin. . . . These age assessments for the marine strata of Burma suggest a Bartonian (late middle Eocene) age for the Pondaung sandstones and fauna based on their intermediate stratigraphic placement. . . . (Holroyd and Ciochon 1995:180–181)

For a faunal or floral study, you may want to limit your discussion of general distribution to that region; for example, in a study of the bryozoans of the Indian River area I mentioned that some species were cosmopolitan, but limited discussion of their range to western Atlantic records. Thus, for *Zoobotryon verticillatum,* a widely distributed fouling species, the entry read as follows:

> DISTRIBUTION: A circumtropical species. Western Atlantic: Beaufort to Brazil. Gulf of Mexico and the Caribbean. (Winston 1982:114)

For the purposes of this regional study, the western Atlantic records were carefully checked, but records from other areas could be ignored. More examples of regional distribution sections from a study of the Mecoptera of a region of South Africa by Jason Londt.

> Afrotropical distribution: South Africa (Transvaal, KwaZulu–Natal); Zimbabwe. The bulk of records are from the Transvaal.

KwaZulu–Natal distribution (Fig. 4): Known from only two localities; both between 300–600 m asl (Table 2). (Londt 1995:174)

And a study on the flora of Hispaniola by Thomas Zanoni and Ricardo García G.

DOMINICAN REPUBLIC: base Cordillera Septentrional: Prov. Puerto Plata, "La Unión," E of new Puerto Plata Airport, 4 km E of Sosúa on highway to Puerto Plata, along mouth of Río Forma, 19°45' N, 70°33' W, sea level to 10 m, 16 Jul 1990 (fl and fr), Mejía & Zanoni 7329 (JBSD, MO, NY, S, TEX). (Zanoni and García G. 1995:262)

For a monograph or complete revision of a taxon, a complete distribution is given. All previously published records should be checked and analyzed and, if possible, the specimens themselves should be examined. The following example is from a study of the sacoglossan mollusk genus *Thuridilla* by Terry Gosliner. In this paper general locations along with citations from all published records of each species are included in each distribution section.

DISTRIBUTION:–Specimens with *T. bayeri* coloration are known from the Maldives (Yonow, 1994), the Marshall Islands (Marcus, 1965; Johnson and Boucher, 1983), Guam (Carlson and Hoff, 1978), ?Fiji (Brodie and Brodie, 1990), Papua New Guinea (present study). . . . (Gosliner 1995:2)

while detailed locality and habitat information on specimens studied is presented in the material examined section:

MATERIAL EXAMINED.–Specimens with coloration more similar to *T. bayeri:* CASIZ 065743, one specimen, dissected, harbor wharf, Madang, Papua New Guinea, 10 m depth, 15 January 1988, T. M. Gosliner, CASIZ 086385, one specimen, barrier reef, wnw of Rasch Passage, 4 m depth, 14 June 1992, T. M. Gosliner, . . . one specimen dissected, Saint Crispin Reef, nw of Port Douglas, Queensland, Australia, 15–20 m depth, 9 December 1984, M. L. Gosliner, CASIZ 087123, one specimen, under rock, Montehage Island, Manado, Sulawesi Indonesia, 3 m depth, 17 May 1990, P. Fiene-Severn. CASIZ 099057, one specimen, radula removed, pinnacle, G. Buoy, Kwajelin Atoll, Marshall Islands, 6 m depth, 6 March 1994, S. Johnson. . . . (Gosliner 1995:2)

When a distribution seems difficult to explain biologically, the distribution section may incorporate discussion of possible explanatory factors, as in this spider example:

DISTRIBUTION: Oklahoma south to northeastern Mexico. A single male (in AMNH) was reportedly collected in Phoenix, Arizona on April 6, 1938; the specimen may simply have been transported by humans (or mislabelled). Juveniles that might belong to this species have been collected in Sonora, Mexico, and the range could conceivably extend that far west, but the Texas records are all from the eastern half of the state. (Platnick and Shadab 1989:9)

Distribution Papers

As was true for ecological information, distributional information may become the focus of the paper. The entire paper may be written around a description or analysis of an organism's distribution, as the following examples show.

Bishop, G. A. and E. C. Bishop. 1992. Distribution of ghost shrimp; North Beach, St. Catherine's Island, Georgia. *American Museum Novitates* 3042, 1–17.

Edwards, C. and S. M. Harvey. 1983. Observations on the hydroids *Coryne pintneri* and *Thecocodium brieni* new to the British list. *Journal of the Marine Biological Association of the United Kingdom* 63: 37–47.

Erséus, C. 1987. Records of *Limnodriloides* (Oligochaeta: Tubificidae) from Venezuela. *Proceedings of the Biological Society of Washington* 100: 272–274.

MacPhee, R. D. E. and A. R. Wyss. 1990. Oligo–Miocene vertebrates from Puerto Rico, with a catalog of localities. *American Museum Novitates* 2965: 1–45.

Segura-Puertas, L. 1992. Medusae (Cnidaria) from the Yucatan shelf and Mexican Caribbean. *Bulletin of Marine Science* 51: 353–359.

Smith, C. L. and T. R. Lake. 1990. Documentation of the Hudson River fish fauna. *American Museum Novitates* 2981: 1–17.

Sources

Topographic Maps

To order topographic maps you need to know the ordering information and reference codes from the index maps and catalogs of topographic and other published maps for each state, which may be obtained free from Map Distribution, U.S. Geological Survey, Box 25286, Building 810, Federal Center, Denver, CO 80225; (303) 236-7477.

Global Positioning System Guides

Ferguson, M. 1997. *GPS Navigation: A Complete Guidebook for Backcountry Users of the NAVSTAR Satellite System.* Boise, Idaho, Glassford.

Hotchkiss, N. 1997. *A Comprehensive Guide to Land Navigation with GPS,* 2d ed. rev. Herndon, Va., Alexis.

Kennedy, M. 1996. *The Global Positioning System and GIS: An Introduction.* Chelsea, Mich., Ann Arbor Press.

Letham, L. 1995. *GPS Made Easy: Using Global Positioning Systems in the Outdoors.* Seattle, Mountaineers.

Biogeography

American Museum of Natural History Systematics Discussion Group. 1981. *Vicariance Biogeography: A Critique.* New York, Columbia University Press.

Bailey, R. G. 1998. *Ecoregions: The Ecosystem Geography of the Oceans and Continents.* New York, Springer.

The Biogeography of the Oceans. 1997. Advances in Marine Biology, vol. 32. San Diego, Academic Press.

Briggs, J. C. 1995. *Global Biogeography.* Amsterdam, Elsevier.

Brown, J. H. and A. C. Gibson. 1983. *Biogeography.* St. Louis, Mosby.

Caley, M. J. and D. Schluter. 1997. The relationship between local and regional diversity. *Ecology* 78: 70–80.

Clark, T. W. and H. H. Seebeck, eds. 1990. *Management and Conservation of Small Animal Populations.* Chicago, Chicago Zoological Society.

Cox, C. B. and P. D. Moore. 1988. *Biogeography: An Ecological and Evolutionary Approach,* 4th ed. Palo Alto, Calif., Blackwell.

Craw, R. C. 1998. *Panbiogeography: Tracking the History of Life.* New York, Oxford University Press.

Edwards, W. and M. Westoby. 1996. Reserve mass and dispersal investment in relation to geographic range of plant species: Phylogenetically independent contrasts. *Journal of Biogeography* 23: 329–338.

Eldredge, N., ed. 1992. *Systematics, Ecology, and the Biodiversity Crisis.* New York, Columbia University Press.

Gaston, K. J. 1994. Geographic range sizes and trajectories toward extinction. *Biodiversity Letters* 2: 163–170.

Gilbertson, D. D. 1995. *Practical Ecology for Geography and Biology: Survey, Mapping, and Data Analysis.* London, Chapman & Hall.

Hallam, A. 1994. *An Outline of Phanerozoic Biogeography.* Oxford, U.K., Oxford University Press.

Harris, L. D. 1984. *The Fragmented Forest: Island Biogeography Theory and the Preservation of Biotic Diversity.* Chicago, University of Chicago Press.

Hengeveld, R. 1992. *Dynamic Biogeography.* Cambridge, U.K., Cambridge University Press.

Krebs, C. J. 1994. *Ecology: The Experimental Analysis of Distribution and Abundance,* 4th ed. Menlo Park, Calif., Benjamin Cummings.

MacArthur, R. H. 1972. *Geographical Ecology: Patterns in the Distribution of Species.* Princeton, N.J., Princeton University Press.

MacArthur, R. H. and E. O. Wilson. 1967. *The Theory of Island Biogeography.* Princeton, N.J., Princeton University Press.

Mauer, B. A. 1994. *Geographical Population Analysis: Tools for the Analysis of Biodiversity.* Oxford, U.K., Blackwell.

Nelson, G. and N. Platnick. 1981. *Systematics and Biogeography: Cladistics and Vicariance.* New York, Columbia University Press.

Nichols, D., ed. 1994. *Taxonomy and Geography: A Symposium.* London, Systematics Association.

Pielou, E. C. 1979. *Biogeography.* New York, Wiley.

Quammen, D. 1996. *The Song of the Dodo: Island Biogeography in an Age of Extinction.* New York, Simon & Schuster.

Rosenzweig, M. L. 1995. *Species Diversity in Space and Time.* Cambridge, U.K., Cambridge University Press.

Species Diversity in Ecological Communities: Historical and Geographic Perspectives. 1993. Chicago, University of Chicago Press.

Tivey, J. 1993. *Biogeography: A Study of Plants in the Ecosphere,* 3d ed. Burnt Mill, Essex, U.K., Longman.

Vermeij, G. J. 1978. *Biogeography and Adaptation: Patterns of Marine Life.* Cambridge, Mass., Harvard University Press.

Veron, J. E. N. 1995. *Corals in Space and Time: The Biogeography and Evolution of the Scleractinia.* Sydney, University of New South Wales Press.

Vicariance Biogeography. 1988. Washington, D.C., Society of Systematic Zoology. (Special issue of *Systematic Zoology.*)

Chapter 15

Material Examined

Citations of specimens examined must always be given as precisely as possible. –HAWKSWORTH (1974:68)

Many years ago people cited material in detail. This was later discouraged or forbidden by many editors because it is expensive in paper and typesetting. It was often replaced by just a spot map, or the name of a place where you could find the documentation.
–R. L. HOFFMAN, ENTOMOLOGIST, PERSONAL COMMUNICATION, 1996

Practical Value

One of the kindest things you can do for biologists who come after you is to include a clear and accurate account of all the material you studied somewhere in your species description. It may seem that including a section on type material (chapter 9) would be enough. However, once you have undertaken some taxonomic research of your own and have searched through paper after paper without finding so much as a note on the deposition of the author's specimens, or arrived at the institution where an author's specimens were deposited only to spend hours or days trying to figure out just which specimens the author used, you will realize the practical value of a material examined section.

Because there are no rules about the inclusion of material examined in taxonomic description papers, use of this section has largely depended on historical tradition among systematists studying different groups and, as pointed out in the second introductory quotation, the need to produce papers more and more economically has worked against inclusion of any such raw data categories. Some workers, such as entomologists (perhaps because they must deal with so much material), traditionally have been conscientious about including information on the specimens they examined whenever it was possible to do so. But for many other groups of animals, including a section on material examined in taxonomic descriptions is still not usual practice.

In Original Descriptions

This section is not needed in a simple description of a new species. If the species has never been collected before, the material you study will stem entirely from your own collecting efforts and will become type or paratype material. A materials examined section may be needed in more complex cases, as when a species has long been misidentified or when a single named species is found to be composed of two or more species. In such cases a large quantity of newly collected and museum specimens may be examined and ascribed to the named or the new species.

Taxonomists sometimes discover new species in museum collections while carrying out phylogenetic studies of a genus or family. For example, Nancy Simmons was conducting a phylogenetic analysis of the Neotropical bat genus *Micronycteris*. This project involved examining and measuring variation in morphological characters of all the *Micronycteris* specimens she could locate in museum collections. In the course of her survey she noticed that one specimen from the National Museum of Natural History's mammal collection that was labeled *Micronycteris schmidtorum* differed in a number of characters from the others given that name. Continuing her survey, she found additional specimens like it in another collection. These had been labeled *Micronycteris minutus,* but, like the first specimen, differed significantly from any known *Micronycteris* and so warranted description as a new species. One of the specimens she discovered became the holotype of the new species. Other specimens from the Carnegie Museum were named as paratypes, but in her paper she noted that further material might still be waiting to be found in other collections:

> TYPE MATERIAL: The holotype of *Micronycteris sanborni* (USNM 555702) is an adult female skin and skull collected by A. L. Gardiner on 26 March 1978 at Sitio Luanda, Itaitera, 4 km S of Crato in the Brazilian state of Ceará (original field number ALG 13745). . . .

> Paratypes include one additional specimen from Ceará (CM 98913) and four from two nearby localities in Pernambuco (CM 98914, 98915, 98916, 98917). These specimens, also skins and skulls, were collected by Michael Willig in 1977. An unknown number of additional specimens may be present in the Museu de Zoologica, Universidade de São Paulo, which received half the collections made by Willig and his colleagues (Mares et al., 1981). (Simmons 1996:6)

It is also surprising how often it happens that once a person has discovered something new in the field and collected specimens to describe, he or she is then able to turn up other specimens already in various museum collections, either mislabeled or still unidentified. In *A Parrot Without a Name,* a book about a search for the last unknown bird species, Don Stap describes how, after specimens of a new species of *Tolmomyias,* a tyrant flycatcher genus, had been collected in Peru in 1984, Louisiana State University ornithologist Ted Parker, who was writing the species description, requested loans of similar-looking specimens from colleagues at museums with large Neotropical bird holdings. He received not only specimens of the other four species of the genus, but also two specimens from the Field Museum that had been questionably identified with another species of the genus (but turned out to belong to the new species), and another specimen from the American Museum of Natural History whose differences had been noticed in 1939 and which had even been labeled as a possible subspecies (it also turned out to belong to the new species). He even found one in his *own* museum's collection. It had been collected in northern Peru in 1979 and labeled as one of the other species of the genus (Stap 1990).

In Other Descriptions

A material examined section is also useful in revisions, redescriptions, and taxonomic analyses of species or other taxa. Large quantities of material may be examined during the course of such work. One way to handle its inclusion efficiently is to plot the specimen locations on a distribution map and include more detailed information in an appendix. Figure 15.1 shows a distribution map for two species of sand lances, *Ammodytes americanus* and *Ammodytes dubius,* examined as part of a study carried out to determine the best way to distinguish the two species morphologically (Nizinski et al. 1990). In this paper the Methods section began as follows:

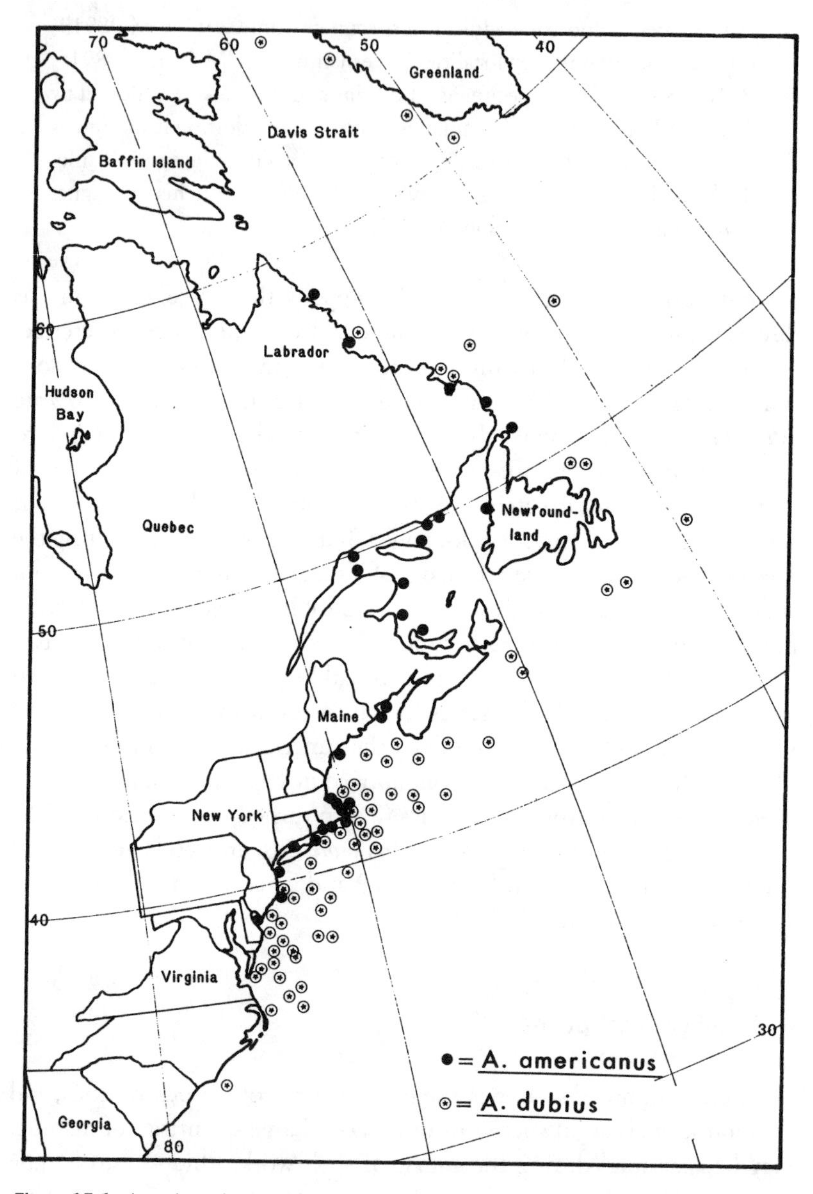

Figure 15.1 Locations showing where specimens of two sand lance species, *Ammodytes americanus* and *Ammodytes dubius,* were collected. Such a map can be an effective way to show a large amount of material examined. *–Source: From figure 2 of Nizinski et al. (1990).*

> Approximately 1500 specimens from a range of locations (North Carolina to Greenland) along the western North Atlantic coast were examined in this study (Fig. 2; Appendix 1). (Nizinski et al. 1990:243)

Their figure 2 (figure 15.1) showed the sites where the two species had been collected. The appendix gave more detailed information, including museum initials and catalog numbers of specimens examined from each site:

> **Ammodytes dubius**
> **Labrador** 14 specimens (121–203 mm SL) from two collections. USNM 165263 (11,121–187) Pack's Harbor, 53°54'N 56°59'W; *Blue Dolphin;* 24 July 1949. USNM 165264 (3,143–203) Hare Harbor; *Blue Dolphin;* 2 July 1950.
> **Newfoundland, Nova Scotia, and Quebec** 72 specimens (107–244 mm SL) from five collections. NMC 64–763 (16, 178–209) off Newfoundland, Grand Banks, 45°06'30''N 49°01'00''W; A.T. Cameron; 14 Oct. 1964. (Nizinski et al. 1990:254)

Material Examined Section

If you examined only a small number of specimens of a species, you can put that information into the main body of the description. More commonly, a small amount of information on material examined is put into the discussion section:

> Our specimen most closely resembles the specimen illustrated by Harmer (1977, pl. LXII, fig. 32) from the *Siboga* Collection (no. 489C). Of the other *Siboga* material of *F. malusii* in the collection of the British Museum, specimen no. 551A-3 is also *F. harmeri,* but specimen 371 G is not. . . . (Winston and Heimberg 1986:28)

The most efficient way to incorporate a large amount of information of this type is in a separate section, variously titled "Material," "Material Examined," "Specimens Examined," or "Additional Material," depending on journal style or author preference. This section lists the specimens, institutions, catalog numbers, and any other information that might be useful to a user. Usually the names of the institutions where the specimens are located are abbreviated to save space. A key to the abbreviations is then given in the introduction, materials and methods or elsewhere in the paper. Catalog numbers are given if the specimens you examined are cataloged specimens. For example, in a study of Indo-Pacific cardinal fishes, Randall et al. (1990) looked at specimens from 14 different museums. The description of each species included a material examined section, giving for each species its geographic location, holding institution, catalog number, and a standard length measurement (which had been defined in their introduction):

> Apogonid material for the present study has been examined at or obtained from the following institutions: Australian Museum, Sydney (AMS), Academy of Natural Sciences of Philadelphia (ANSP); British Museum (Natural History), London (BMNH); . . . Lengths given for specimens are standard length (SL), measured from the front of the upper lip to the base of the caudal fin. . . . (Randall et al. 1990:41)

Thus, for *Apogon aureus:*

> *Material examined.*–East Africa: RUSI 3205, 54 mm. Kenya: Mombasa, BMNH 1913.4.7.51, 92 mm. Zanzibar: BMNH 1867.3.7.555, 84 mm. . . . (Randall et al. 1990:49)

Sometimes, you may have examined uncataloged material. In that case you give the accession number (if there is one) and any identifying label material as well as any other information (e.g. The Osburn Collection, the Hincks Collection) that might help a future user locate those specimens in the institutional collection where they are maintained.

The following examples illustrate more ways of presenting information on material examined. Illustrated specimens may be listed separately from other material examined, as in this entry for a fossil plant, *Sphenolepis sternbergiana:*

> Number of Specimens Examined: 30.
> Illustrated Specimens: USNM 446026; FMNH PP43801, PP43809.
> Other Identified Specimens: FMNH PP43846–PP43868. (Upchurch et al. 1994:16)

In another fish taxonomy paper, this one on eelpouts of the eastern Pacific (Anderson 1995), information on station locations and depths is included when available, in addition to length measurements, as in the entry for *Lycenchelys jordani:*

MATERIAL EXAMINED–Alaska, off Sitka Sound: SU 20014 (paratype; 208 mm); Cape Edgecumbe S 84°, E 21 miles; *Albatross* sta. 4267; 1686 m; 1731 hrs.; 31 July 1903; U.S. Fish Commission. British Columbia: NMC 68–1783 (1; 320 mm); 50°54.5'N, 130°06.0'W; *G. B. Reed;* 2103–2196 m; 11 Sept. 1964. . . . (Anderson 1995:81–82)

In a study of two new Aegean frog species, where electrophoresis was used as one of the means to distinguish the species, the material included frozen tissue specimens.

SPECIMENS EXAMINED. We examined a total of 41 individuals (including the type series): Type locality, 13 (3 females, 1 male, 9 juveniles); Petros river, 2 (females); Lavris, 6 (4 females, 2 males); Iraklion (Ιακλειον; river in town), 7 (juveniles); Elaphonissi (extreme southwest corner of Crete, near sea shore, collected in April 1982 by H. Maier), 13 (juveniles). Frozen tissue samples of the type series and other specimens examined are stored at the Zoologisches Museum, Universität Zurich. (Beerli et al. 1994:2)

As mentioned previously, if information on specimens examined is extensive, it may be placed in an appendix, as in this taxonomic review of the North American freshwater snail genus *Probythinella:*

Appendix: Material Examined

Lots containing anatomical material (all of which were dissected during the course of this study) are highlighted with an asterisk. The following abbreviations for institutions are used:
ANSP: Academy of Natural Sciences, Philadelphia
BMNH: The Natural History Museum, London
FMNH: Field Museum of Natural History, Chicago . . .

Probythinella protera PILSBRY

United States. Alabama. *Mobile County:* Fowl River, 1.9 km west of Mon Lewis, 1.6 km northeast of Bellingrath (*UF 224042).–west coast Mobile Bay, just south of mouth of Dog River (USNM 680257). **Louisiana:**

> Lake Pontchartrain, at mouth of Tchefuneta River (*USNM 874156) (Tangipahoa–St. Tammany Parishes). . . . (Hershler 1996:139–140)

Material Examined: Botanical Taxonomy

Botanists also include a section on specimens examined; in fact, they seem to have been much more consistent about including this information than have zoologists. The *Mycologist's Handbook* (Hawksworth 1974) includes a section on material examined, with examples. Most of his recommendations are applicable to other botanical material as well. Host, substratum, and locality (e.g., country, state, province, county, map grid reference), dates, and (when applicable) collector's numbers should all be included. Locality information should be verified and spellings checked in an atlas or gazetteer before they are used in your paper. Where spellings are incorrect or names outdated, he recommends using the currently accepted spelling first and the actual herbarium packet spelling in brackets. Dates should be written out unambiguously, e.g., 8 March 1986, or 8 MAR 1986, not 3/8/86 or 3-8-86.

The following example comes from a paper on the algal genus *Colpomenia:*

> SPECIMENS EXAMINED.–*Gulf of California.* Sonora: Punta Pelicano, vicinity of Puerto Peñasco, 17 March 1973, JN-3811 (UC); 19 Apr 1973, JN-3966 (MEXU); 9 Mar 1974, JN-4982 (US), (leg. JN and KB); Feb 1965, Dawson 27558 (ARIZ). Playa Arenosa, vicinity of Puerto Peñasco, 26 Dec 1972, Wynne 3754 (MICH, TEX); 17 Apr 1973, JN-3911 (Mich). (Wynne and Norris 1976:2)

Collection dates are given in a standard format, along with collection number and herbarium abbreviation. In this paper the full names of the herbaria where the studied specimens are held are given in the introduction, along with the corresponding abbreviations. Full names of collectors identified by initials are given in the introduction also.

In this excerpt from a description of a Micronesian morning glory, geographic information is combined with that on specimens examined. The collector's name and number are italicized. If the specimen was examined by the authors, the name of the herbarium where it is deposited is given, using the standard herbarium initials as given in the *Index Herbariorum* (Holmgren et al. 1990):

GEOGRAPHIC RECORDS AND SPECIMENS EXAMINED

MARIANAS ISLANDS–Pagan: Fresh Water Lake, *Fosberg 31393* (US, BISH, Fo).

Saipan: *Cameron* in 1944 (GISH); *Momose* in 1930 (TI); Kagman Peninsula, *Courage 20* (US); S of Mt. Tapotchau, in swamp, 10 ft [3 m], *Hosaka 2912* (US, BISH); marshes and ponds about 1 km S of Lake Susupe, SE of Chalankoa, *Fosberg 25275* (US, BISH).

Tinian: *Okatani 66* (FU), *67* (FU).

Rota: Slopes above As Malote, S side of island, 250 m, *Fosberg 31884* (US, BISH, Fo, NY, L). . . . (Fosberg and Sachet 1977:8)

The Material Examined Paper

It is possible to devote an entire work to the examination of specimens, or even a single specimen. Such work may need to be carried out to settle the identity of species that were described long ago, or for which material has been lost and is later rediscovered, or with which a problem of some kind is found. This kind of project is unlikely to be undertaken by anyone except a professional systematist, as it demands a good deal of experience with the taxonomy of the group involved. Examples of such titles are as follows:

Riedl, H. 1995. On two doubtful species of *Cynoglossum* (Boraginaceae) described by J.G.C. Lehmann. *Annalen des Naturhistorischen Museums in Wien* 97B:509–512.

Olson, S. L. 1989. Two overlooked holotypes of the Hawaiian flycatcher *Chasiempis* described by Leonhard Stejneger (Aves: Myiagrinae). *Proceedings of the Biological Society of Washington* 102: 555–558.

Musser, G. G. and J. L. Patton. 1989. Systematic studies of oryzomyine rodents (Muridae): The identity of *Oecomys phelpsi* Tate. *American Museum of Natural History Novitates* 2961: 1–6.

This last paper describes a professional systematist's nightmare, a case in which the holotype specimen of a Venezuelan rodent turned out to be a composite, made up of a skin and jaw from one genus mixed with the cranium of another. Straightening out this mismatch in a way that preserved information and clarified the use of the scientific name for future workers was no doubt gratifying for the taxonomists involved, but it represents the kind of problem most other biologists would be glad to avoid.

Sources

See *Sources* in chapters 5 and 9 for reading and references applicable to this chapter also.

Chapter 16

Publication

The goal of scientific research is publication. Scientists, starting as graduate students, are measured primarily not by their dexterity in laboratory manipulations, not by their innate knowledge of either broad or narrow scientific subjects, and certainly not by their wit or charm; they are measured, and become known (or remain unknown), by their publications. —Day (1988:vii)

New knowledge is not science until it is made social. The scientific culture can be defined as new verifiable knowledge secured and distributed with fair credit meticulously given.
—Wilson (1995:210)

Once your description has been written, you are ready for the final step: publication. But publication of a taxonomic description has two aspects. As with any other form of scientific research, you must deal with the logistics of getting your paper accepted and published in a scientific serial. You must also make sure that you satisfy the nomenclatural rules that govern the publication of the names of new taxa. This chapter covers both aspects of publication. First, it reviews what makes a new species name available (zoology) or valid (botany). Second, it describes the steps in the publication process and lists some of the journals that publish taxonomic work.

Criteria of Publication: Zoology

To be published, a new species must first be given a properly composed genus and species name, as described in chapter 8 of this book. To become available for use, the name must be published, accompanied by certain descriptive information, in an appropriate medium. The Third ICZN (1985, Art. 8) states that to be regarded as published a work must

"Be issued publicly for the purpose of providing a permanent scientific record"

"Be obtainable, when first issued, free of charge or by purchase"

Be "produced in an edition containing simultaneously obtainable copies by a method that assures numerous identical copies." The fourth draft of the code adds one more criterion for availability. The new name must be explicitly indicated as being new by use of a term such as *new species, sp. nov.,* or *new genus, gen. nov., fam. nov., nom nov.,* or the equivalent term identifying it as a new term in the language in which the paper is written.

The rules formerly specified that the publication must have been produced "by conventional printing," such as letter press, offset printing, hectographing, or mimeographing, *in ink on paper.* Works produced since 1986 may be published by other means, if

They "contain a statement by the author that any new name or nomenclatural act within it is intended for permanent, public scientific record."

They include a statement *in the work itself* that "it was produced in an edition containing simultaneously available copies."

So you *could* use photocopies, microcards, microfiche, CD-ROM, or desktop-published and computer-printed media as your means of publication of a species description, provided that work carried the statements that any new name within it was "intended for permanent, public, scientific record," had the date of publication on it, and had the number of copies produced on that date on it, as well as a price (or a statement that it was available without charge) and the address from which it could be obtained.

But although the rules allow such forms of publication, their authors recommend against using them. Because the main goal of publishing the description is to give your new species a name that you and others can use, you should attempt to get your work published in the best available medium, a legitimate refereed scientific publication. That way, your work will have the maximum degree of scientific credibility and your new species name will achieve the broadest possible distribution. In other words, if you want your new name to be accepted and used you should do your best to ensure that your description appears in a serial publication, a scientific journal or monograph series (e.g., *American Museum Novitates* or *Bulletin, Smithsonian Contributions in Zoology, Proceedings of the Biological Society of Washington*), or the proceedings volume of a scientific society meeting.

In fact, that is what most biologists do. It is rare to see a privately published or marginal publication labeled as the code suggests. It is much more common to see a disclaimer in a society newsletter or popular scientific publication in which taxonomic matters are discussed, but where people want to avoid having their preliminary discussions of a new entity taken as the actual description:

> *The ISC Newsletter* is not issued for permanent scientific record, and thus, for the purposes of zoological nomenclature, does not fulfill the criteria for publication as defined in the International Code of Zoological Nomenclature. (1990, 9:11)

As noted in chapter 9, all new species descriptions published after 1999 must include the designation of a name-bearing type (holotype or a syntype series) and must state where that type or types have been deposited.

The ICZN also deals with what does not constitute publication, and in doing so shows how the code has developed over time. Early descriptions were often not good at all by today's standards, and succeeding versions of the ICZN show a slow phasing out of bad practices. Although we may still accept an early description that would not meet today's requirements, we must adhere to current rules in the descriptions we publish today.

For example, a handwritten work (even if the handwritten pages were reproduced in large numbers by some process) produced after 1930, microfilms, photographs (except for microcard or microfiche in works after 1986), proof sheets, computer printouts, acoustic recordings, the limited circulation of a note (even if printed) in explanation of an accompanying illustration, labeling of a specimen, mention at a meeting, or deposit of a single document (e.g., a thesis or dissertation) in a library or archive is not acceptable as published (ICZN Art. 9). The fourth edition of the ICZN adds electronically distributed text or illustrations, including down-loaded copies or printouts and abstracts of pages, posters, lectures given at a meeting but not otherwise published to the list of what does not constitute valid zoological publication.

In addition, the ICZN rules state that every new scientific name published after 1930 must be accompanied by a written description or bibliographic reference to a published description (as described in chapters 10 and 11 and in Art. 13 of the ICZN). Before 1950 such a new name might be considered available even if it were published anonymously. This is no longer allowed (Art. 14). Before 1960, a new name might also be declared to be available if published conditionally. This is no longer allowed either (Art. 15).

Another change has to do with the availability of species names that were originally proposed as varieties or forms; those proposed before 1960 might be considered available. Those published after 1960 are excluded from zoological nomenclature as infrasubspecies categories (Art. 16). This change in rules is significant because often what were considered varieties by the original nineteenth-century authors have turned out to be valid species when examined by modern workers. We can turn those old names into species names, but we cannot do the same with new ones (fortunately, form and variety names are seldom used anymore).

Criteria of Publication: Botany

To be considered validly published in botany, a new species name must be properly constructed according to the ICBN (chapter 8, this book). It must also be published in a publication that is printed and is made adequately available to those interested and it must be accompanied by adequate descriptive information (chapters 10 and 11). One of the most important differences between botany and zoology is the requirement for a Latin diagnosis in botany regardless of the language of the rest of the description (ICBN, Art. 36.1, and chapter 11, this book). This rule has been in effect since 1958 for algae and since 1935 for all other nonfossil plant groups. The only exception is that for fossil plant taxa and even for fossil plants, new descriptions (those made after 1996) must be accompanied by a Latin *or* English diagnosis or description (ICBN 36.3).

Chapter 4 of the ICBN covers effective and valid publication. The ICBN is even clearer than the ICZN about publication being "effected . . . only by distribution of printed matter . . . to the general public or at least to botanical institutions with libraries accessible to botanists generally"

(ICBN, Art. 29.1). Like the zoological code, the ICBN no longer allows publication of handwritten descriptions (even if the handwriting is reproduced in some printed form).

Valid publication in botany (equal to ICZN availability) requires a "full and direct reference" to "author and place of valid publication with page or plate reference and date" (ICBN, Art. 33.2). For names of new taxa published after January 1, 1958, it also requires that the nomenclatural type be indicated (Art. 37). In fact, for names of a new species or genus, not only must the type be indicated by the words **typus** or **holotypus** or a modern language equivalent, but citation of corroborating details (collecting locality, collector's name or number, herbarium where deposited) is also required (Art. 37.3). The description must be accompanied by an illustration showing the essential characters or a reference to such an illustration (Art. 38).

Preparation of the Manuscript

This section is not intended to be a detailed course in scientific publication, but merely an outline of the steps in the process, especially as they apply to taxonomic papers. Many books are available on writing in general and on scientific writing in particular. Some of those books were discussed in chapter 11 and listed in the *Sources* for that chapter. Anyone who is planing a career in science should have at least one or two of them as desk references.

The First Draft

By the time you read this chapter you should have a pretty clear idea of what you need to write. You may have found some good papers by specialists in the group to serve as models. In fact, you may have already written the first draft of the introduction, methods, description, and discussion of your results. *Writing* the first draft of a paper is not so bad. We all complain about it, but once we get started we usually enjoy the creative process. However, even the first draft will go more smoothly if you first choose a journal and follow its instructions to authors. That way, you won't risk having to rewrite a whole section (because you combined results and discussion and the journal guidelines required that them to be separate sections) or, even worse, having your paper rejected because it was out of that particular journal's scope.

Revision

Once you've gotten through the first draft, check your paper for obvious grammatical and typographical errors (don't forget to use your computer's spell checker). Then, put it aside for a few days or even a couple of weeks. At that point, you should be able to see it more objectively and notice errors and omissions that you missed in the first round. Reread it, and revise it as you think necessary. Then send it to the your most critical friend. This is not a required step in the process, but it is highly recommended because you will have critical feedback from a friendly source before it is read by colleagues and competitors. By the time you get it back from your friend, you should be able once again to look at your work objectively and revise it some more, if necessary, before getting it ready for submission. Check the headings, synonymy, and description for accuracy. Review the nontelegraphically written sections to make sure that you have followed the general guidelines for readable writing. Those who teach and study scientific writing say that often what makes scientific writing seem unreadable is not the difficulty of concepts or even the scientific terminology or jargon, but failure by the authors to follow general rules of good writing. These include putting information where the reader expects to find it, checking sentences to make sure that the subject is followed immediately by the verb in most of them, that only one point is being made per sentence, and that what is being emphasized in the sentence is placed in the stress position at the end of the sentence (Gopen and Swan 1990).

After you have made your writing as clear as possible, you should go back over the paper again, looking very critically at two sections: the title and the abstract.

TITLE Once you've responded to your friend's criticism and put together a third draft, you probably think your paper looks pretty good, and you hope it will have wide readership. But these days, with ever more literature to search and ever less time to search even for the articles of greatest significance to your own research, you should realize that most people will read only the title of your paper. If the title is meaningful to them, they will read the abstract. Only those most interested in what you say in the abstract will read the entire paper and request a reprint from you or make a copy for themselves.

But before you feel too bad about carefully prepared but unread text, remember that your goal with a species description is to get the information that the species exists–its scientific name, what its distinguishing characteristics are, and (usually) where it can be found–out to as many people as possible. The same technology that has given us the ability to produce greater quantities of information has made it possible to spread that information rapidly across the world. Your title and your abstract can do that for you, but only if your title clearly describes your paper's content and your abstract includes its most important findings.

What makes a good title? According to Day (1988), a title should be as short as possible without leaving out vital information. However, it should also be *specific,* even if that makes it a little longer. For example, "*Galathea coralliophilus,* a new decapod crustacean (Anomura: Galatheidae) from Singapore, Gulf of Thailand and West Irian" (Baba and Oh 1990) is a much more informative and useful title (in the information retrieval sense) than "Notes on Some New or Little-Known Marine Animals" (Gosse 1855).

When only a few hundred papers were published in a year, titles such as Gosse's did not present an information retrieval problem. Many scientists still seem to enjoy the nineteenth-century flavor of titles beginning, "Observations on . . . ," "Studies on . . . ," "Investigations of . . . ," and so on. But most editors will consider these wasted words, so think carefully about whether you need them. Remember, your title is a label and a data field for automated information retrieval, and it should contain the key words that indexing and abstracting systems need to classify it correctly.

ABSTRACT The most useful kind of abstract summarizes the paper. The abstract should give the objectives, methods, results, and conclusions of the paper as briefly as possible. Usually, this means in about 100–250 words. The maximum allowable number varies from publication to publication, and if you can do it in less than the maximum number, no editor will complain. If possible, those words should be clear and jargon free in order to persuade the reader to read the whole paper (Day 1988). Although many journals still ask you to give them several key words, the older automated retrieval systems in which only key words could be searched have largely been replaced by databases in which the full text of titles and abstracts is retrievable; every word in them is potentially a key word.

An abstract for a single species description can be quite short, but it should use standard terminology for taxonomic description in that particular group of organisms, a recommendation that seems to conflict somewhat with the recommendation just given for using jargon-free words. However, as was mentioned in chapter 11, standard terminology can aid comprehension and information retrieval, especially for those who are reading or writing in a language other than their native one. This English abstract from a Spanish paper on a new nematode species does the job adequately in just 80 words; the Spanish version is slightly longer, but just as technical. However, even a specialist who speaks neither language, armed with a scientific dictionary, could make an identification based on this abstract:

ABSTRACT

Description of *Vanderlindia hispanica* sp. n. (Nematoda: Tylencholaimidae)

In the present article one new species of *Vanderlindia* Heyns, 1964, named *V. hispanica* sp. n., is described from Spain. It is distinguished by the body length (L = 3.08–4.04 mm), body width (a = 39–50), esophagus length (b = 4.5–5.6), amphid width (13–15 µm), odontostyle only 1.5–1.7 lip regions long, guiding ring 0.9–1 lip region widths from anterior end, spicules length (82–93 µm), lateral guiding pieces structure, renal supplements location and by the tail shape. (Jiménez-Guirado 1990:3)

Another example of a brief but informative abstract is given below. In 113 words, the author has given diagnostic, geographic, and phylogenetic information on the species, as well as a bit of functional morphology.

ABSTRACT

Pycnoclavella belizeana new species is described from mangrove lagoons on the Barrier Reef in Belize, Central America. The species is characterized by its small size, simple branchial sac with only three rows of stigmata, and white or orange pigment in the thorax. The last and sometimes the first rows of stigmata may be deflected along the dorsal mid line of the thorax. Deflection of the stigmata is probably due to reduction in the size

> of the branchial sac and a need to fold the posterior and anterior rows of stigmata during contraction. *P. belizeana* is probably more closely related to *P. stanleyi* from the California coast than it is to *P. minuta* from West Africa. (Goodbody 1996:590)

Students may be surprised to discover that many (if not most) scientists write the abstract last, but writing the abstract last gives you additional time and perspective to think about what the most important points of the paper are and what its significance is, especially to those in other fields. This does not mean that you should exaggerate its significance. But even a simple species description can often be placed in a more general context with just a connecting sentence or two. When the description is combined with an ecological or biological study, this is even easier to do. In a note to fellow limnologists, Nelson Hairston (1990) suggested two simple rules of cross-communication that we would all do well to apply in our work. First, write abstracts, introductions, and discussions that connect your observations and conclusions to research in other systems, and second, describe your organism so that a nonspecialist can understand what kind of creature it is.

Submission of the Manuscript

Your final job before mailing your manuscript is to write a cover letter to accompany it. You use this cover letter to tell the editor of the journal the title and subject of the manuscript, as well as where you (and your coauthors) may be reached (addresses, phone and fax numbers, e-mail addresses, etc.). You may want to mention that it is original research and is not being submitted elsewhere (but you shouldn't be sending it to more than one journal anyway). It is also a good idea to state why you think the manuscript is appropriate for that particular journal, why it will interest its readership. Journal editors maintain files of potential reviewers for different subjects, so you don't usually suggest reviewers unless the instructions to contributors ask you to do so. However, if only one or two people could legitimately review the topic (as can be the case for the taxonomy of some groups of organisms), you may want to give the editor their names, addresses, e-mail addresses, and telephone numbers.

Pack the cover letter, along with originals of the illustrations (unless the instructions state that you are to hold the originals until requested) along with the specified number of copies in a snugly fitting manila or padded envelope or mailing box, and reinforce its seal with bands of tape. The best way to send the manuscript are by a courier service such as Federal Express or by first-class or express mail. Overseas packages should be sent airmail, express mail, or one of the courier services, not by surface mail.

Most journals send you a postcard or letter acknowledging receipt of your manuscript quite promptly, but don't expect much action for quite awhile after that. The editor sends it out to at least two reviewers and waits for their reports before contacting you again. It takes a minimum of 1 month to get copies out to reviewers and to get their reviews back, and 2–3 months is usually more realistic. But if you don't hear within that time, don't be afraid to write or call, just in case the manuscript has been lost somewhere in the process.

Final Revision and Publication

If your manuscript is rejected, you have no choice but to try again, either with the same journal or with another journal after dealing with the reasons for rejection. However, with the kind of paper you are submitting, outright rejection is unlikely (unless you have submitted it to a journal that does not publish taxonomic work). But you seldom will get a letter saying that your manuscript is perfect and can be published just as it is. You are most likely to get a letter accepting your paper for publication, with the provision that you make certain revisions, major or minor. Usually reviewers' comments point out areas where your meaning was not clear. Incorporating their suggestions, which often means rewriting unclear sentences and adding sentences or paragraphs of further documentation or explanation, will increase your work's value. Once you've made the revisions, you should return the revised manuscript by the deadline given, or if none was given, as soon as you possibly can.

Sometimes, however, a reviewer has made a mistake, misread something, or made an unfair criticism. If you don't think you can make the suggested changes, you need to make that clear to the editor by politely rebutting the reviewer's comments in the cover letter that you send along with the revised manuscript. If you can clearly and objectively point out why you are correct, the editor will probably accept your rebuttal (Day 1988).

Proofs and Proofreading

Once your paper has been accepted, you will receive an acceptance letter from the editor telling you approximately when you can expect proofs. Then you will wait for several months (sometimes even a couple of years) until the issue for which your paper is scheduled is set up to be printed. Unless you have submitted the manuscript on a diskette or electronically, as many journals now request that you do, the manuscript will be rekeyboarded at the editorial office. It will also be copyedited, the spelling and punctuation corrected, and typefaces and page layout codes inserted for the typesetting system (Day 1988). You may receive the copyedited manuscript for correction; some publications include this step, others don't. You will receive printed galley proofs, the draft version of your printed article, to correct. Proofs will be marked with the copyeditor's corrections and marginal queries from the copyeditor to you. The copyeditor will use standard proofreading marks (as found in the CBE style manual), and you should attempt to use the same system (some journals send you an explanation of proofreaders' marks along with your proofs).

Proofreading is probably more important in taxonomic papers than in most other scientific manuscripts because of the problems errors in scientific names, synonymies, and references can cause: Any mistake may lead to information being lost, and a mistake in spelling a new species name may cause long-lasting nomenclatural hassles. Even if your manuscript had no errors when you submitted it, some might have appeared during the typesetting process, so you still need to proofread carefully. The most foolproof way to proofread a manuscript requires two readers. One person reads through the corrected manuscript copy sentence by sentence, speaking the punctuation out loud: "The cheilostome *comma italics Conopeum tenuissimum comma* is one of the most common bryozoans encrusting sea *hyphen* grasses in the Indian River *period.*" The second person follows along on the proof copy, checking the text against the version being read for errors and making the necessary corrections on the proofs.

If you must read proofs by yourself, Day (1988) suggests that you read them through one time to catch any errors of omission (and one disadvantage to computer manuscript production is that words, paragraphs, and even pages can disappear with a touch of the wrong control key). Then you should go through the manuscript a second time, this time carefully examining each word, especially the scientific names and terms, and number by number. As Day points out, proofreaders and copyeditors can pick up most of the spelling errors, but only you can check for errors in your numbers. Check references and numbers, particularly in tables, synonymies, and references cited, for transpositions and other errors (1890 is not 1980 and 29 mm is certainly not the same as 92 mm or 2.9 mm); such small errors can greatly decrease the usefulness of your work to others.

Reprints

Most journals send a reprint order card or form along with your proofs. Return your order with the proofs if so requested, or send it in as soon as possible so that you don't miss the opportunity for good copies. Reprints are usually printed in multiples of 100. The first hundred are expensive to cover setup costs; the next hundred are much cheaper (Day 1988). Order enough copies. Although a taxonomic paper doesn't draw the number of immediate requests that an ecological paper might, people will be using it for years to come, and you may still get requests for it 20 or even 30 years later. The quality of a reprint copy is way above that of a photocopy when half-tone illustrations are involved, and any serious user will want to see the best version.

Journals That Publish Taxonomic Papers

The prestigious international journals do not publish routine taxonomic descriptions. Don't expect to publish in *Science* or *Nature* unless you find a new fossil *Homo,* the earliest dinosaur, or something equally notorious. But you don't need to publish in that kind of journal to give your new species adequate recognition. In deciding where to publish, you should consider which publications the people you hope to reach are most likely to read, of course, but even if the journal you choose is not the first source your audience looks at, the abstracting and indexing services of today are quite complete and thorough. To give your new species name and description worldwide distribution, all you really need to do is to publish in a journal or other scientific publication that is routinely indexed. These include the general journals in a field, the regional, national, and international scientific society journals, and museum publication series, as well as journals that publish all kinds of research (including taxonomy) on a particular group of organisms, or those from a particular geographic region.

It is surprising to beginners in systematics that the journals whose titles indicate that their subjects are systematic or evolutionary in scope don't publish species descriptions. Instead, they publish theoretical phylogenetic reviews and revisions of whole groups of organisms. Such journals include *Evolution, Systematic Biology,* and *Plant Systematics and Evolution.* Other journals with more obscure titles, or titled with the name of the sponsoring society, do publish species descriptions as a minor or major part of their coverage.

Many museums can afford to publish work only by in-house authors (curators and sometimes research associates). However, this is not always the case, and it is worth asking about, particularly if you have worked with a curator, used the museum's collections, or contributed to the collections or their curation. Some museum and other journals publish articles in several languages; you might not have to write a paper in Spanish to publish in a Spanish journal. In fact, a number of European journals now accept only papers in English (e.g., *Tropical Zoology,* an Italian journal). Society journals usually, but not always, require at least the senior author to be a member of the society. Page charges may also be an important point of consideration. If you don't have funds to pay the cost of publication in a journal that charges them, you will probably need to consider only the journals that have no charges or will waive them for people without funding.

The following list of journals to consider is not an exhaustive one. Doing the library research necessary for your species description will have shown you the journals that workers on that group of organisms have used to publish their descriptions. You may also find other journals that would suit your purpose right in your institutional library. Check with a reference librarian to make sure the journal you select is indexed and abstracted by at least one of the major services (*BIOSIS, Zoological Record, Aquatic Sciences and Fisheries Abstracts,* etc.). These organizations publish guides listing the publications they cover.

The following list is divided into several categories, but they are not necessarily exclusive. Also, journal titles don't always accurately reflect the range of acceptable topics: Some journals with regional-sounding titles or regional society sponsorship are actually quite international in scope, whereas some journals with very general titles are quite narrowly focused.

To make certain, you need to scan titles and subjects of several issues of the journal and read the journal's statement of purpose and the instructions to contributors.

General Journals

Biological Bulletin
Bulletin of Marine Science
Journal of the Marine Biological Association of the United Kingdom
Journal of Natural History
Netherlands Journal of Zoology
Proceedings of the Biological Society of Washington
Zoologica Scripta
Zoological Journal of the Linnaean Society
Zoologisches Anzeiger

Journals Covering a Particular Group of Organisms

Acarologia
Annals of the Entomological Society of America
Aquatic Insects: International Journal of Freshwater Entomology
Ardea (journal of the Netherlands Ornithologists' Union)
The Auk: A Quarterly Journal of Ornithology
Biological Journal of the Linnaean Society
Bulletin of the British Arachnological Society
Canadian Entomologist
Coleopterists Bulletin
Copeia
Crustaceana: International Journal of Crustacean Research
Entomological News
Herpetologica
The Herpetological Journal
International Journal of Primatology
Invertebrate Biology

Journal of Arachnology
Journal of Crustacean Biology
Journal of Entomological Research
Journal of Eukaryotic Microbiology
Journal of Fish Biology
Journal of Foraminiferal Research
Journal of Herpetology
Journal of Mammalogy
Journal of Molluscan Studies
Journal of Nematology
The Journal of Parasitology
Journal of the Australian Entomological Society
Journal of the Kansas Entomological Society
Journal of the Lepidopterists' Society
Journal of Plankton Research
Malacologia
Mammalia: Morphologie, Biologie, Systématique des Mammifères
Marine Mammal Science
Mycotaxon
Myriapodologica
The Nautilus
Nematologica
Primates: A Journal of Primatology
Proceedings of the Entomological Society of America
Proceedings of the Entomological Society of Washington
Psyche: A Journal of Entomology
Systematic Entomology
Systematic Parasitology
Transactions of the American Entomological Society
Transactions of the American Fisheries Society
Transactions of the New York Entomological Society
The Veliger
The Wilson Bulletin: A Quarterly Magazine of Ornithology

Journals Covering a Particular Geographic Area

American Midland Naturalist

Annales Zoologici Fennici (North European and other boreal regions)

Canadian Field Naturalist

Canadian Journal of Aquatic Sciences and Fisheries

Canadian Journal of Entomology

Canadian Journal of Zoology

Caribbean Journal of Science

Entomologist's Gazette: A Journal of Palearctic Entomology

Entomon

Florida Entomologist

Florida Scientist

Great Lakes Entomologist

Invertebrate Taxonomy (Indo-Pacific)

Ohio Journal of Science

Ophelia

Oriental Insects (insects and other land arthropods of the old-world tropics)

Pacific Science

Pan-Pacific Entomologist (primarily western coastal U.S.)

Revue Zoologie Africaine

Sarsia

Southwestern Entomologist (southwestern U.S. and Mexico)

Tropical Zoology (tropical areas with particular attention to Africa)

Virginia Scientist

Molecular Systematics

Biochemical Systematics and Ecology

Journal of Molecular Evolution

Marine Ecology Progress Series

Molecular Biology and Evolution

Museum and University Publications

American Museum Novitates

Anales del Instituto de Biologica Universidad Nacional Autónoma de México

Annalen des Naturhistorischen Museum in Wien

Annals of the Carnegie Museum of Natural History

Annals of the Natal Museum

Annals of the South African Museum

Beaufortia (Institute for Systematics and Population Biology, University of Amsterdam)

Brenesia. Museo Nacional de Costa Rica

Brimleyana. The Journal of the North Carolina State Museum of Natural History

Bulletin du Museum National d'Histoire Naturelle, Paris

Bulletin of the Natural History Museum: Botany Series

Bulletin of the Natural History Museum: Entomology Series

Bulletin of the Natural History Museum: Palaeontology Series

Bulletin of the Natural History Museum: Zoology Series

Bulletin. Zoölogisch Museum Universiteit van Amsterdam

Contributions in Science, Natural History Museum of Los Angeles County

Fieldiana, Geology

Fieldiana, Zoology

Jeffersoniana (Virginia Museum of Natural History)

Mitteilungen aus dem Hamburgischen Zoologisichen Museum und Institut

Notulae Naturae (Academy of Natural Sciences of Philadelphia)

Proceedings of the California Academy of Sciences

Records of the Western Australian Museum

Royal Ontario Museum Life Sciences Contributions

Smithsonian Contributions in Zoology

Spixiana (Bavarian State Zoological Collection)

Paleontology

Acta Palaeontologica Polonica
Journal of Paleontology
Palaeontology
Lethaia

Botany

American Journal of Botany
Annals of Botany
Annals of the Missouri Botanical Garden
Botanical Journal of the Linnaean Society
Brittonia
Bryologist
Bulletin of the Torrey Botanical Club
Journal of Bryology
Journal of Phycology
Mycologia
Mycological Research
Mycotaxon
Phycologia
Systematic Botany
Taxon

Part Four

BEYOND SPECIES DESCRIPTION

Chapter 17

Subspecies

Much has been written in recent decades about subspecies and their use in taxonomy. There are strong feelings that they are usable, useful and desirable. There are also strong feelings that they are not really relevant to taxonomy and are an unnecessary encumbrance to classification and nomenclature.

–Blackwelder (1967:171)

The species is the lowest rank which it is essential to recognize for general taxonomic purposes. It is not, however, the lowest rank which it is desirable and useful to recognize.

–Stace (1989:192)

Why Are Subspecies Important?

This chapter was not part of the original outline for this book. An academic background in ecology and evolutionary biology led me to think of subspecies as flexible, responsive, and changing populations, unlikely to fit well into the hierarchical categories of formal biological nomenclature. Recent texts on evolutionary biology (e.g., Grant 1981; Otte and Endler 1989; Avise 1993) do not even mention the term *subspecies.* Research on marine invertebrates made me even more wary of giving names to subspecies in groups where we still have little idea of the limits of most species. So why write about describing subspecies at all? There are several reasons.

The first reason is that so much of biology is done at or below the species level. From molecular biology, genetics, population genetics, population biology, population ecology, and speciation studies to much of agriculture and horticulture, populations of plants and animals provide the great battleground or harvest ground of modern biology. Even though most of this work doesn't lead to the naming of subspecies in a formal way, the sheer volume of work being produced dictates that some discussion of the biological concepts involved and nomenclatural rules that may be applied would be useful here.

A second reason is that a moderate amount of taxonomic literature *is* being produced on subspecies. Over the last 28 years (as figure 6.1 showed), 3 percent of all taxonomic papers have included one or more new subspecies, their description thus following description of new species (33 percent) and new genera (8 percent) in popularity (Winston and Metzger 1998). Many of these papers represent thesis and dissertation work by graduate students, who may dream of traveling the world for their research but must usually settle for working on projects nearer home, projects that cover only part of a species range and one or a few populations of a species.

Then, too, there are the groups of vertebrates such as birds for which species limits are well known, most species are already described, and most new discoveries relate to subspecies (Blackwelder 1967). This is also true for certain flowering plant groups, and even some invertebrates, such as mollusks and butterflies.

Finally, there is another very practical reason for naming infraspecific taxa. Naming attracts attention to a taxon, and these days, recognition and attention can mean the difference between continued survival and extinction of a population or geographic race. It can even affect survival of a species, if further study proves a "subspecies" or "variety" to actually be a cryptic species, as happened with the tuatara (Daugherty et al. 1990; chapter 12), for example. From a conservation point of view, giving names to infraspecific taxa can be valuable because it points out unique commercially or evolutionarily valuable genetic resources within a species, resources that might vanish even if other populations of the species survive (Stace 1989).

Infraspecific Variation

Blackwelder (1967) makes the point that subspecies can't be known until the species limits are known, until that particular species has been well enough studied that variants that might possibly be described as subspecies can be recognized. What is a subspecies? As was true for species, the answer to that question depends on the group of organisms being studied. Modes of inheritance and biology make a subspecies in one group quite different from a subspecies in another.

Molecular analyses of geographic population structure have now been carried out on hundreds of species. Populations of most animal species show some amount of genetic differentiation from one locality to another. These differences result from the interaction of many evolutionary processes: dispersal ability and pattern, the physical basis of gene flow, migration, random genetic drift, natural selection, mutation, genetic recombination, mating systems, habitat structure, and geographic isolation (Avise 1993). It is easy to describe how to write a subspecies description. However, to evaluate the desirability of naming a subspecies within a particular species, a biologist must have a strong understanding of how these processes might be working in the group being studied, and preferably some hard evidence for the scope of action of at least some of them.

The most commonly used definition for subspecies is the one that applies to sexually reproducing animals. As defined by Mayr (Mayr and Ashlock 1991:430), a *subspecies* is "an aggregate of local populations of a species inhabiting a geographic subdivision of the range of the species and differing taxonomically from other populations of the species."

In botany, a *subspecies* has been defined as "a population of several biotypes" (given genotypes expressed in one or more individuals, according to Jeffrey 1989) "forming a more or less distinct regional facies of a species" (Du Rietz 1930, as quoted in Stace 1989:193) or a geographic race. Like the zoological definition, it is most applicable to sexually reproducing species.

Even among sexually reproducing organisms, not all species have a population structure showing a pattern of variation compatible with the naming of subspecies. Some species, although they contain considerable genetic variability, are taxonomically homogeneous throughout their range. Others show patterns such as clinal variation, in which one or more characters vary gradually over a series of contiguous populations (Mayr and Ashlock 1991). Such species cannot be divided into subspecies in any logical way. On the other hand, species in which *step* or *stepped clines* (those with sharp breaks in the rate of change in the clinal character over a short geographic distance) might usefully be divided into subspecies (Hubbell 1954; Blackwelder 1967).

Plant populations also show variation in genetic composition, sometimes over very short distances (a few meters or kilometers; Avise 1993).

Local races differ statistically in gene pool composition and genetically determined phenotype characters from other populations of the species. Plant dispersal is carried out by pollen, fruits and seeds. In most cases dispersal falls off with distance from the parent, so related plants live close together, with a considerable degree of inbreeding developing, even in an outcrossing plant population. The balance between intensity of selection and rate of gene flow determines whether local racial differentiation will occur (Grant 1981).

Climatic factors usually vary gradually, resulting in clinal variation patterns rather than local race differentiation, except in special environments such as high mountains where extreme changes over very short distances are possible. Edaphic (soil-related) factors often change greatly over short distances, and may result in the development of many geographic races in a species. Small isolated or semi-isolated outliers or colonies favor action of genetic drift. Each colony or group of colonies may develop its own distinctive characteristics (Grant 1981). Although some of these local variants might be recognized taxonomically as subspecies, they are much more commonly described and used in population studies, but not formally named.

Local variation in plant populations usually grades into geographic variation, but for convenience Grant (1981) classified nonlocal races of plants into three categories: *continuous geographical races, disjunct geographical races,* and *ecological races.* Species with fairly continuous populations over ranges large enough to show environmental gradients often show clinal changes in one or more characters (continuous geographical races). On the other hand, when populations of a species inhabit only small and disjunct areas within their range, they become isolated or semi-isolated from each other, and those living in different environments diverge (disjunct geographical races). Geographical races are allopatric (occupy different geographic areas). In patchy environments ecological races may also exist. In such cases, two or more races live in different habitats in the same geographic area. They may interbreed and intergrade in contact zones, but they maintain their distinctive racial character in their preferred habitats (Grant 1981).

Clonal animals and plants reproduce by asexual or sexual and asexual means. A local population may consist of one or a few clones. Even a regional population would have fewer genetic individuals than genetic individuals in a geographically similar, unitary sexually reproducing species. Parthenogenic asexually reproducing animals are much less common than asexually reproducing plants, but they do occur in a number of invertebrates, mostly arthropods (e.g., aphids, gall wasps, cladocerans, and rotifers), and a few vertebrates. Some show cyclical patterns in which sexual and parthenogenetic reproduction alternate.

There are more than 300 taxa of apomictic (reproducing by means of asexual seed formation or agamospermy) angiosperms. Some algae, mosses, and ferns also reproduce asexually by apogamy (development of an embryo from a cell other than the egg and without fertilization) or apospory (production of the gametophyte gamete from a somatic cell of the sporophyte without spore formation). At and below the species level these plants cause particular problems for taxonomists because they are properly categorized as lineages rather than populations, and subspecies as groups of related populations cannot apply (Asker and Jerling 1992). Luckily for most of us, 75 percent of angiosperm apomicts are found in only three plant families: Poaceae, Asteraceae, and Rosaceae. Unfortunately, these families include 10 percent of angiosperm species, including some very important and interesting groups, such as cereal grains, roses, blackberries, raspberries, and hawthorns (Asker and Jerling 1992; Zomlefer 1994).

In both plants and animals, asexual taxa show distinct ecological and geographic patterns. Some evidence indicates that they have broader geographic ranges than their sexual relatives, and plant apomicts, at least, are both more common at higher altitudes and better colonizers of disturbed areas (Asker and Jerling 1992).

Rules of Infraspecific Nomenclature: Zoology

In both the ICBN and the ICZN these categories are covered under species group rules. The ICZN includes only two categories: the species and the subspecies.

In zoology subspecies names are *trinomial,* consisting of the binomial species name plus an additional subspecies name, which may add further descriptive information, tell where the subspecies occurs, or honor the collector or some other person. Like the species name, the subspecies name begins with a lowercase letter. The subspecies names is also constructed like a species name (chapter 8) and is considered an additional modifier of that name grammatically. For example:

Leiocephalus personatus elattoprosopon, a Haitian subspecies of a Hispaniolan iguanid lizard (Gali et al. 1988)

Zygodontomys brevicauda microtinus, a coastal savanna-dwelling subspecies of Neotropical muroid rodent (Voss 1991)

Dryocopus javensis cebuensis, the Cebu Island subspecies of the white-bellied woodpecker (Kennedy 1987)

Rallina eurizonoides alvarezi, the Batan subspecies of the slaty-legged crake (Kennedy and Ross 1987)

The creation of a subspecies within a species actually creates two subspecies: the nominotypical subspecies and the new subspecies. The *nominotypical (nominate) subspecies* is the one that contains the name bearing type of the species. Its subspecies name is the same as the species name, and it has the same date and author as the species does. The new subspecies has a separate species name and its date and author are taken from the publication in which it was first described. Usually that author is the person who first described it as a subspecies. However, this is not so in cases where it was originally described as a species, but later recognized as a subspecies. For example, the nominotypical subspecies of the neotropical mouse *Zygodontomys brevicauda* (J. A. Allen and Chapman) 1893 is *Zygodontomys brevicauda brevicauda* (J. A. Allen and Chapman) 1893. The additional subspecies mentioned above has its own author and date: *Zygodontomys brevicauda microtinus* (Thomas 1894). Note that the author and date are the date and author of its original description. In this particular case the animal was originally described as a species: *Oryzomys microtinus* by Thomas in 1894. Much later another mammalogist (Hershkovitz 1962) realized that it represented subspecific variation of a broadly ranging species; he then called it *Zygodontomys brevicauda microtinus,* but his name and the date of his paper making it a subspecies are not used.

The authors of the ICZN must have realized that including a trinomial subspecies name in a paper might be construed by some zoologists to invalidate the whole work because after all, the code is based on the principle of *binomial* nomenclature: "The author must have consistently applied the Principle of Binomial Nomenclature [Art. 5 a] in the work in which the name or nomenclatural act was published" (1985, Art. 11 c). They forestalled this argument with the following statement: "The scientific name of a subspecies, a trinomen [Art. 5 b], is accepted as consistent with the principle of Binomial Nomenclature" (1985, Art. 11 c, ii).

In papers published recently (after 1960), the author must also have stated that the taxon is considered a subspecies. ICZN (1985, Art. 45 f–g), deals with establishing subspecies rank in papers from the older literature. Varieties, forms, and aberrations are now considered infrasubspecific and are not covered by the code. Those found in publications before 1961 are considered subspecies unless the paper's content makes it clear that they were meant as infrasubspecific. That is not particularly important to know unless you encounter a species that was once considered a "variety" or "morph," but looks to you like a good species. Then what matters nomenclaturally is whether anyone else cited it as a subspecies in the literature before 1985. If it was cited as a subspecies, that name is available to be used as a species name. If not, you will need to give it an entirely new species name.

Rules of Infraspecific Nomenclature: Botany

The ICBN (Section 5, Art. 24–27) recognizes five infraspecific categories: subspecies, variety, subvariety, form, and subform. Only three of them, *subspecies, variety,* and *form,* are commonly used (Stace 1989). However, because the botanical code includes several categories below the species rank, their names must be qualified by a term for the category (Jeffrey 1989), such as *Atriplex longipes* subsp. *praecox, Chrysothamus viscidiflorus* ssp. *humilis, Ameria maritima* var. *sibirica,* or *Aloe macrocarpa* var. *major.*

A *variety* is "a population of one or several biotypes, forming a more or less distinct local facies of a species," or a local or ecological race (Du Rietz 1930, quoted in Stace 1989:194). These divisions are useful to botanists because one or more of them may be recognized in many plant species. However, because they are hierarchical their use is limited by the fact that they do not accommodate the many nonhierarchical patterns of variation in plant species. Many botanists have limited themselves to using only subspecies, and using that as a dumping ground for all levels of infrasubspecific variation (Stace 1989). But as Stuessy (1990) explains, the situation has been made even more complicated in U.S. botany, where *variety* and *subspecies* have been used in a totally confusing way. Historically, there have been two schools in U.S. botany, an Eastern and a Californian group. The Californian group used *subspecies* as their infraspecific category of choice, whereas the Eastern school used *variety,* but pretty much in the sense of *subspecies* (as defined by Du Rietz) rather than according to his definition of *variety.* The Eastern group used *subspecies* only for groups of morphologically similar "varieties."

No such problems occur with the lowest commonly used botanical category, *form* (**forma**). *Form* is defined as a minor genetic variant, "a population of one or several biotypes occurring sporadically in a species population in one or several distinct characters" (Du Rietz 1930, quoted in Stace 1989:194). Botanists seem to agree that this category refers to plants with unusual morphological features, which occur scattered throughout a population. They are genetically controlled, usually by small genetic mutations or recombinations (as opposed to changes in ploidy). The question is, rather, whether it is worth giving these variations formal names at all.

The ICBN (Art. 24.4, note 2) specifically states that the same final epithet (in this case, subspecies, variety, or form name) cannot be used within the same species for names with different types. Infraspecific taxa within different species may have the same epithets. It also recommends (Rec. 24B.1) that authors avoid using the same epithets for infraspecific taxa in species of the same genus (only common sense because taxonomic reassignments could then cause nomenclatural chaos).

Art. 25 states that a species or any taxon below "is regarded as the sum of its subordinate taxa." Again, this is just common sense. If you are discussing a species with more than one subspecies, you are assumed to be referring to all of them unless you clearly state otherwise.

Art. 26 states that the nominotypical subspecies, variety or form repeats the specific epithet as its final epithet (just as for the nominotypical subspecies in zoology) but is not followed by an author's name. It is called an *autonym.* Just as in zoology, the publication of an infraspecific taxon name automatically creates a corresponding *autonym,* an equivalent subdivision of the name-bearing type (e.g., the publication of *Xxxxa yxxxa* var. *zxxxa* automatically establishes also *Xxxxa yxxxa* var. *yxxxa*).

In older botanical literature you may notice that the nominotypical variety was designated by a epithet such as *typica, genuina,* or *originaria,* but this was not done after 1950 (Benson 1962).

Deciding When to Name an Infraspecific Taxon

Taxonomists usually must base the decision whether to represent infraspecific variation nomenclaturally on morphological and geographic data, but any genetic and reproductive data that can be added will improve the result. This is certainly one of the areas of systematics in which an ecologist or geneticist, who has studied a species from this point of view, can make an important contribution. The point of reviewing species differences again was to point out how an organism's biology figures in the decision to name an infraspecific taxon. You should also remember that although genetic and morphological differentiation often correspond, they do not always do so, a fact that can be critical in cases in which the preservation of an organism is at stake. Two recent cases point up the problem.

After the extinction of the last dusky seaside sparrow, *Ammodramus maritimus nigrescens,* in 1987, an mtDNA sequencing study of almost all of the whole seaside sparrow complex (nine named subspecies) was carried out. This study showed that that particular subspecies was almost identical genetically to other Atlantic coast populations. However, all the Atlantic coast subspecies were phylogenetically distinct from Gulf Coast populations. In this case, taxonomy based on morphology had given emphasis to an unimportant character (coloration), whereas many greater differences, important to the species's future, were being overlooked (Avise and Nelson 1984; Avise 1993).

Determining actual population structure can be just as important to invertebrate conservation. Eastern oysters (*Crassostrea virginica*) have a variable shell morphology that is known to be influenced by the environmental conditions under which they live and grow (Groue and Lester 1982). They are impossible to divide into genetically differentiated populations on the basis of outward morphology. Electrophoretic studies of nine populations in the Laguna Madre area of the Texas coast showed that genetically differentiated groups were present. The main Laguna Madre group was genetically divergent from populations of eastern and central Texas as well as those from the rest of the Gulf of Mexico and the Atlantic coast. The authors of the study concluded that they should be treated as discrete management units and, if further investigation warranted, protected as unique genetic resources (King et al. 1994).

Once you believe that you have sorted out these kinds of problems and that naming an infraspecific taxon might be useful, these guidelines from Stuessy (1990) may help you decide. (For animals, of course, only subspecies are relevant.)

Subspecies should be characterized by several conspicuous morphological differences, cohesive geographic allopatric or parapatric population patterns, and usually marked multigenic genetic divergences, with natural hybridization possible along contact zones and with markedly reduced fertility in hybrids. *Varieties* should be characterized by one to a few conspicuous differences, geographic patterns that are cohesive and largely allopatric (with some overlap), genetic divergences (multigenic or simple), natural hybridization probable in overlapping populations, and hybrids with reduced fertility. A *form* is characterized by a single conspicuous difference, sporadic, sympatric geographical pattern, genetic divergence usually controlled by a single gene, natural hybridization, and complete fertility between hybrids (Stuessy 1990).

Writing Infraspecific Descriptions

As indicated earlier, taxonomists are most likely to give subspecies names to geographically isolated and morphologically distinctive populations of a species. The description is carried out just as for a new species, with the exception that the characters used in the description section may be limited to characters that can be compared with other subspecies of the species rather than covering all the taxonomic characters that may be found in the species. Type material is described, as are specimens examined, and comparisons are made with other subspecies of the species. The range of the subspecies is included, as is an etymology section defining the subspecies name. For example, the slaty-legged crake, *Rallina eurizonoides,* is found from India through southern Asia to the lesser Sunda Islands and from Japan through the Philippines to Celebes, and several subspecies have been named. The nominate subspecies, *R. e. eurizonoides,* is restricted to the Philippines, where it occurs on many islands. Recently specimens were collected by Robert S. Kennedy from the Batan Islands and named as a new subspecies on the basis of morphological and plumage differences by Kennedy and Charles A. Ross:

Rallina eurizonoides alvarezi, new subspecies

Holotype.–PNM 16301, adult male in breeding condition (largest testis–7 × 14 mm), 31 May 1985, western slope of Mt. Iraya, 180 m elevation, Sitio Nacamaya, 3 km NE of Basco, Batan Island, Batanes Province, Philippines, R.S. Kennedy and party (collector's no. 1048).
Paratypes.–LSUMZ 105082 and USNM 582810–582812, Batan Island, 1.4 km N or NE of Basco. . . .

> *Subspecies characters.*–Differs from *R. e. eurizonoides* and *R. e. formosana* by having: the upperparts, and outer edge of wing and tail feathers darker olive brown; the top and sides of head, hind neck, lower throat and breast darker chestnut; the white barring of underparts narrower and less pronounced; the throat, in males, chestnut, not pale rufous as in *R. e. eurizonoides* or white as in *R. e. formosana*. . . . Differs further from *R. e. formosana* in having a longer tail . . . , shorter culmen . . . , and a shorter tarsus. . . . Like *R. e. minahasa* in being of similar size, in having dark olive brown underparts and in having narrow and less pronounced white barring on the underparts. Differs from *R. e. minahasa* in having the top and sides of head, hind neck, lower throat and breast darker chestnut and throat, in males chestnut and not pale rufous. . . .
> *Range.*–Batan Island. . . . The species was not encountered on nearby Ivojos Islands and is unknown to the inhabitants. . . .
> *Etymology.*–Named in honor of Jesus B. Alvarez, Jr., for his untiring efforts to conserve Philippine fauna and flora.
> *Remarks.* . . . It is most similar to *R. e. minahasa* of Celebes and least similar to *R. e. formosana* and the other races of *R. eurizonoides* owing to its more richly colored upperparts and narrower abdominal barring.
> *Specimens examined.*–1 (PNM), 3 (USNM). *R. e. amauroptera;* 4 (USNM., *R. e. eurizonoides:* Bohol, 1 (FMNH); Cebu, 3 (DMNH). . . . (Kennedy and Ross 1987:459–460)

Those who work on other vertebrates also use subspecies to describe distinctive isolated populations of a species. For example, Gali et al. (1988) described a new subspecies of the iguanid lizard *Leiocephalus personatus*, *Leiocephalus personatus elattoprosopon,* on the basis of its geographic isolation from the next closest population of the species by a mountain range, the Massif de la Hotte.

In their remarks section they made clear the similarities and differences between their new subspecies and the nominotypical subspecies, *L. p. personatus:*

> *Remarks.–Leiocephalus p. elattoprosopon* is similar to *L. p. personatus* in having a head scutellation formula of 3–5–4 (prefrontals–medians–frontoparietals). Both subspecies have masks, but they differ in that the mask in *L. p. elattoprosopon* usually is restricted dorsal to the ear opening, unlike that of *L. p. personatus* in which the mask extends half way to the forelimb insertion, including the eyes, and the loreal region, and as far anteriorly as the tip of the snout. (Gali et al. 1988:2)

Their subspecies name describes the chief diagnostic characteristic of the new subspecies:

> *Etymology.*–The name *elattoprosopon* is from the Greek elatton (less) and prosopon (face, mask), in reference to the reduced dark mask in this subspecies in contrast to that in *L. p. personatus.* The name is used as an appositional noun. (Gali et al. 1988:2)

They also use the new subspecies as evidence for a preliminary biogeographic hypothesis:

> . . . it is intriguing that *Ameiva chrysolaema evulsa* Schwartz (1973) and *L. p. elattoprosopon* occur in much the same general area. . . . Intervening areas, which have been intensively sampled, seem not to harbor populations of either lizard. It seems that the Aquin-Morne Dubois "peninsula" region has a peculiar relictual fauna that was once more widespread or continuous with relatives farther to the east and north. (Gali et al. 1988:2)

Those who have made an intensive study of a single widely distributed genus or species have often found patterns of subspecific variation useful in explaining its evolutionary history. For example, in his study of *Zygodontomys,* Robert Voss (1991) found that the geographic patterns of morphological variation in *Zygodontomys brevicauda* best fit a vicariance hypothesis based on paleoclimatic evidence. The subspecies *microtinus* currently occurs in discontinuous coastal savannahs from Guyana to French Guiana. However, the presence of derived cranial characters indicates that the currently separated populations of the subspecies form a historically related group, reflecting the original contiguity of the nonforest habitats before invasion of lowland gaps in the mountains by forests.

Sources

As was true for species (chapter 3), the literature on infraspecific variety is so vast that it is impossible to do more than list some of the relevant books and a sampling of articles that demonstrate the variety of research topics falling under this category.

Books

Avise, J. C. 1993. *Molecular Markers, Natural History and Evolution.* New York, Chapman & Hall.

Endler, J. A. 1977. *Geographic Variation, Speciation, and Clines.* Princeton, N.J., Princeton University Press.

Ferraris, J. D. and S. R. Palumbi, eds. 1996. *Molecular Zoology: Advances, Strategies, and Protocols.* New York, Wiley.

Grant, V. 1981. *Plant Speciation,* 2d ed. New York, Columbia University Press.

Grant, V. 1991. *The Evolutionary Process: A Critical Study of Evolutionary Theory,* 2d ed. New York, Columbia University Press.

Hillis, D. M., C. Moritz, and B. K. Mable. 1996. *Molecular Systematics,* 2d ed. Sunderland, Mass., Sinauer.

Williams, G. C. 1992. *Natural Selection: Domains, Levels, and Challenges.* New York, Oxford University Press.

Articles on Subspecies

Stace, C. A. 1986. The present and future infraspecific classification of wild plants, pp. 9–20 in B. T. Styles, ed. *Infraspecific Classification of Wild and Cultivated Plants.* Oxford, U.K., Clarendon Press.

Starrett, A. 1958. What is the subspecies problem? *Systematic Zoology* 7: 111–115.

Wilson, E. O. and W. L. Brown, Jr. 1953. The subspecies concept and its taxonomic application. *Systematic Zoology* 2: 97–111.

Examples of Infraspecific Level Research

Bolger, D. T. and T. J. Case. 1994. Divergent ecology of sympatric clones of the asexual gecko, *Lepidodactylus lugubris. Oecologia* 100: 397–405.

Bowen, B. W., N. Kamezaki, C. J. Limpus, G. R. Hughes, A. B. Meylan, and J. C. Avise. 1994. Global phylogeography of the loggerhead turtle (*Caretta caretta*) as indicated by mitochondrial DNA haplotypes. *Evolution* 48: 1820–1828.

Cronin, M. A., S. Hills, E. W. Born, and J. C. Patton. 1994. Mitochondrial DNA variation in Atlantic and Pacific walruses. *Canadian Journal of Zoology* 72: 1035–1043.

Floate, K. D., G. W. Fernandes, and J. A. Nilsson. 1996. Distinguishing intrapopulational categories of plants by their insect faunas: Galls on rabbitbrush. *Oecologia* 105: 221–229.

Graves, G. R. 1991. Bermann's rule near the equator: Latitudinal clines in body size of an Andean passerine bird. *Proceedings of the National Academy of Science* 88: 2322–2325.

James, F. C. 1991. Complementary descriptive and experimental studies of clinal variation in birds. *American Zoologist* 31: 694–706.

Johannesson, B. and K. Johannesson. 1990. *Littorina neglecta* Bean, a morphological form within the variable species *Littorina saxatilis* (Olivi)? *Hydrobiologica* 193: 71–87.

Kilar, J. A. and M. D. Hanisak. 1989. Phenotypic variability in *Sargassum polyceratium* (Fucales, Phaeophyta). *Phycologia* 28: 491–500.

Morafka, D. J., L. G. Aguirre, and R. W. Murphy. 1994. Allozyme differentiation among gopher tortoises (*Gopherus*): Conservation genetics and phylogenetic and taxonomic implications. *Canadian Journal of Zoology* 72: 1665–1671.

Pagani, M., C. Ricci, and A. M. Bolzern. 1991. Comparison of five strains of a parthenogenetic species, *Macrotrachela quadricornifera* (Rotifera, Bdelloidea). II. Isosenzymatic patterns. *Hydrobiologia* 211: 157–163.

Sperling, F. A. H. 1990. Natural hybrids of *Papilio* (Insecta: Lepidoptera): Poor taxonomy or interesting evolutionary problem? *Canadian Journal of Zoology* 68: 1790–1799.

Takabayashi, M. and O. Hoegh-Guldberg. 1995. Ecological and physiological differences between two colour morphs of the coral *Pocillopora damicornis. Marine Biology* 123: 705–714.

Thorpe, R. S. 1991. Clines and cause: Microgeographic variation in the Tenerife gecko (*Tarentola delalandii*). *Systematic Zoology* 40: 172–187.

Tibayrenc, M., P. Ward, A. Moya, and F. Ayala. 1986. Natural populations of *Trypanosoma cruzi,* the agent of Chagas disease, have a complex multiclonal structure. *Proceedings of the National Academy of Science, USA* 83: 115–119.

Chapter 18

Genus-Level Description and Revision

The genus is a mandatory category to which every species must belong if binomial nomenclature is to be preserved.

–WILEY (1981:205)

It is by the arbitrariness of definition that all higher categories differ from the species category.

–MAYR AND ASHLOCK (1991:135)

Despite the vagaries of systematists, there are, in a taxonomic sense, many dense clusters of biotic diversity.

–MINELLI (1993:185)

You are not likely to revise a whole classification unless you get very serious about systematics, but you may well face the problem of describing a new genus. Eight percent of taxonomic papers published during the last 25 years have included descriptions of one or more new genera (Winston and Metzger 1998). If you are working on a group in which there are many undescribed species, it is quite possible that the particular one you have studied won't fit into any described genus, either. This chapter covers generic concepts, the rules of nomenclature dealing with publication of new genera, and the construction and typification of generic names, and gives examples of generic description for several groups of organisms, as well as introducing some of the common problems that may be encountered in generic revisions.

The Genus Concept

Although no one species concept applies to all groups of organisms, biologists do at least seem agreed that species represent biological realities.

The same cannot be said for the higher-level categories of the Linnaean hierarchy; their definitions are arbitrary. Mayr and Ashlock put it this way: "The species category signifies singularity, distinctness and difference, while the higher categories have the function of grouping and ordering by emphasizing affinities among groups of species." However, even if there are no nonarbitrary criteria, "a taxon correctly delimited and placed in a higher category is natural provided it is consistent with the theory of common descent" (Mayr and Ashlock 1991:134).

As the first term of the Latin binomial that makes up an organism's scientific name, the noun to which the species adjective is appended, it could be claimed that the genus is more essential to nomenclature and classification than the species. The fact that ancient and folk taxonomies are genus based (chapter 2) seems to indicate that this category represents some biological reality, describing broadly recognized clusters or discontinuities. The genus is useful to modern-day scientists as well, as the category on which many paleontological and biological macroevolutionary and systematic studies are based. Considering how much has been written about the species category (see chapter 3 *Sources,* for example), a large literature on the genus might be expected also.

But much less has been written on the genus concept. In contrast to the dozens of authors who have published on the biological species concept, only a few workers have attempted to provide a biologically based genus concept. Alain Dubois, a French amphibian systematist, reviewed work on the genus concept in zoology up to 1988, including the evidence for morphological, ecological, and genetic discontinuities, and suggested using a biological criterion–the ability to produce viable adult hybrids–to determine whether two species can be included in the same genus (Dubois 1988).

This criterion may be valid, but like the biological species definition, it is difficult to put into practice. In addition, it does not take into account the whole range of organisms, both plant and animal, in which other modes of inheritance may control the patterns of diversity observed.

Evolutionary and phenetic taxonomists have placed emphasis on the gaps between clusters in defining genera: "A genus (generic taxon) is a monophyletic group composed of one or more species that are separated from other generic taxa by a decided gap" (Mayr and Ashlock 1991:135). Cladistic taxonomists do the opposite, rejecting gaps and defining the genus as follows: "A mandatory category to which every species must belong and which contains one species or a monophyletic group of species" (Wiley 1981:205). Or you can take the pragmatic approach, as Quicke (1993:268) does, and define the genus as a grouping of "one or more species that are (usually) believed to be closely related."

The idea of the genus as a "kind" of organism implies a collection of species, and systematists have long been concerned about the number of species to be grouped in a genus, with some taxonomists being lumpers and some splitters.

However, large genera are a real phenomenon. Some kinds of organisms are just more diverse than others. Alessandro Minelli (1993) devotes a whole chapter of his *Biological Systematics* to a survey of the unequal distribution of taxonomic diversity among organisms. Some large genera are certainly "dumping ground" or "wastebasket" taxa. Heterogeneous and sometimes clearly polyphyletic, they are still waiting for someone to sort out their relationships. Other large genera (e.g., fungi) may be artificially large because of large numbers of described "species" that really represent the products of hybridization. Still others represent the results of apomictic reproduction, such as the blackberries, *Rubus,* mentioned in chapter 17. But other large genera do share one or several strongly derived characters, (e.g., the Staphylinid genus *Stenus,* with more than 1800 species described so far) and do seem to represent true monophyletic groups (Minelli 1993).

Minelli lists some of the largest genera for plants, insects, birds, and mammals. He also points out that few marine animal genera contain above 100 species; *Conus,* with 500 species (Kohn 1991), may be the largest. The largest marine fish genera have just over 100 species, but some freshwater fish genera have more than 300; 28 percent of all amphibians so far described belong to just four genera (*Eleutherodactylus, Rana, Hyla,* and *Bufo*). *Large* is relative, of course. A "large" genus in mammals might have 50 species, whereas in some flowering plants, 500 species might not be considered a large genus.

At the other extreme from large genera are *monotypic genera,* those in which only one species has been described. Some of these may represent situations in which taxonomic knowledge is still incomplete; as research on the group is continued, more related species will eventually be found. Yet experience with plants demonstrates the reality of monotypic genera (Stace 1989). In botany the predominance of monotypic genera in groups from algae to angiosperms seems to have been accepted for years. A 1922 work estimated that 39 percent of angiosperm genera were monotypic, another 13 percent had only two species, and only a very few had large numbers of species (Willis and Yule 1922). The characters that define these monotypic genera represent fundamental differences, and further study only adds more evidence of their uniqueness. In plants some monotypy may result from polyploidy. However, monotypy is found in numerous sexually reproducing plant and animal groups as well. No one has carried out a study similar to that of Willis and Yule for the major animal groups, but it seems likely that many would show a similar pattern. In bryozoans, for example, where understanding of genera is quite good, there are many monotypical genera and only a few genera that are really speciose.

It seems that there is still much to learn about the meaning of genera in plants and animals. But there is no doubt that genus-level research figures importantly in studies of living and fossil organisms. It is the level most commonly used in macroevolutionary studies (Allmon 1992). Inter- and intragenetic comparisons figure in a considerable amount of ecological, genetic, and physiological research as well, as the following sample titles show:

Baird, R. W., P. A. Abrams, and L. M. Dill. 1992. Possible indirect interactions between transient and resident killer whales: Implications for the evolution of foraging specializations in the genus *Orcinus. Oecologia* 89: 125–132.

Brock, V. 1990. Intergeneric distances between *Ostrea, Crassostrea,* and *Saccostrea,* studied by means of crossed immuno-electrophoresis. *Marine Ecology Progress Series* 68: 59–63.

Bulnheim, H.-P. and S. Bahns. 1996. Genetic variation and divergence in the genus *Carcinus* (Crustacea, Decapoda). *International Revue des gesampte Hydrobiologie* 81: 611–619.

Jensen, G. C. and D. A. Armstrong. 1991. Intertidal zonation among congeners: Factors regulating distribution of porcelain crabs *Petrolisthes* spp. (Anomura: Procellanidae). *Marine Ecology Progress Series* 73: 47–60.

Newell, S. J., O. T. Solbrig, and D. T. Kincaid. 1981. Studies on the population biology of the genus *Viola.* III. The demography of *Viola blanda* and *Viola pallens. Journal of Ecology* 69: 997–1016.

Wendel, J. F. 1995. Bidirectional interlocus concerted evolution following allopolyploid speciation in cotton (*Gossypium*). *Proceedings of the National Academy of Science USA* 92: 280–284.

When to Describe a New Genus

If you can write a species description, you can certainly do the same for a new genus. The same procedures that you follow to describe a new species (background research, composing a name, writing a description, selecting a type specimen, publishing in accordance with the codes of nomenclature) also apply to the description of a new genus. Before you start to write, though, you must decide whether the species you are describing requires a new genus or can be best handled by being placed in an existing genus.

Unfortunately, there is no such thing as a "generic character," a taxonomic character whose presence would clearly place your species in a certain genus. Nor is there any character that is always useful at the generic level rather than at the species or some other level, although in a particular group of organisms there may be characters that are important in defining genera or species. The essential thing here, as always, is knowing the organism and the standards of taxonomic practice in the group to which it belongs.

Genera are defined not by one character, but by a group of carefully chosen characters. Usually at least some of them are correlated either functionally or genetically and so are present in all the members of the groups. The members of a genus do not have to share all characters, however. One or more species in a genus may lack one or more of the diagnostic characters, or may have a character present, but in a modified form. In cases where modifications are slight it is probably better to revise the genus diagnosis than to create a completely new genus (Mayr and Ashlock 1991). However, when most of the taxonomic characters found in a new species do not correspond well with those in other species of any known genus and cannot, by a reasonable amount of modification of the generic diagnosis, be made to fit into any described genus, then a new genus should be created.

Generic Names

Unlike the species name, which has to be unique only within that genus, the genus name must be unique within the kingdom. A zoological genus can refer to only one kind of animal, a botanical genus to only one kind of plant. Therefore, it is even more important to invent a good genus name than a new species name. A large number of people will have to use, write, and pronounce it, particularly if the organism is common or experimentally or commercially important.

The basic characteristics of the genus name are that it is singular, *uninomial* (one word, even if a compound word), Latinized, spelled with an initial capital letter, and italicized. According to the ICZN (1985, Art. 11g), "A genus-group name must be or be treated as a noun in the nominative singular."

Genus names may be any of the following:

Latin nouns (substantives):

Convolvulus, the morning glory genus = Latin word for bindweed, morning glory

Cornus, the dogwood genus = Latin word for dogwood

Latinized Greek nouns:

Coryphaena, the dolphin fish = Latinized from Greek **Koryphaina** = dolphin fish

Compound Greek words in which the (usually) first part is a modifier:

Mesoglossum, an orchid genus, from **meso-** = Greek, middle, plus **glossum** = Greek, tongue

Asterocampe, Emperor butterfly genus, from **Aster** = Greek, star, plus **campe,** Latinized version of Greek **Kampe** = caterpillar

Compound Latin words (in which the first part is a modifier):

Recurvirostra, avocet, from **re-** = back, **curvus** = bent, **rostrum** = beak

Greek or Latin *derivatives* (words used to show comparison, diminution, resemblance or possession):

Lactucella, "little lettuce," a plant genus = dim. of Latin **lactuca** = Lettuce

Fratercula, "little brother," puffin genus = dim. of Latin **frater** = brother

Personal (classical) or mythological names (if not Latin to begin with, they receive a Latin ending, as follows:

FI *Dryas, Daphne, Peristylus, Ione*

Modern last names (patronyms) with a generic ending, as follows:

For names ending in a consonant the endings **-ius, -ia,** and **-ium** are used:

Lamarckia, Cunninghamia

For names ending in *e, i, o, u,* or *y* the endings **-us, -a, -um** are used:

Blainvillea

For those ending in *a* the ending *-ia* is used:

Danaia

They could also include expeditionary names:

Challengeria

Names from modern languages. These may or may not have Latin endings:

Idahoa, a plant genus, with ending Latinized

Cabraca, a milliped genus, named after a figure from indigenous Mexican mythology, ending unchanged

Names made up of an arbitrary combination of letters (which, like species names constructed in this manner, must be pronounceable):

Yalpa and *Acauro,* South American moth genera

Names formed by anagram from a genus name in the group

Dacelo, anagram of *Alcedo*

As with species group names, diacritical marks and hyphens are not used in zoology (ICZN 1985, Art. 27, 31d). In botany a generic name cannot be made up of two separate words unless they *are* joined by a hyphen (Art. 20.3). However, as in zoology it can be a compound word (unhyphenated).

Determining the gender of the word is important and can be difficult. A generic name ending in a Greek or Latin word is given the gender assigned to it in a Greek or Latin classical dictionary. This is also true of names taken from a modern language having genders for nouns (e.g., German or French). In zoology a genus group name that is of variable gender is to be treated as masculine unless the author has stated it is feminine or uses it as feminine in combination with a species name (that is, gives the species name a feminine ending) (ICZN 1985, Art. 30). In botany it is recommended (ICBN Rec. 20A.1 (i)) that a feminine ending be given to all personal generic names whether they honor a man or a woman.

The simplest kind of generic name is one that already exists in Latin (i.e., the name that the ancient Romans used to describe a particular plant or animal). Unfortunately, because most of them refer to common European plants or animals, taxonomists used them up as genus names long ago. Some examples are *Bufo,* a toad, *Hirudo,* a leech, and *Quercus,* an oak. Other names were taken from the Greek names for the animal or plant and their spelling and their endings Latinized: from **Klematis** to *Clematis,* from **Geranion** to *Geranium* (Savory 1962).

The next step in complexity is to take an ordinary Greek or Latin noun and give it as a genus name to an organism that possesses some feature related to that word. For example, *Unio* (pearl in Latin) became the name for a genus of freshwater mussel (which sometimes produces pearls). *Helix* (spiral) became the name of a snail genus, and *Gladiolus* (little sword) became the name of a genus of plants with sword-shaped leaves (Savory 1962).

Other Greek and Latin words that can be used as generic names include participles and adjectives used as nouns (substantives). **Primulus** means first in Latin, but standing alone it can mean "the first man"; thus *Primula,* "the first woman," became the name for a plant genus, the primrose, which flowers early in spring. Generic names derived in a similar way include the horseshoe crab *Limulus* (= askew) and the flower genus *Impatiens* (= impatient) (Savory 1962).

Classical names, such as those of gods and goddesses, heroes and monsters, historians and historical localities, have all been borrowed as genus names. As Savory (1962:34) points out, "*Nereis, Argiope* and *Dryas* were classes of nymphs; now they are also worms, spiders and insects." But with almost 2 million organisms already named, the chances of finding a simple unclaimed classical name are slim. Most generic names created today are compound names. They may be compound Greek words, compound Latin words, or modern language names or words given a generic ending as shown earlier. "Nomina hybrida," compound names with components taken from different languages (e.g., Greek and Latin), used to be considered uncultured. However, the necessity of composing a *unique* name seems to be eroding that bias.

It is still possible to come up with Latin combinations that will describe the organism:

Saginospongia for a new genus of Permian sponge (Rigby and Senowbari-Daryan 1996), from **sagino** (Latin for fishnet) in reference to its netlike skeleton and **spongia,** sponge

Other methods of constructing a unique name are to use the name of the country or locality where the organism was found in combination with a descriptive term or a name commonly used in that group of organisms:

Chilehexops, a six-eyed spider from Chile (Coyle 1986)

Sinoniscus, the first cave-dwelling terrestrial oniscidean isopod from China (Schultz 1995)

Taxonomists also commonly construct genus names in honor of another worker on the group

Verseveldtia, an octocoral genus named for Jakob Verseveldt, an octocoral systematist (Williams 1990)

Taxonomists may use the person's last name or first name, or combine either name with a generic ending already in use in a group to create a unique compound name whose taxonomic relationships will still be readily recognizable to other students of the group:

Olgalepidonotus, a new genus of scale worm named in honor of Olga Hartman, a polychaete systematist (Pettibone 1995)

This procedure has a strong advantage for identification and placement, as anyone with a knowledge of the group will immediately recognize where in the classification the organism belongs. For example, many murid genera end in *-mys* from **mys,** the Greek word for mouse. Should you describe *Xxxxxomys,* mammalogists will know right away that you're talking about a mouselike rodent and not a rabbit or a cat, let alone a sweet pea or sea cucumber.

Before publication you must determine whether the name is unique. Reference works that list all genera named so far (up to their publication date) should be consulted to determine whether the name has already been used (see this chapter's *Sources*).

You also want to avoid giving your genus a name that is too similar to that of another existing genus; for example, *Desmazieria* and *Desmazeria* and *Tichothecium* and *Trichothecium* are two bad examples from botany (Hawksworth 1974).

Publication of Generic Names

Like a new species name, a new generic name must be published in an acceptable manner to be valid (legitimate). The following guidelines apply in both zoology and botany:

Make it clear (in the heading) that you are describing a new genus (e.g., *Xxxxxus,* new genus," or *Xxxxxus,* gen. nov.).

Coin an acceptable generic name that doesn't violate the rules and regulations.

Check the proper authorities to make sure that the name is unique and does not already exist either as a homonym (a name that can't be used because it is already in use in some other group) or a synonym (a name that is already published for the same group of species).

Include a diagnosis with a clear statement of the characters by which the genus differs from other members of the group.

Clearly designate a type species.

Give the etymology and gender of the name.

Generic Types

To become available, every generic name must have a type species chosen. This used to be called the "genotype" by taxonomists (and you will see this in the older taxonomic literature). But that usage was confusing to most biologists, who normally use the term *genotype* to describe the genetic makeup of an organism. Taxonomists have pretty much abandoned the term and replaced it with "type species of the genus," "type species," or "generitype" (Jeffrey 1989). The type specimen of that species is the voucher, the biological reality, that links the author's concept of the genus with a particular specimen. Generic limits may be modified, but if that species as defined by its type does not fit some specimens or species, then the ones that do not fit are split off into another genus. The name of the genus *must* stay with the portion that is compatible with the type species.

If you are describing an organism new as to both genus and species you won't have any problem because the only species so far in the new genus–

yours–will become the generic type, but you should state this directly, just in case. If you are proposing a new genus for which a number of species have been described, then you must choose a type. (In botany this is required. In zoology, if you don't chose a type, the next person who studies that genus can do so, and it's too bad for you if you don't agree with their decision.) The same recommendations apply in choosing a type specimen for a genus as for a species (chapter 9).

Zoology

There are three ways to *fix* (determine which one will be used as) the type species of a genus (ICZN 1985, Art. 67–69): by original designation, by indication, and by subsequent designation.

The simplest way is for you as the author to clearly state in your description that your new species is the type, e. g., for a new genus, new species:

Type Species: *Yourgenus yourspecies* new genus, new species

If you are describing a new genus, either for species that are already known or for a group of known species, plus a new one that you have found, you need to choose a type species for the group. Pick the species that most closely represents your conception of the group, whether it is a new species you are describing in that paper, or one that someone else had described previously, but that you have decided belongs in your new genus. You should also try to pick a type species that is well described and illustrated (in your paper or elsewhere) and for which type material exists (ICZN, Rec. 69A):

Type Species: *Yourgenus yourspecies* new genus, new species

Or if you decide that a species originally called *Somebodygenus somebodyspecies* Somebody really fits your new genus and should be the type species of it, then use

Type Species: *Yourgenus somebodyspecies* (Somebody)

Note that author's name is now in parentheses because you have assigned a new genus name to his or her species.

Although just mentioning a structure as typical of a genus or a species or mentioning a species as an example of a genus is not enough to fix it as the type species, naming it *typicus* or *typus* is considered by the ICZN to do so by indication (ICZN, Art. 68c).

A genus group established with only one included species takes that nominal species as its type species by indication (called type by monotypy).

If the species group name or cited synonym included in a genus group taxon is identical to the genus group name, that name becomes the type of the genus by "absolute tautonymy" (ICZN, Art. 68a).

If the authors established a genus but didn't fix its type species, the first author to designate one of species originally included in the genus to be the type is considered to have validly designated the type species. No later designation is valid.

Botany

The application of the rules in botany differ because names in botany are not coordinate. To be validly published in botany the rank must be made clear, there must be a description or diagnosis in Latin, and a type must be cited. Another difference is the existence of many conserved names in botany. Changes in a conserved generic name can be made only by the same process of committees and voting at International Botanical Congresses that led to the conservation of the name in the first place (ICBN, Art. 14.12).

As in zoology, the name of each genus is based on a type species, actually on its type specimen or an illustration (ICBN Art. 10.1). "Designation of the type is achieved only if the type is definitely accepted as such by the typifying author, and if the type element is clearly indicated by direct citation including the term 'type' or an equivalent" (ICBN, Art. 7.11).

In botany you should give a feminine form to all personal names when used for a genus, subgenus, or section, whether the person is a woman or a man. The following endings apply. If the personal name ends in a consonant use *-ia,* but when the name ends in *-er,* then either of the terminations (*-a* or *-ia*) is appropriate (ICBN, Rec. 60B(b)). If the name ends in a vowel, add the ending *-a* to the name (Benson 1962).

The ICBN also states that a name cannot coincide with a technical term used in morphology. Few technical terms are exactly the same in modern languages as in Latin, but botanists, because they must use the Latin technical terms in the diagnosis or description of new taxa, may run into this problem (ICBN, Art. 20.2).

Examples of Generic-Level Description

There are three basic reasons for writing papers on genera:

In conjunction with the description of a new species (new genus and species)

In conjunction with intensive research on a known species

In conjunction with a taxonomic review or revision

The double description (new genus, new species) is actually the simplest case scenario for this chapter, and as the one you are most likely to face, it is covered first here.

New species from poorly known habitats often turn out to require new genera as well–once you manage to catch them. For example, in November 1971, during a dive to 635 m in the Tongue of the Ocean, off the Bahama Islands, the pilot of the D.S.R.V. *Alvin* observed a green and yellow powder-puff–like object lying on the muddy bottom. However, when he tried to pick it up with the sub's mechanical arm, it escaped, swimming away with poorly coordinated wriggles of its segmented body. A few days later, at nearby site at 597 m, the submariners were able to capture two of the creatures with a spring-powered box corer before they could swim away. The animals turned out to be terebellid polychaetes; the powder-puff effect was created by their masses of tentacles, which completely surrounded the worms' bodies. Further study showed them to be related to members of the non–tube-dwelling subfamily Polycirrinae, but they were distinct enough that a new genus as well as a new species seemed warranted. In this case, the genus was diagnosed very briefly and the etymology for both species and genus, as well as the discussion of relationships and ecology, were given in the species description:

Terebellidae Grube, 1850; Polycirrinae Malmgren, 1865;
***Biremis,* new genus; gender: feminine**

Type species *Biremis blandi,* n. sp.
Diagnosis: Tentacular lobe enlarged, bearing numerous, long, grooved tentacles in marginal groove; eyes absent; branchiae absent; two longitudinal muscular ventral ridges separated by midventral groove; thoracic segments achaetous; abdominal segments with bilobed uncinigerous

pinnules bearing avicular uncini supported by long bristles; ventral glandular scutes indistinct. (Polloni et al. 1973:170).

[And from the species description]

Biremis (Latin: a ship with two banks of oars) recognizes the swimming capability of the terebellid, with its segmental pairs of bilobed uncinigerus pinnules. The species is named for Edward L. Bland, Jr., "Alvin" pilot who caught the specimens. (Polloni et al. 1973:172)

On the other hand, a new genus and species can turn up even in an area that has received a good deal of study, especially if it is an organism that is small or cryptic in habit. When entomologist Thomas Henry saw a few specimens of a puzzling seed bug that had been collected in a pitfall trap at Seashore State Park in southeastern Virginia, three things about the bug struck him immediately: its very small size, the fact that its hemelytra were oddly modified and constricted, and the fact that its male genitalia were quite different from those of other U.S. members of the subfamily. Examination of more specimens convinced him that this combination of characters was sufficient to show that it was not only a new species, but a new genus of the Bryocorinae, a group of mirid plant bugs scarce in the Nearctic regions but abundant in the Neotropics (Henry 1993). Excerpts show how he handled its description:

Pycnoderiella, new genus

Type species: *Pycnoderiella virginiana,* new species.

Diagnosis: Bryocorinae: Eccritotarsini. Characterized by a combination of the small size (ca. 2.00 mm for male; 1.70 mm for females); the greatly modified male hemelytra (Figs. 1, 3) that are constricted on the basal third of the corium, flared just before the base of the cuneus, and again narrowed from the base to the apex of the slender cuneus; the modified female hemelytra having the cuneus and membrane narrowed to give an overall rounded appearance; and the unique male genital capsule and parameres (see species description).

Description: oblong oval. Head: broader than long, uniformly set with short, semierect setae, basal margin straight, contiguous with anterior margin of pronotum, eyes small. . . .

Name: This name is constructed from the existing generic name *Pycnoderes* and the diminutive suffix "ella" to imply a smaller, superficially similar taxon. The gender is feminine.

Remarks: Relationships in the Eccritotarsini are not sufficiently well known to place *Pycnoderiella* in a phylogenetic context. . . . *Pycnoderiella* differs externally from these genera in having swollen, apically narrowed metafemora and peculiarly modified hemelytra and from all, except *Halticotoma,* in having the rostrum extending to the metacoxae. Additionally it differs in male genital characteristics. . . . The male genitalia of *Pycnoderiella* are most like those of *Sixeonotopsis* Carvalho and Schaffner (1973) in lacking a tubercle on the genital capsule and having a shortened, more curved, right paramere. . . . (Henry 1993:2–6)

After his comparison of the characteristics of the new genus with those of other genera within the subfamily, he went on to describe the new species itself.

The two preceding examples are typical of cases in which the newly found species differs sufficiently from its relatives that a new genus must also be described. Three more excerpts from descriptions of new genera in conjunction with new species follow. They show the variability in style, content, and order of presentation of sections that can occur even in a single publication series. Some differences in the descriptions reflect descriptive traditions within a taxonomic speciality, others, the amount of information available or the author's own preferences. However, all the descriptions cover the basics, presenting *diagnostic information, etymology,* and *naming a type* for that genus.

A New Spider Genus

CHILEHEXOPS, new genus

TYPE SPECIES: *Chilehexops platnicki,* new species.

ETYMOLOGY: The generic name is derived from the country (Chile) where both species have been collected and the Greek words *hex* (six) and *ops* (eye); it is masculine in gender.

DIAGNOSIS: *Chilehexops* can be distinguished from other diplurid genera by the following character states: (1) the absence of posterior median eyes (figs. 17,18); (2) heavily sclerotized spermathecal trunks (figs. 28–34); (3) a male palpal tibia nearly as long as the femur (PTL (100/PFL

= 90–100), figs. 22, 25), and (4) the absence of mating spurs, apophyses, and keels on the legs of adult males.

DESCRIPTION: Small mygalomorph spiders. . . . (Coyle 1986:4)

A New Genus of Cardinalfish

Cercamia, new genus

Type species *Cercamia cladara*
Randall and Smith

DIAGNOSIS: Apogonid fishes with 9 + 15 vertebrae, two pairs of epipleural ribs, no predorsal bones, and hypural 1–4 fused into a double fan-shaped structure. Included species: *Cercamia cladara* and *Rhabdamia eremia* Allen, 1987.

DESCRIPTION: Both species are transparent and without obvious markings. Hypurals 1–4 are fused; hypural 5 is a separate bone. There are 3 epurals, no uronurals, and no autogenous hemal spines. Epipleural ribs are present on the first two pairs of pleural ribs. There are one to two spinules on the posttemporal and spines on both flanges of the preopercle.

RELATIONSHIPS: The genus appears to be related to *Rhabdamia* (as defined by Fraser, 1972) on the basis of its caudal skeleton, but it differs from *Rhabdamia* in having spines on the posttemporal and preopercle and in the absence of predorsals. It also resembles *Pseudamiops* in having epipleural ribs only on the first two pairs of pleural ribs rather than five to seven pairs as in *Rhabdamia.* The vertebral count of 9 + 15 occurs in no other species of the family Apogonidae.

ETYMOLOGY: *Cercamia* is from the Greek *Kerkos,* tail, and *Amia,* a Greek fish name that has been applied to apogonid fishes. This name alludes to the elongated tail resulting from the unique vertebral count with one fewer abdominal and one more caudal vertebrae than other cardinalfishes. It is feminine. (Randall and Smith 1988:7)

A New Fossil Shark Genus

CHONDRICHTHYES
ELASMOBRANCHI
PLESION HYBODONTIFORMES
SUBORDER HYBODONTOIDEI
Hamiltonichthys, g. nov.

Diagnosis: Hybodontoid sharks of approximately 300 mm total length, with low-crowned teeth, each with a single asymmetrical peak; teeth traversed by numerous cristae, with a lingual swelling and continuous shoulder; lower dentition comprising a symphyseal series, six anterior lateral series, a single elongate lateral series, and a single, short, posterior series; upper dentition comprising parasymphyseal series; six anterior lateral series, a single long lateral series, and a single short posterior series; jaws broad, lacking postorbital articulation; four simple multicuspid cephalic spines in males, each cusp with an open pulp cavity; pharyngeal dentition consisting of several multicuspid tooth-whorls; pelvic girdle a continuous puboischiadic bar only in males; posterior dorsal fin with metapterygial-like axis; anal fin supported by series of small cartilage plates.

Etymology: After the type locality.

Type Species: *Hamiltonichthys mapesi,* g. nov., sp. nov. (Maisey 1989: 11)

Sometimes more extensive study, taxonomic or biological, can lead you to discover that a particular species does not belong in the genus to which it was originally assigned by its authors. To give a taxonomic example, Guy Musser had undertaken a long-term study of the morphology, distribution, and natural history of Philippine murids. Although there was an impression among mammalogists that this fauna was known, his research showed that the diversity of the fauna was greater than had been realized, both because recent exploration had added samples of new species from various islands and because many specimens of new taxa had been hidden under misidentifications in museum collections or placed under other genera. One of the latter was *Rattus latidens:*

ABDITOMYS, NEW GENUS

TYPE SPECIES: *Rattus latidens* Sanborn (1952, p. 125); based on the holotype, an adult female (FMNH 62347), collected April 29, 1946, from Mount Data, 7500 feet, Mountain Province, Luzon, the Philippines. An adult male (USNM 357244) obtained on March 14, 1967 from Los Baños, Laguna Province, Luzon (Barbehenn, Sumangil, and Libay, 1972–1973) is the only other known specimen of the species.

KNOWN DISTRIBUTION: The island of Luzon in the Philippine Islands (fig. 2).

ETYMOLOGY: *Abditomys* is formed by combining the Latin *abditus,* meaning hidden or concealed, with the Greek, *mys,* for mouse or rat. I

> allude to the past concealment of *latidens*–so distinctive in features of skin, skull, and teeth–within the morphological confines of the genus *Rattus,* to which it is not closely related.
>
> DIAGNOSIS: A genus of arboreal murid distinguished from all other species by the following combination of features: large body size, long, monocolored tail; four pairs of mammae; a large nail instead of a claw on each hallux. . . .
>
> DESCRIPTION: *Abditomys latidens* is a large-bodied rat with a tail longer than the combined lengths of head and body. . . . (Musser 1982:3)

The rest of the paper consists of a thorough description of *latidens* morphology and a detailed comparison of its features with those of members of the genus *Rattus* (which have only claws, never nails), as well as with other murines having nails, as evidence for the decision to create a new genus for the species.

There is one other taxonomic problem that may arise in the study of a known species: Your literature research may show that the species fits the genus in which it was described, but that the name for the genus either is not unique or is a junior synonym of some other genus. For example, while studying two Australasian ground spider genera, arachnologists Vladimir Ovtsharenko and Norman Platnick realized that one of the genera was a junior synonym of the other. To resolve the problem they created a new synonomy, rediagnosed and described the senior genus, and in their synonymy section explained the history of the problem:

> SYSTEMATICS
>
> *Anzacia* Dalmas
>
> *Anzacia* Dalmas, 1919: 249 (type species by original designation *Drassus perexiguuus* Simon).
>
> *Adelphodrassus* Rainbow, 1920: 235 (type species by monotypy *Adelphodrassus inornatus* Rainbow).
>
> NEW SYNONYMY. . . .
>
> SYNONYMY: According to the table of "Contents" of volume 1 of the Records of the South Australian Museum, volume 1 (number 3) of that journal, containing Rainbow's description of *Adelphodrassus,* was published on June 30, 1920. Bonnet's (1955) catalog indicates that Dalmas' description of *Anzacia* also appeared in 1920, raising the question of the relative priority of these two names. However, Roewer's (1955) catalog indicates that Dalmas' paper was published in 1919, and this

> appears to be one of the rare cases in which Bonnet's catalog is in error. We have found no internal evidence indicating that volume 25, number 4, of the Bulletin of the Muséum National d'Histoire Naturelle (for 1919) was not actually published in 1919, and Bonnet's catalog itself lists a paper by Berland, published in a later number (6) of volume 25, as having appeared in 1919. It is possible that Bonnet replicated an error from the Zoological Record for 1920, which erroneously listed Dalmas' paper as having appeared in the Museum's Bulletin for 1920 (perhaps confusing it with another paper by Dalmas that appeared in the Bulletin de la Société Entomologique de France for 1920). In any case, it seems clear that Dalmas' paper appeared before that of Rainbow, and that the name *Anzacia* has priority over *Adelphodrassus.* (Ovtsharenko and Platnick 1995:10–11)

Other generic descriptions are written in the course of reviewing or revising a genus or higher category.

When you review the species of a genus, either worldwide or from a particular region, you compile a synonymy that clears up any old mistakes. Then you update the diagnosis and description of the genus if necessary, or give a reference to a paper in which the information is available. You may also include a section to mention species included or species that do not belong. This is followed by the systematic descriptions of the species of the genus you found or studied. For example, in their review of zelotine spiders of China, Platnick and Song (1986) first gave a brief review of the genus *Zelotes:*

> *ZELOTES* GISTEL
>
> *Melanophora* C. L. Koch, 1833, pt. 120 (type species by original designation *Melanophora subterranea* C. L. Koch); preoccupied by *Melanophora* Meigen, 1803 (Diptera).
>
> *Zelotes* Gistel, 1848, p. xi (nomen novum for *Melanophora* C. L. Koch).
>
> *Prosthesima* L. Koch, 1872, p. 139 (superfluous nomen novum for *Melanophora* C. L. Koch).
>
> DIAGNOSIS: Specimens of *Zelotes* can be distinguished from all other gnaphosids by the combined presence of a preening comb on metatarsi III and IV . . . and an intercalary sclerite on the male palp. . . .
>
> DESCRIPTION: See Platnick and Shadab (1983, p. 104).
>
> MISPLACED SPECIES: Two species belong elsewhere. The holotype of "*Zelotes ?joanisi*" Schenkel (1963), from Nanking in Jiangsu Province,

lacks metatarsal preening combs and is not a zelotine. Similarly, the holotypes of the Japanese species *Zelotes pallidipatellis* (Bösenberg and Strand), which was recorded from China by Zhu (1983), also lacks a preening comb; Japanese specimens of both sexes kindly loaned to us by Dr. Takahide Kamura have genitalia and abdominal patterning typical of members of the *Poecilochroa* group of genera. (Platnick and Song 1986:2–3)

The last category is the taxonomic revision. You may intentionally take on a revision of a genus, usually with the goal of working out the phylogenetics of the species in that genus. You may also revise all the genera of a family (see chapter 20 also), studying the phylogenetic relationships among them.

For example, Stonedahl and Cassis (1991) revised the species of the plant bug genus *Fingulus.* They began their paper with an introduction giving the history of the genus, a section on taxonomic characters used, and material and methods used in the study. The systematics (results) section began with a redescription of the genus:

SYSTEMATICS
Fingulus Distant

Fingulus Distant, 1904: 275 (n. gen.).–Carvalho, 1955a: 27 (key); 1955b: 221 (syn.); 1957: 86, 87 (cat.).–Linnavuori, 1975: 11 (descr.).–Akingbohungbe, 1981: 182 (descr.). Type species by monotypy: *Fingulus atrocaeruleus* Distant.

Ix Bergroth, 1916: 234, 235 (n. gen.).–Carvalho, 1955b: 221 (syn.). Type species by monotypy: *Ix porrecta* Bergroth.

Anchix Hsiao, 1944: 377, 378 (n. gen.).–Carvalho, 1955b: 221 (syn.). Type species by monotypy: *Anchix atra* Hsiao.

DIAGNOSIS: Distinguished from other genera of Deraeocorinae by the porrect head with strongly developed necklike postocular region, usually with transverse constriction or furrow posteriad of eyes (fig. 1, 13, 29); broad, flattened or rarely weakly rounded pronotal collar (figs. 1, 11, 13, 15, 29); strongly protruding ostiole of metathoracic scent efferent system (fig. 2); deep cuneal incisure marking strongly deflexed hemelytra (figs. 13, 15); and structure of the male genitalia, especially the vesica with gonopore opening into sclerotized, tooth-lined cavity (see description of genitalia).

REDESCRIPTION: *Male.* Length from apex of tylus to cuneal incisure 1.90–4.33; width across humeral angles of pronotum 1.01–2.04; castaneous to dark fuscous general coloration, sometimes tinged with red especially on cuneus.; body elongate to broadly ovate; pronotum and hemelytra coarsely punctate, shining, glabrous, or with very short, fine setae protruding from punctures. . . . (Stonedahl and Cassis 1991:5–6)

This section was followed by description of all the species, including several new species they included in the genus, a cladistic analysis, and a biogeographic discussion.

Bieri (1991) revised the genera of the chaetognath family Sagittidae to create more homogeneous groupings based on use of the same suite of taxonomic characters throughout the family:

Sagittidae Tokioka, 1965

***Adhesisagitta,* new genus**
Figure 1

Type species: *Sagitta hispida* Conant, 1895, by monotypy and present designation.

Name: From the Latin "adhaesus," clinging to, referring to the habit of this "quasi-planktonic" species of clinging to the substrate.

Definition: Corona ciliata long, extending from just anterior to eyes to well posterior on trunk but not reaching ventral ganglion, intestinal diverticula present; lateral fins completely rayed; body rigid; seminal vesicles slightly swollen at anterior end when mature, but not forming distinct knob; ovaries moderately long and narrow with ova irregularly spaced rather than in a single row; mouth appearing sutured or buckled by 4 rod-like structures or bars about 1/3 as long as distance between them; species can cling to vertical substrate.

Species included: One, *Sagitta hispida.*

Discussion: Published drawings and descriptions of this species are not consistent with one another in several critical features. . . . Although the number of chaetognath species examined with SEM now exceeds 17 (Thuesen et al 1988), *A. hispida* is the only species found to have well developed bars or buckles across the mouth. . . . *Adhesisagitta* is distinguished from the possibly closely related genera *Sagitta* (sensu Tokioka) and *Ferrosagitta* not only by the mouth buckles but also by its clinging ability, shape of the seminal vesicles, and the irregular arrangement of the ova. (Bieri 1991:221)

Note that in the format used here the "definition" is a description, not a diagnosis, and the "diagnosis" of the genus comes in the discussion section that follows.

Problems Caused by Generic Revision

Sometimes careful study of described species of a genus can lead to the necessity of creating a new genus for one or more of them, or splitting genera. For example, as part of a long-term study of meiolaniids, or horned turtles, a fossil group with skulls characterized by development of bony hornlike protuberances, Eugene Gaffney restudied a fossil skull and tail club from Australia that were housed in the British Museum's collections. The species had originally been named by Victorian paleontologist Richard Owen, who had mistakenly described it as a giant horned lizard. This mistake was later corrected and the species placed in the genus *Meiolania.* However, other work that Gaffney and colleagues had already done on *Meiolania* led them to conclude that that genus should be restricted only to species that shared a certain group of derived characters and were clearly monophyletic (Gaffney and Meylan 1988). Contrasting characters in the British Museum specimen indicated to him that it should be separated from *Meiolania* and placed in a new genus.

Ninjemys, new genus

Megalania Owen, 1881 (in errore).
Meiolania Owen, 1886 (in part).

TYPE SPECIES: *Meiolania oweni* (Woodward).

ETYMOLOGY: *Ninja,* in allusion to that totally rad, fearsome foursome, epitomizing shelled success; *emys,* turtle.

KNOWN DISTRIBUTION: Pleistocene of southern Queensland, Australia.

DIAGNOSIS: A meiolanid known only from skull and tail, characterized by unique possession of laterally projecting B horns and anterior extension of the nasals beyond rest of skull: A scale area large and forms posterior shelf as in *Niolamia* but A scale not significantly larger than B scale; D scales probably meet in midline, X scale small as in *Meiolania;* D scale area raised as in *Niolamia,* not flat as in *Meiolania;* Y scale relatively large as in *Meiolania;* apertura narium interna partially divided

> as in *Meiolania* but in contrast to *Niolamia;* well-developed second (more medial) accessory ridge on triturating surface of palate reaching nearly to midline in contrast to *Meiolania* in which it is lacking anteriorly and *Niolamia* in which it is absent; tail ring enclosed ventrally, as in *Niolamia* but in contrast to *Meiolania;* tail club formed from two segments, rather than four as in *Meiolania.* (Gaffney 1992:5)

At other times new information means that two genera should be merged. For example, botanists Thomas Lammers, Thomas Givnish, and Kenneth Sytsma used phylogenetic analyses based on molecular data to determine that species of the endemic Hawaiian plant genus *Rollandia* actually formed a monophyletic group with the members of another Hawaiian genus, *Cyanea.* They merged the two genera under *Cyanea:*

> In order to provide a more natural classification of the baccate Hawaiian lobelioids, *Cyanea* and *Rollandia* are here merged. Although Gray (1861) transferred some species of *Rollandia* to *Cyanea,* the species that was later designated as the lectotype of the genus (*R. lanceolata*) was assigned to *Delissea.* Because no previous author has effectively merged just *Cyanea* and *Rollandia,* and because the two generic names have equal priority, we have the prerogative of choosing the name to be retained. We here choose *Cyanea,* as doing so will require far fewer nomenclatural innovations. . . . (Lammers et al. 1993:437–438)

In their paper they followed this step with a description of the merged genus and information on the species included in it.

"Resurrection" of a Genus

Sometimes new material can make it necessary to take a new look at a genus that has been put into synonymy. This happened to Richard Zweifel and Fred Parker when they examined a collection of microhylid frogs they had made in New Guinea in 1987. In 1917, Fry had described *Aphantophryne pansa,* a new genus and species of frog from the Owen Stanley Mountains of New Guinea. Later workers disagreed about the morphological characters used to distinguish the genus, and it was put into the synonymy of other genera. However, in their 1987 collections from the same region Zweifel and Parker not only found two new species related to *pansa,* but were able to identify a new skeletal character shared by all three, evidence that they did indeed make up a monophyletic group. On that evidence they resurrected the genus *Aphantophryne* to be used for all three species.

> *Aphantophryne* Fry, 1917
>
> *Aphantophryne* Fry, 1917: 172 (type species by designation, *A. pansa* Fry). van Kampen, 1923. Parker, 1934.
>
> *Asterophrys:* Loveridge, 1948: 417 (*Aphantophryne* considered a junior synonym of *Asterophrys*).
>
> *Cophixalus* Zweifel, 1956: 41 (*Aphantophryne* considered a junior synonym of *Cophixalus*). Zweifel and Allison 1982.
>
> DIAGNOSIS: A genus of genyophrynine (Burton, 1986: Zweifel, 1972, as Sphenophryninae) microhylid frogs that differs from other genera of Genyophryninae in having seven rather than eight presacral vertebrae. Additional distinctive characteristics include short hind legs, digits with rounded tips lacking flattening or terminal grooves . . . and absence of clavicles and procoracoid cartilages. . . .
>
> CONTENT: *Aphantophryne pansa* Fry and two new species described herein, *A. sabini* and *A. minuta.*
>
> DISTRIBUTION: Mountains of eastern Papua New Guinea, from Mt. Amungwiwa southwest of Wau, Morobe Province, southeastward at least to Myola Guest House, Northern Province, northeast of Port Moresby (fig. 1).
>
> DISCUSSION: The presence of seven presacral vertebrae noted in *Aphantophryne* is decidedly unusual among genyophrynine microhylids and appears to be unique to this genus as a typical condition. Combining counts made on cleared-and-stained specimens and on radiographs, we find seven vertebrate in all of 11 *A. pansa,* 17 *A. sabini,* and 1 *A. minuta.* . . . (Zweifel and Parker 1989:3)

Infrageneric Categories and Names

In zoology only one category is recognized between the genus and species: the subgenus. Like genera, subgenera are given names that are singular uninomial Latinized nouns or noun substitutes, written with an initial capital letter. Their taxonomic use varies greatly from one group of organisms to another. In some groups they are never used. In others, especially in large genera of well-studied organisms (e.g., the amphibian genus, *Rana,* or some large insect or molluscan groups) they are in common use by taxonomists. Even where they are used, however, they are primarily used in taxonomic studies and seldom seen in nontaxonomic, nonmonographic work (Blackwelder 1967).

As far as nomenclatural regulations go, subgeneric names are treated like generic names, but with these differences:

When used with a particular species name, subgenus names are always used with the genus name, too. That is, although you can discuss the subgenus by itself, you cannot use *Subgxxxus spxxus*; you must use *Gxxxus (Subgxxus) spxxus.*

Subgenus names are always written in parentheses between the genus and the species name, as in *Drosophila (Hirtodrosophila) pleothoracis* (Grimaldi 1987)

They must be published later than the name of the genus in which they are included.

They can't be used as the basis of family group names unless they have first been made into genera.

The name of the nominate subgenus must be identical to that of the genus. (Blackwelder 1967)

A new subgenus is described just as a new genus is described, as in this excerpt from a paper on the subgenera of the squid family Enoplotheuthidae by K. Tsuchiya and T. Okutani:

Subgenus ***Paraenoploteuthis*** nov.

Type species: *Enoploteuthis chuni* ISHAKAWA, 1914.

Diagnosis: Right ventral arm is hectocotylized. . . . The ventral flap of tentacle is present. . . .

Included Species:

1. *Enoploteuthis chuni* ISHKAWA, 1914
2. *Enoploteuthis galaxias* BERRY, 1918
3. *Enoploteuthis anapsis* ROPER, 1964
4. *Enoploteuthis jonesi* BURGESS, 1982
5. *Enoploteuthis higginsi* BURGESS, 1982.

Geographical distribution: warm to cool temperate waters of the Atlantic and Pacific.

Remarks: On the evaluation of the characters to split the genus *Enoploteuthis*, BURGESS (1982) considered that the presence or absence of the ventral flap of the carpus is one of the important criteria. RIDDELL (1985) described the presence of the ventral flap in *E. galaxias* from New Zealand waters. The papillation on the oral surfaces of arms characterizing the subgenus *Enoploteuthis* is not present in this subgenus. (Tsuchiya and Okutani 1988:120–121)

In botany several infrageneric categories are used: subgenera, sections, subsections, series, and subseries. They have names similar to those of genera, but with different endings. Their chief use is to divide some extremely large plant genera (Hawksworth 1974).

Article 21 of the ICBN covers generic names. "The name of a subdivision of a genus is a combination of a generic name and a subdivisional epithet connected by a term (subgenus, sectio, series, etc.) denoting its rank." So in writing about a genus and its subgenus, you would write "*Costus* subg. *Metacostus*" or, in writing about a section, "*Ricinocarpos* sect. *Anomodiscus*." But in writing about a particular species, it is recommended that the infrageneric name be placed in parentheses between the genus and species names (as in zoology), with its rank indicated if desired (e.g., *Astralagus (Cycloglottis) contortuplicatus*).

The infrageneric name either has the same form as the generic name or is a plural adjective agreeing in gender with the generic name and written with a capital initial letter:

Costus subg. *Metacostus*

Euphorbia subsect. *Tenellae*

The ICBN recommends that the name of a subgenus or section be a noun or substantive (word acting as a noun) and that of a subsection or below be a plural adjective, but if other names already exist in the genus, it is recommended that an author choose a name that parallels the existing names in construction.

ICBN Art. 21.3 states that an infrageneric name "is not to be formed from the name of the genus by adding the prefix Eu-." **Eu-** in Greek means well, good, truly, originally. According to Stearn (1983), botanists used to use it in subgeneric and sectional names to form the name of the subdivision to which the type of the name belonged. Now, they, like the zoologists, use a name identical to the genus name for the nominotypical subdivision.

Sources

Genus Concept

Allmon, W. D. 1992. Genera in paleontology: Definition and significance. *Historical Biology* 6: 149–158.

Bartlett, H. H. 1940. History of the genus concept in botany. *Bulletin of the Torrey Botanical Club* 67: 349–362.

Cain, A. J. 1956. The genus in evolutionary taxonomy. *Systematic Zoology* 5: 97–109.

Clayton, W. D. 1972. Some aspects of the genus concept. *Kew Bulletin* 27: 281–287.

Dubois, A. 1988. The genus in zoology: A contribution to the theory of evolutionary systematics. *Mémoires du Muséum National d'Histoire Naturelle, Zoologie* 140: 1–122.

Lemen, C. A. and P. W. Freeman. 1984. The genus: A macroevolutionary problem. *Evolution* 38: 1219–1237.

Van Gelder, R. G. 1977. Mammalian hybrids and generic limits. *American Museum Novitates* 2635: 1–25.

Williams, C. B. 1951. A note on the relative sizes of genera in the classification of animals and plants. *Proceedings of the Linnaean Society of London* 162: 171–178.

References Listing Generic Names Already in Use

Zoology

Zoologists rely on this series of reference works for the older literature:

Neave, S. A., ed. 1939. *Nomenclator Zoologicus: A List of Names of Genera and Subgenera in Zoology from the Tenth Edition of Linnaeus, 1758 to the End of 1935,* 4 vols. London, Zoological Society of London.

Neave, S. A., ed. 1939. *Nomenclator Zoologicus,* vol. V, 1936–1945. London, Zoological Society of London.

Edwards, M. A. and A. T. Hopwood, eds. 1966. *Nomenclator Zoologicus,* vol. VI, 1946–1955. London, Zoological Society of London.

Edwards, M. A. and G. Vevers, eds. 1975. *Nomenclator Zoologicus,* vol. VII, 1956–1965. London, Zoological Society of London.

From 1966 on, generic names must be checked by a search of *Zoological Record* (as described in chapter 4).

The ICZN has published the *Official Lists and Indexes of Names and Works in Zoology* 1987. This volume lists all the names on which the ICZN has ruled from its beginning through 1985. To catch up with rulings after that date, it is necessary to do a taxon search or check the *Bulletin of Zoological Nomenclature.* The *Journal of Paleontology* carries summaries of rulings that apply to fossil taxa.

Botany

One of the most important references for botany is the *Index Nominum Genicorum (Plantarum).* This project, the result of a joint effort by the International Association for Plant Taxonomy and the Smithsonian Institution, is an attempt to compile the generic names of all organisms covered by the ICBN. It is available in print form:

Farr, E. R., J. A. Leussink, and F. A. Stafleu, eds. 1979. *Index Nominum Genicorum (Plantarum). Regnum Vegetabile* 100–102: 1–1896.

Farr, E. R., J. A. Leussink, and G. Zijlstra, eds. 1986. *Index Nominum Genicorum (Plantarum).* Supplementum I. *Regnum Vegetabile* 113: 1–126.

It is also available online in a searchable draft form (**http://www.nmnh.si.edu/ing/**). This electronic version is current through 1990, and is continually being updated through the efforts of many collaborating botanists.

Greuter, W., et al. 1993. *NCU-3. Names in Current Use for Extant Plant Genera. Regnum Vegetabile* 129: 1–1464. (Covers generic names currently in use for flowering plants, ferns, algae, mosses, and fungi.)

Botanists have a number of other references available, but their scope is limited to certain plant groups. They were also listed in chapter 4:

Index Filicum. 1906–. (Gives binomial names of all ferns from Linnaeus on, and of all pteridophytes from 1961 on.)

Index Kewensis. (Print and CD-ROM versions. All flowering plant names at rank of family and below from 1971 on; all binomial and generic names from Linnaeus [1753] on. Useful guide to tracing first publication of a name.)

Crosby, M. R. 1997. *Index of Mosses: A Catalog of the Names and Citations for New Taxa, Combinations, and Names of Mosses Published During the Years 1993 Through 1995, Inclusive, with Citation of Previously Published Basionyms and Replaced Names Together with a Bibliography of the Publications in Which These Nova Appeared.* St. Louis, Missouri Botanical Garden.

Index Muscorum. 1959–. Updates published in journal *Taxon.* (Names of mosses with basionyms [the name on which a new combination is based], synonyms, and places of publication.)

Index of Fungi. 1940–. Kew, U.K., CAB International Mycological Institute.

Bonner, C. E. B. 1962–78. *Index Hepaticarum,* Parts 1–9. Weinheim, J. Cramer. (For Liverworts, gives names at all ranks, with places of publication, typification, and geographic distributions.)

Brummitt, R. K. 1992. *Vascular Plant Families and Genera.* Kew, U.K., Royal Botanic Gardens. (Alphabetically lists accepted generic names and selected synonyms, with families. Also lists accepted families alphabetically, with included genera. Lists several classification schemes.)

Crosby, M. R., et al. 1992. *Index of Mosses: A Catalog of the Names and Citations for New Taxa, Combinations, and Names for Mosses Published During the Years 1963 Through 1989 with Citations of Previously Published Basionyms and Replaced*

Names Together with Lists of the Names of Authors of the Names and Lists of Names of Publications Used in the Citations. St. Louis, Missouri Botanical Garden.

Wielgorskaya, T. 1995. *Dictionary of Generic Names of Seed Plants.* New York, Columbia University Press. (Most recent compilation for seed plant genera. Gives detailed geographic distribution, updated lists of species within genera, and corrected names and dates of publications.)

For fossil plants see various volumes of the *Index of Generic Names of Fossil Plants:*

Andrews, H. N. 1970. *Index of Generic Names of Fossil Plants, 1820–1965.* Washington, D.C., U.S. Geological Survey.

Blazer, A. M. 1975 *Index of Generic Names of Fossil Plants, 1966–1973.* Washington, D.C., U.S. Geological Survey.

Watt, A. D. 1982. *Index of Generic Names of Fossil Plants, 1974–1978.* Washington, D.C., U.S. Geological Survey.

Chapter 19

Keys

The use of keys in identification is old indeed. Much of Aristotle's classification of animals was presented in the form of simple dichotomous alternatives. –MAYR (1969:276)

The purpose of a key is to enable identification, and it should not be a vehicle for expressing systematic opinions. First, good systematic characters are very often poor or even unusable key characters, and second, classifications are all too frequently subject to modification. –QUICKE (1993:99)

Keys in Taxonomy

From the user's point of view, one of the most important functions of taxonomy is to provide a fast and accurate way to identify organisms (Quicke 1993). A taxonomic key, sometimes called a *diagnostic key, determinator,* or *artificial key,* is basically a series of questions about the characteristics of an unknown organism. By working through it and answering each question in order, the user is led to a correct identification of the unknown organism.

As Mayr pointed out in the quotation at the beginning of this chapter the use of dichotomous choices in identifying taxa goes back to classical times. However, biological keys as we know them were first used in the seventeenth and eighteenth centuries and became increasingly popular during the nineteenth and twentieth (Voss 1952, Pankhurst 1978). They are still an important means of identifying organisms and are found in a variety and level of sophistication far beyond that of the simple dichotomous keys you may have learned to construct in beginning biology. They can be grouped into two general categories: *single-access* or sequential *keys* and *multiaccess keys.*

Taxonomic keys are used in biology for two purposes: recognition of major divisions within a groups and detailed identification of taxa to genus or species name. General keys, such as *How to Know the Living Things* (Winchester and Jacques 1981) and *How to Know the Beetles* (Arnett et al. 1980), help a beginner identify members of the chief taxonomic subdivisions of a group of organisms. Figure 19.1 gives an example of a such a key. They are usually heavily illustrated and designed to work even if the user has no previous experience with the systematics of the group. They are the ones you may have used at the beginning of your search for an unfamiliar species (chapter 4). Keys seem most extensively developed in plant taxonomy and entomology, where groups of related taxa can be very large, and field guides often include keys to orders, families, and genera.

Those who work on groups of organisms for which taxonomic knowledge is less complete (such as most groups of marine invertebrates) have fewer keys to turn to and must remain skeptical of those that do exist. When the geographically closest monograph or paper you can use to attempt an identification of an unknown is for an area hundreds or even thousands of miles from the locality where the species you are trying identify lived, you soon realize that trying to use a key from one of those sources will at best help to identify your unknown as "none of the above." At worst, it will lead you to a misidentification if you try to force your unknown species to fit one of the available possibilities. In such cases keys are not a good means of identification. You will more likely be able to identify a species by using a matching strategy, that is, by comparing your unknown with figured specimens and written descriptions in published literature (chapter 4). That way, you are still attempting to match as many of the characteristics as possible, but without the pressure of trying to come out with one definitive answer. For instance, from descriptions in a paper from the nearest available geographic region, you might be able to place your undetermined species in a genus. However, slight differences in morphology might indicate that it does not belong to any of the species named in that paper. Then your task becomes tracing and, if necessary, examining all described species of that genus, continuing the matching process until you either confirm your unknown as belonging to one of them or decide that it must be undescribed.

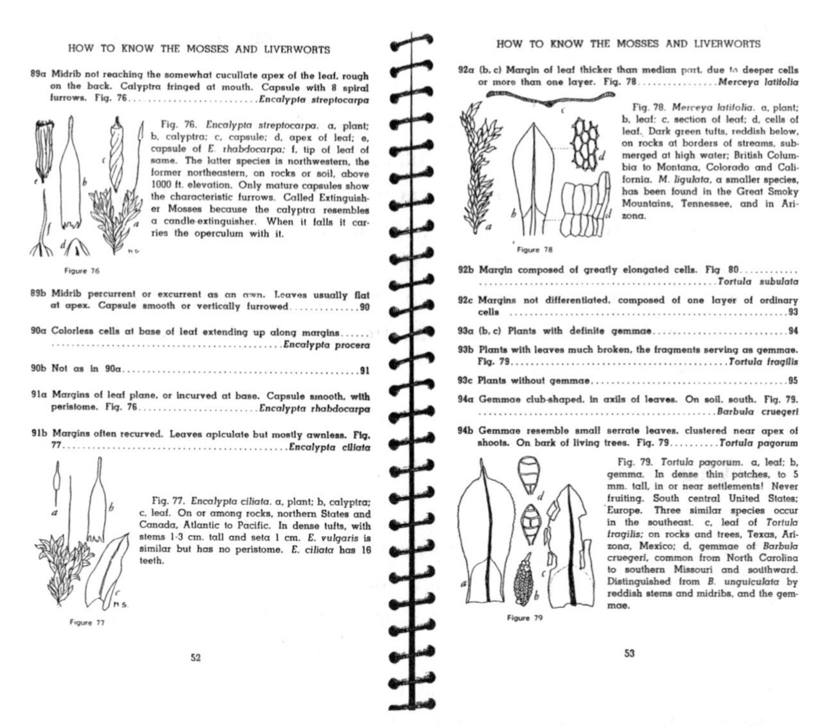

HOW TO KNOW THE MOSSES AND LIVERWORTS

89a Midrib not reaching the somewhat cucullate apex of the leaf, rough on the back. Calyptra fringed at mouth. Capsule with 8 spiral furrows. Fig. 76 .... *Encalypta streptocarpa*

Figure 76

Fig. 76. *Encalypta streptocarpa*. a, plant; b, calyptra; c, capsule; d, apex of leaf; e, capsule of *E. rhabdocarpa;* f, tip of leaf of same. The latter species is northwestern, the former northeastern, on rocks or soil, above 1000 ft. elevation. Only mature capsules show the characteristic furrows. Called Extinguisher Mosses because the calyptra resembles a candle-extinguisher. When it falls it carries the operculum with it.

89b Midrib percurrent or excurrent as an awn. Leaves usually flat at apex. Capsule smooth or vertically furrowed 90

90a Colorless cells at base of leaf extending up along margins .... *Encalypta procera*

90b Not as in 90a 91

91a Margins of leaf plane, or incurved at base. Capsule smooth, with peristome. Fig. 76 .... *Encalypta rhabdocarpa*

91b Margins often recurved. Leaves apiculate but mostly awnless. Fig. 77 .... *Encalypta ciliata*

Figure 77

Fig. 77. *Encalypta ciliata*. a, plant; b, calyptra; c, leaf. On or among rocks, northern States and Canada, Atlantic to Pacific. In dense tufts, with stems 1-3 cm. tall and seta 1 cm. *E. vulgaris* is similar but has no peristome. *E. ciliata* has 16 teeth.

52

HOW TO KNOW THE MOSSES AND LIVERWORTS

92a (b, c) Margin of leaf thicker than median part, due to deeper cells or more than one layer. Fig. 78 .... *Merceya latifolia*

Figure 78

Fig. 78. *Merceya latifolia*. a, plant; b, leaf; c, section of leaf; d, cells of leaf. Dark green tufts, reddish below, on rocks at borders of streams, submerged at high water; British Columbia to Montana, Colorado and California. *M. ligulata*, a smaller species, has been found in the Great Smoky Mountains, Tennessee, and in Arizona.

92b Margin composed of greatly elongated cells. Fig 80 .... *Tortula subulata*

92c Margins not differentiated, composed of one layer of ordinary cells 93

93a (b, c) Plants with definite gemmae 94

93b Plants with leaves much broken, the fragments serving as gemmae. Fig. 79 .... *Tortula fragilis*

93c Plants without gemmae 95

94a Gemmae club-shaped, in axils of leaves. On soil, south. Fig. 79. .... *Barbula cruegeri*

94b Gemmae resemble small serrate leaves, clustered near apex of shoots. On bark of living trees. Fig. 79 .... *Tortula pagorum*

Figure 79

Fig. 79. *Tortula pagorum*. a, leaf; b, gemma. In dense thin patches, to 5 mm. tall, in or near settlements! Never fruiting. South central United States; Europe. Three similar species occur in the southeast. c, leaf of *Tortula fragilis;* on rocks and trees, Texas, Arizona, Mexico; d, gemmae of *Barbula cruegeri*, common from North Carolina to southern Missouri and southward. Distinguished from *B. unguiculata* by reddish stems and midribs, and the gemmae.

53

Figure 19.1 Portion of a key to mosses and liverworts. This type of general key is designed to help identify members of the chief taxonomic subdivisions of a group of organisms, even when the user has had no previous experience with taxonomy of the group. *–Source: From Conard (1956), How to Know the Mosses and Liverworts* in the Pictured-Key Nature Series.

This example points out two of the pitfalls in using keys:

They do not work if the species to which the unknown belongs is not included in the key.

They may lead to misidentification if an unknown shares some of the attributes of species which are included.

Keys also fail if

The specimen is a hybrid between two species (zoologists may not have to face this problem too often, but it is very common in some plant groups).

The specimen is fragmentary (as is often the case in paleontology) or lacks the characters needed (e.g., flowers, fruit) needed to follow the key.

The specimen falls on the borderline of a character supposedly separating two species.

But under the right circumstances, keys can work perfectly and be great time savers. Even long keys (up to 200 couplets) rarely require the user to make more than a dozen choices to get to any given species. In the marine example given earlier, for example, if you used a matching strategy to figure out the genus and then were able to find a monographic treatment with a key that covered all the known species, you could rely on that key to give you either a positive identification or reasonable assurance that you had a new species to describe.

Key Characters

You probably have already used keys, either as a student or in the identification search for your potential new species. Even if you never need to write one, as a user of keys you should understand the kinds of characters used in their construction. To begin with, a key character in taxonomy has nothing to do with a key character in ecology. In ecology the term *key character* has been used in the sense of an evolutionary innovation, a character that has allowed a group of organisms to make use of a previously unavailable niche (Van Valen 1971; Wiley 1981). A key character in taxonomy is one that provides some clearly and conveniently observable distinguishing feature that helps identify of an organism.

Key character choice should be strictly practical. The characters used in keys should be the most clear-cut and distinctive diagnostic characters. They do not need to be the most phylogenetically significant characters. In fact, they can be quite superficial, which is why taxonomic keys are sometimes called artificial keys (Stace 1989), as long as they lead to the desired result: a correct identification in as few steps as possible.

One of the best ways to get an argument going among taxonomists is to start discussing what characters are useful in keys. A character that seems useful to the person making up a key may be obscure to a reader who is not a taxonomist, or who just happens to have found the wrong sex of the organism, or has a specimen collected at the wrong time of year. But there are some reasonable guidelines, the foremost being "whatever works."

Taxonomists (e.g., Benson 1962; Mayr 1969; Quicke 1993) agree that the best key characters

Apply to all members of the taxon (not just to males, females, juveniles, adults, etc.)

Are qualitative and absolute

Are observable without special equipment, dissection, or histological preparation

Are fairly constant, don't show a lot of individual or temporal variation

Are fairly indestructible, that is, not easily broken or lost (in life, preparation, or storage)

It is not always possible to use such ideal characters. Obviously, microscopic organisms must be keyed out under a microscope, for example. But good keys rely on the most distinctive and convenient characters that can be found. Continuously variable characters (e.g., measurements of size of body parts) are difficult to handle in keys. However, they give valuable identification for many groups and can't always be eliminated. For example, "4–6 cm in length" versus "8–15 cm in length" is much more useful than "shorter" versus "taller" (especially to a user who has no idea what short or tall means for that group of organisms). Mayr (1969) recommends that differences in quantitative key characters should be nonoverlapping, but this can't always be done either. Hall (1970) recommends that differences involving continuous characters should be greater than 0.1 of the total range of the character. But it is still wise to remember Murphy's Law of Keys (courtesy R.L. Hoffman), "If a couplet contrasts 'length 8–11 mm' versus 'length 13–15 mm,' you will arrive at the couplet with a 12 mm long specimen."

Single-Access (Analytical or Sequential) Keys

Almost all single-access keys are dichotomous keys. *Dichotomous* means dividing into pairs. *Dichotomous keys* consist of pairs (couplets) of opposing statements (leads). Each statement leads to another pair of choices, sequentially narrowing the user's search down until a final statement identifies the organism in question (usually by its scientific name). Dichotomous keys may be designed in two styles: *bracketed* (open) or *indented* (closed). In most keys the couplets are numbered to help the user follow the sequence.

An example of a bracketed key is given in figure 19.2. In a bracketed key the two leads of the couplet are set together and each lead indicates

Key to Amphibians and Reptiles

A dichotomous key is a mechanical way of identifying an individual amphibian or reptile in hand. It consists of a series of couplets or steps. Start with number 1 and determine which of the two choices contains the description that best fits your specimen. Look to the end of the couplet to see whether a species or another number is indicated. If the latter, then proceed to that step and repeat the procedure until you come to a species name.

Selected body features used in the following couplets are illustrated in Figures 3 and 4 for frogs, Figure 5 for turtles, and Figure 6 for snakes.

1a. Skin smooth or bumpy, without scales or hard shell 2
1b. Skin with scales or hard shell, may be smooth or rough 8

2a. Body elongated, long tail present, legs very short . *Plethodon cinereus*
2b. Body short, without tail, legs as long or longer than body . . . . Frogs and toads: 3

3a. Skin with bumps, light middorsal stripe, dark patches on dorsum containing several bumps, legs shorter than body . . *Bufo woodhousii*
3b. Skin smooth and without distinct light middorsal stripe, legs longer than body 4

4a. Body uniform green with golden flecks or body with three dark longitudinal stripes, toe discs present 5
4b. Body color green to brown without golden flecks or longitudinal stripes, toe discs absent 6

5a. Tips of toes distinctly expanded (discs large), body uniform green with golden flecks, no dark triangular patch on head . *Hyla cinerea*
5b. Tips of toes slightly expanded (discs small), body gray to brown with three dark longitudinal stripes, dark triangular patch on head . . . . *Pseudacris triseriata*

6a. Dorsolateral ridge absent on trunk . . . . *Rana catesbeiana*
6b. Dorsolateral ridge present on trunk 7

16 *Amphibians and Reptiles*

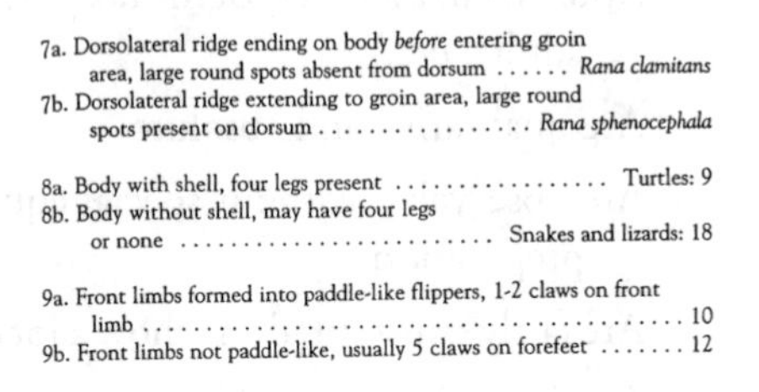

7a. Dorsolateral ridge ending on body *before* entering groin area, large round spots absent from dorsum . . . . . . *Rana clamitans*
7b. Dorsolateral ridge extending to groin area, large round spots present on dorsum . . . . *Rana sphenocephala*

8a. Body with shell, four legs present . . . . Turtles: 9
8b. Body without shell, may have four legs or none . . . . Snakes and lizards: 18

9a. Front limbs formed into paddle-like flippers, 1-2 claws on front limb 10
9b. Front limbs not paddle-like, usually 5 claws on forefeet 12

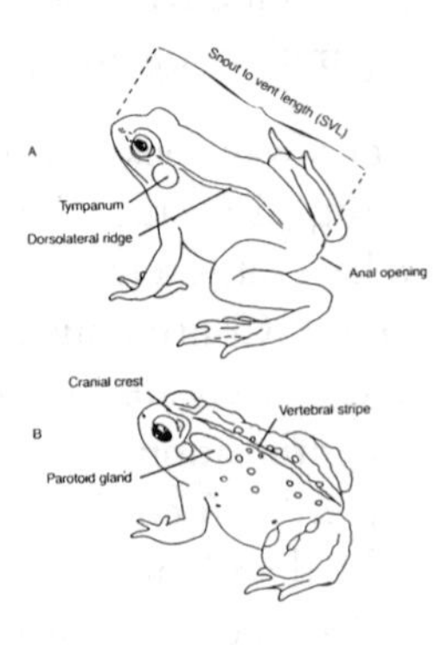

Figure 3. A generalized frog (A) and toad (B) showing selected morphological features of anurans.

Key to Species 17

Figure 19.2 A bracketed key. The two leads of the couplet are set together and each lead indicates what couplet the user should go to next. . *–Source: From Mitchell and Anderson (1994)*

what couplet the user should go to next. Bracketed keys are probably the easiest for a beginner to follow.

An example of an indented key is shown in figure 19.3. In an indented key the possibilities that follow from answering *yes* to the first lead follow that lead before the second lead is dealt with at all. An indented style has advantages in a short key because, like an outline, it shows the logical pattern of the key clearly and it tends to cluster related forms together. But in a long indented key, answering *no* to the first lead of the beginning couplet can send you searching through several pages to find the right spot to continue with the second lead, and you may more easily get lost or discouraged in that search (Quicke 1993). A long indented key has another disadvantage: The indents keep moving further toward the right margin, leaving very little space for describing the final choices.

Because many nonspecialist users find it difficult to get through these

20 ANALYTICAL KEY.

Ovary compound, 1-celled, with two or more parietal placentæ.
Calyx caducous; juice milky or colored. . . . PAPAVERACEÆ, 57
Calyx deciduous, of 4 sepals. . . . . CAPPARIDACEÆ, 74
Calyx persistent, of 3 or 5 sepals. . . . . CISTACEÆ, 76
Ovary compound, several-celled.
Calyx valvate in the bud, and
Persistent; stamens monadelphous; anthers 1-celled. MALVACEÆ, 96
Deciduous; anthers 2-celled . . . . TILIACEÆ, 101
Calyx imbricated in the bud, persistent.
Shrubs; stamens on the base of the petals. TERNSTRŒMIACEÆ, 95
Aquatic or marsh herbs; ovaries many,
On 5 placentæ in the axis. . . . SARRACENIACEÆ, 57
On the 8 – 30 partitions. . . . . NYMPHÆACEÆ, 54

2. *Calyx more or less coherent with the surface of the (compound) ovary.*

Ovary 8 – 30-celled; ovules many, on the partitions; aquatic. NYMPHÆACEÆ, 54
Ovary 10-celled; cells 1-ovuled. . . . Amelanchier, in ROSACEÆ, 166
Ovary 2 – 5-celled.
Leaves alternate, with stipules. . . . Pomeæ, in ROSACEÆ, 151
Leaves opposite, without stipules. . . Some SAXIFRAGACEÆ, 168
Leaves alternate, without stipules. . . . . STYRACACEÆ, 333
Ovary 1-celled, with the ovules parietal.
Fleshy plants with no true foliage; petals many. . CACTACEÆ, 186
Rough-leaved plants; petals 5 or 10. . . . . LOASACEÆ, 193
Ovary one-celled, with the ovules rising from the base. PORTULACACEÆ, 90

B. *Stamens of the same number as the petals and opposite them.*

Pistils 3 – 6, separate; flowers diœcious; woody vines. MENISPERMACEÆ, 51
Pistil only one.
Ovary one-celled; anthers opening by uplifted valves. BERBERIDACEÆ, 52
Ovary one-celled; anthers not opening by uplifted valves.
Style and stigma one; ovules more than one. . PRIMULACEÆ, 328
Style 1; stigmas 3; sepals 2; ovules several. PORTULACACEÆ, 90
Style twice or thrice forked; flowers monœcious.
Crotonopsis, in EUPHORBIACEÆ, 458
Styles 5; ovule and seed only one. . . . PLUMBAGINACEÆ, 327
Ovary 2 – 4-celled.
Calyx-lobes minute or obsolete; petals valvate. . . VITACEÆ, 112
Calyx 4 – 5-cleft, valvate in the bud; petals involute. RHAMNACEÆ, 111

C. *Stamens not more than twice as many as the petals, when of just the number of the petals then alternate with them.*

1. *Calyx free from the ovary, i. e. the ovary wholly superior.*

* *Ovaries 2 or more, separate.*

Stamens united with each other and with a large and thick stigma common to the two ovaries. . . ASCLEPIADACEÆ, 338
Stamens unconnected, on the receptacle, free from the calyx.
Leaves punctate with pellucid dots. . . . . . RUTACEÆ, 106

ANALYTICAL KEY. 21

Leaves not pellucid-punctate.
Tree, with pinnate leaves. . . Ailanthus, in SIMARUBACEÆ, 107
Low shrub, with pinnate leaves. Xanthorrhiza, in RANUNCULACEÆ, 48
Herbs, not fleshy. . . . . RANUNCULACEÆ, 34
Herbs, with thick fleshy leaves. . . . . CRASSULACEÆ, 176
Stamens unconnected, inserted on the calyx.
Just twice as many as the pistils (flower symmetrical). CRASSULACEÆ, 176
Not just the number or twice the number of the pistils.
Leaves without stipules. . . . . . SAXIFRAGACEÆ, 168
Leaves with stipules. . . . . ROSACEÆ, 150

* * *Ovaries 2 – 5, somewhat united at the base, separate above.*

Leaves punctate with pellucid dots. . . . . . RUTACEÆ, 106
Leaves not pellucid-punctate.
Shrubs or trees with opposite leaves. . . . SAPINDACEÆ, 115
Terrestrial herbs; the carpels fewer than the petals. SAXIFRAGACEÆ, 168

* * * *Ovaries or lobes of ovary 3 to 5, with a common style.* GERANIACEÆ, 102

* * * * *Ovary only one, and*

+ *Simple, with one parietal placenta.* LEGUMINOSÆ, 122

+ + *Compound, as shown by the number of cells, placentæ, styles, or stigmas.*

Ovary one-celled.
Corolla irregular; petals 4; stamens 6. . . . FUMARIACEÆ, 59
Corolla irregular; petals and stamens 5. . . . . VIOLACEÆ, 78
Corolla regular or nearly so.
Ovule solitary; shrubs or trees; stigmas 3. ANACARDIACEÆ, 118
Ovules solitary or few; herbs. . . Some anomalous CRUCIFERÆ, 61
Ovules more than one, in the centre or bottom of the cell.
Petals not inserted on the calyx. . . . CARYOPHYLLACEÆ, 82
Petals on the throat of a bell-shaped or tubular calyx. LYTHRACEÆ, 184
Ovules several or many, on two or more parietal placentæ.
Leaves punctate with pellucid and dark dots. HYPERICACEÆ, 92
Leaves beset with reddish gland-tipped bristles. DROSERACEÆ, 178
Leaves neither punctate nor bristly-glandular.
Sepals 5, very unequal or only 3. . . . CISTACEÆ, 76
Sepals and petals 4; stamens 6. . . Anomalous CRUCIFERÆ, 61
Sepals and petals 5; stamens 5 or 10.
Ovary and stamens raised on a stalk. PASSIFLORACEÆ, 194
Ovary sessile. . . . . . . SAXIFRAGACEÆ, 168
Ovary 2 – several-celled.
Flowers irregular.
Anthers opening at the top,
Six or eight and 1-celled; ovary 2-celled, 2-ovuled. POLYGALACEÆ, 120
Ten and 2-celled; ovary 5-celled. Rhododendron, in ERICACEÆ, 286
Anthers opening lengthwise.
Stamens 12 and petals 6 on the throat of a tubular inflated or gibbous calyx. . . . Cuphea, in LYTHRACEÆ, 186

Figure 19.3 An indented key. The possibilities that follow from answering *yes* to the first lead follow that lead before the second lead is dealt with at all. *–Source: From Gray (1889)*

traditional keys successfully, taxonomists have tried to come up with alternative layouts to make the user's task easier.

Pictorial keys (figure 19.4) are designed especially for field identifications by nonspecialists or even nonbiologists. They may illustrate the diagnostic characters or even the entire organism. Examples can be found in field guides to birds, mammals, wildflowers, seashore invertebrates, and other organisms. Even the division of wildflowers into groups by color, as found in some field guides, is a pictorial key of a sort.

Branching keys (figure 19.5) make it easy to see the choices, but users must keep in mind that the branching patterns they show are not really phylogenetic (even though they resemble the cladograms generated in phylogenetic analyses). The branch choices in branching keys may be based on superficial characters and are intended only to lead you quickly

ILLUSTRATED KEY TO SEASHORE ANIMALS

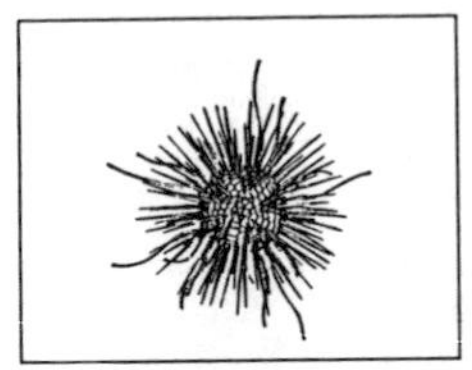

Fig. 58
Sea Urchins 3″ (7.6 cm)
Echinodermata (p. 69)

Fig. 59
Sand Dollars 3–5″ (7.6–13 cm)
Echinodermata (p. 69)

SNAIL-LIKE

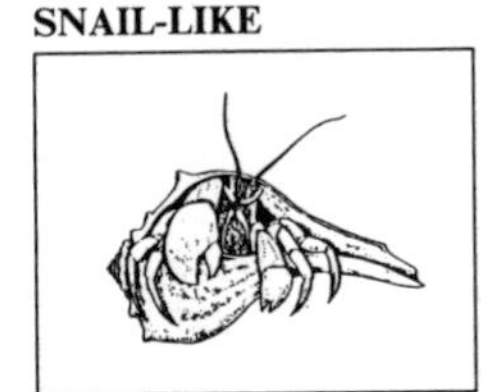

Fig. 60
Hermit Crabs 1–7″ (2.5–18 cm)
Arthropoda (p. 223)

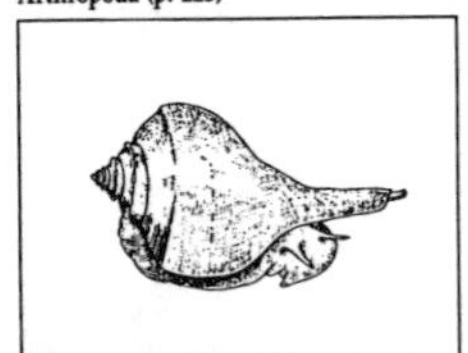

Fig. 61
Snails 0.5–24″ (1.3–61 cm)
Mollusca (p. 92)

Fig. 62
Snails 0.1–4″ (0.3–10 cm)
Mollusca (p. 92)

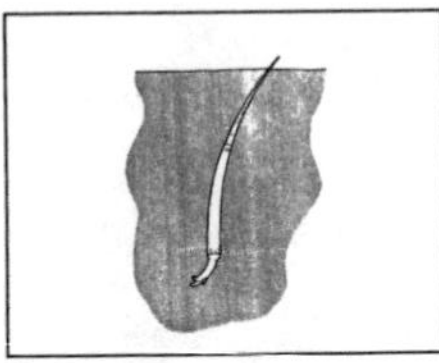

Fig. 63
Tusks 2″ (5 cm)
Mollusca (p. 92)

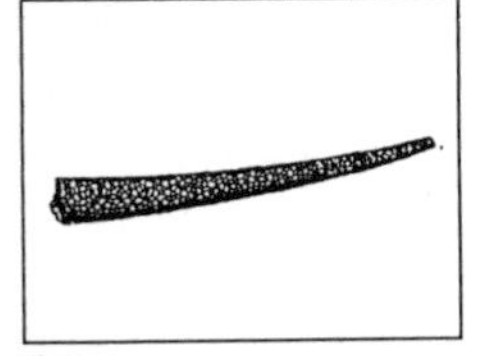

Fig. 64
Ice Cream Cone Worms 2″ (5 cm)
Annelida (p. 185)

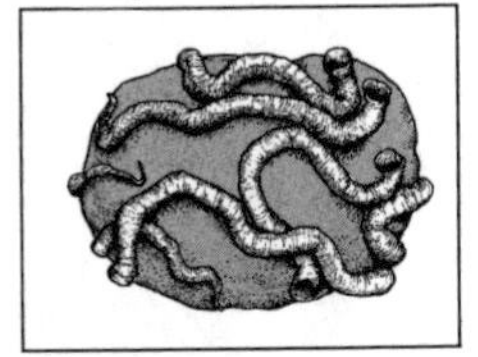

Fig. 65
Segmented Worms 2″ (5 cm)
Annelida (p. 185)

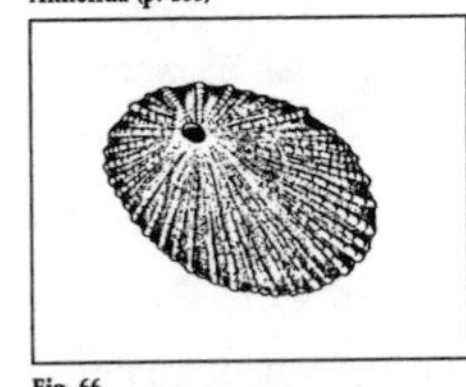

Fig. 66
Keyhole Limpets 1″ (2.5 cm)
Mollusca (p. 92)

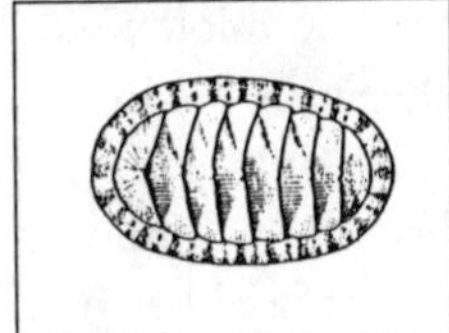

Fig. 67
Chitons 1″ (2.5 cm)
Mollusca (p. 92)

CLAMLIKE

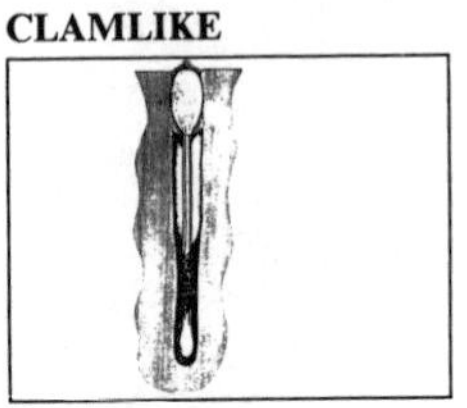

Fig. 68
Lamp Shells 4″ (10 cm)
Brachiopoda (p. 81)

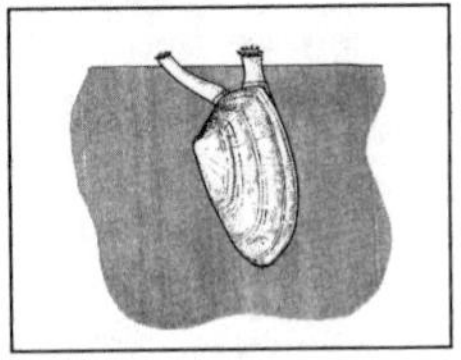

Fig. 69
Clams 0.5–5″ (1.3–13 cm)
Mollusca (p. 92)

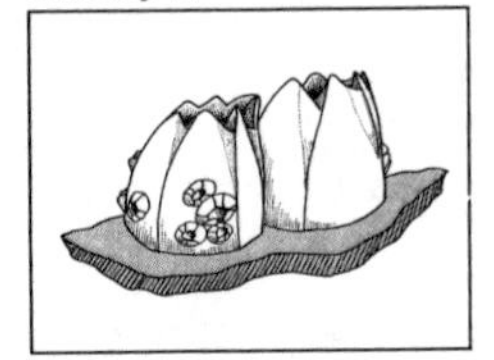

Fig. 70
Barnacles 0.4–1.6″ (1–4 cm)
Arthropoda (p. 223)

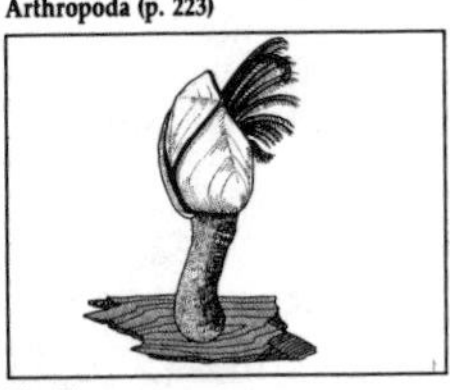

Fig. 71
Goose Barnacles 1.5″ (4 cm)
Arthropoda (p. 223)

SPIDERLIKE

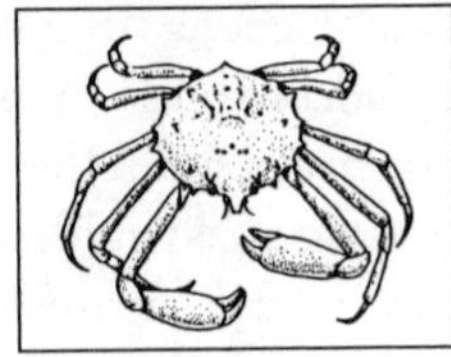

Fig. 72
Spider Crabs 4″ (10 cm)
Arthropoda (p. 223)

Figure 19.4 (opposite) A pictorial key. Pictorial keys are designed especially for field identifications by nonspecialists or even nonbiologists. *–Source: E. Ruppert and R. Fox (1988), Seashore Animals of the Southeast,* p. 13.

to an identification. They are not necessarily based on synapomorphies, the shared derived characters on which phylogenetic branching diagrams are based (chapter 22).

Multiaccess Keys (Polyclaves)

Unlike the inflexible single-access keys, in which you must begin the identification process with the first entry and continue in sequence until the end (and in which progress comes to a halt when a character that can't be determined is reached), multiaccess keys can be entered at any point. There are three kinds of multiaccess key in use: punched card keys, tabular keys, and computerized keys.

Punched card keys are the mechanical precursors of computerized keys. They were invented in the 1930s and today seem to be used mostly by

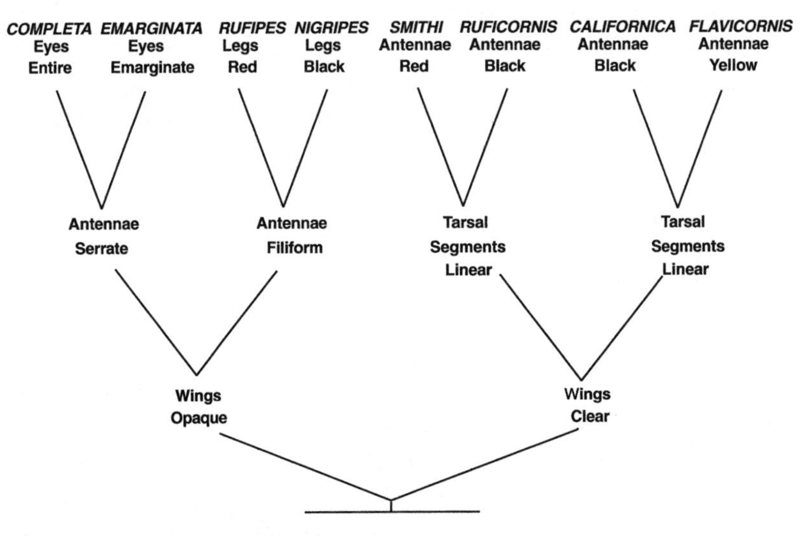

Figure 19.5 Example of a branching key. Unlike those in dendrograms and cladograms, branching points in keys do not necessarily represent relationships of the organisms, but are set up merely to facilitate identification of the unknown. Diagrams could be used instead of words at nodes and terminations. *–Source: Redrawn from figure 13 of Mayr et al. (1953).*

botanists (see Stace 1989). They are operated either by thin rods passed through punched holes or by cards being held to the light to show holes. They can be either edge-punched or center-punched (also called body-punched or peek-a-boo) keys.

In edge-punched systems there is a card for each species or other taxon. Holes in the edge of the card identify the key characters. All the characters *not* displayed by the taxon have their holes clipped open into a U. A thin rod or needle can be passed through the stack of cards to find all the taxa having that particular character (they will be the cards that stay on the rod). The sorting is continued character by character until only one taxon card is left.

The center-punched system sorts in the opposite way. Each card represents a character. All possible taxa the key is meant to determine are printed in a standardized way on each card. The user decides what characteristics the species in question shows, then takes the cards for those characters and stacks them neatly together so that the holes are in alignment. Species that have all those characters will show as holes all the way through the stack. Additional characters (cards) are added until only one hole shows through, representing the species to which the unknown belongs.

In addition to allowing the user to start anywhere, punched card keys have another advantage. If there is an unusual character present in the unknown, most of the possibilities can be eliminated at the first sorting of the cards and identification can be very fast. But these card systems are also expensive to produce and awkward to carry and use in the field.

Of course, punched card keys are a primitive version of the kind of sorting that can most rapidly be done electronically by computer, and in the future we may see most keys in electronic form (Fortuner 1993). But in the meantime, people have tried to come up with printed versions of multiaccess keys called matrix keys. In a matrix key, data matrices (tables) are arranged so that they can be sorted by eye or by using some kind of masking or ruling device to guide the eye or mark out the characters present in the unknown. They are intended to provide some the advantages of multiaccess keys, but in a form cheap and portable enough to use in the field. For example, Westfall and colleagues (1986) devised a set of keys to identify plants of South Africa's Transvaal Waterberg that could be used on site and were based on vegetative characters (because it would have been impossible for the biologists carrying out the census to find all species in flower at any one time). Their keys (part of one is shown in figure 19.6) consist of printed data matrices with the characters in rows and the

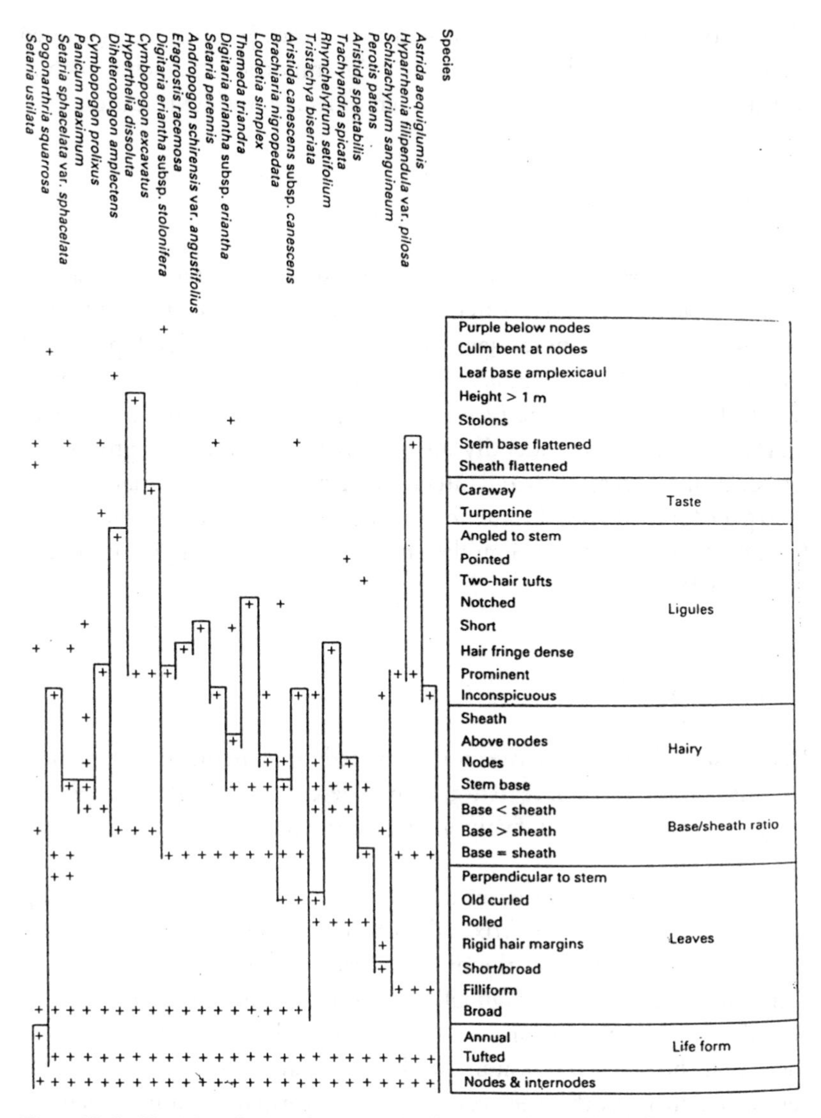

Figure 19.6 Matrix key. Portion of a key to identify plants of South Africa's Transvaal waterberg constructed so it could be used on site and based on vegetative characters (because it would have been impossible for the biologists carrying out the census to find all species in flower at any one time). Key is a table or matrix, having characters on one axis and taxa on the other.

taxa in columns. They used a computer program called PHYTOTAB (Westfall et al. 1982) to order each matrix in a way that would make identification easy. The minimal number of characters for identification of each taxon were grouped to the right in the table and corroborating characters (ones that were variable or just not always present) were grouped to the left. A species that was variable in an important character (such as leaf shape) could appear twice to prevent possible misidentification.

Interactive Identification

Even as cumbersome sets of cards or eye-straining tables, multiaccess keys have advantages over single-access keys in terms of flexibility and ease of use. The computer age, particularly the recent development of affordable CD-ROM technology, is finally bringing them into their own. CD-ROM guidebooks are beginning to appear that enable the user to quickly and accurately identify an unknown. A number of groups are working on expert systems and interactive identification. For example, the Web page for the Expert Center for Taxonomic Identification Biodiversity Center listed 44 CD-ROM guides either completed or in preparation as of March 1998.

Key Construction

Key construction is one of the more advanced topics in practical systematics. It is not one you'll have to deal with in describing a single new species, but you may encounter it if your interest leads you to undertake a revision of a genus or family.

The first step in constructing a key for the taxa you have been working with is to set up a data matrix (table) based on key characters. On one axis you list taxa, and on the other characters, then fill in the blanks. This will mean going through all the taxonomic characters you used in your descriptions and picking out the best, most diagnostic key characters (note that the characters you use in the data matrix for your key may be very different from the ones you used if you carried out a phylogenetic analysis). If you are going to create a printed tabular key, you will want to order your matrix so that the users' eyes will be led to the correct identification. For large numbers of taxa this is best done by computer. For small numbers, it could easily be done using the cut-and-paste features of your word processor, or by hand.

The next step is to attempt to make the wording of the key as unambiguous as possible. For example, the same noun should be used to begin each pair of opposed couplets, as in "Flowers blue . . ." "Flowers pink . . ." (Benson 1962). Quicke (1993) also recommends that the person constructing the key try to find character differences that won't be too subtle for the less experienced user and use illustrations and diagrams of the alternatives to make terminology and morphological structure clear to the user.

If you choose to construct a dichotomous key, Pankhurst (1978) recommends beginning with an easily observable character that will divide the species listed into two nearly equal groups. This will make the shortest key (in numbers of steps) and thereby give the user the fewest chances to make mistakes. He also recommends putting the shorter lead (in terms of number of steps following it) first so that the eye doesn't have to travel as far down the page.

If you have one or more unusual species, species that are really distinct from the rest, you can choose to take those first if that seems most convenient.

Often a few of the species for which the key is being constructed will be common species and the rest will be rare. In such cases you may want to set up the key so that the path to those species is as short as possible. This approach is more difficult to carry out than separating unusual species first because the common species may not have any especially distinctive characters, but it can help your users considerably (Pankhurst 1978).

A final step (often left out in key construction) is testing. Keys should be tried out before publication, preferably by someone other than the author, using additional specimens (Pankhurst 1978).

Computerized Key Construction

In the last 20 years a number of key construction programs have been designed. One of the most popular is that devised by Dallwitz (1980) and colleagues as part of their DELTA (Descriptive Language for Taxonomy system). Their KEY program can be used to generate conventional keys, which can be weighted for ease of use and character stability. They have also created a program called INTKEY for interactive identification. Information on DELTA is available on the Internet (see this chapter's *Sources*). Programs and instructions can be downloaded for use. PANKEY, developed by Richard Pankhurst, is another program using DELTA. A number of other packages have been developed for computerized identification of different groups of plants and animals: CABIKEY, Flora, KEY, ONLINE, ONLINEKEY, XPER (see Edwards and Morse 1995), XID, and LucID.

Computerized key generation has its own problems. These have to do with coding the data and the way the program orders and operates on the data. Each program must be studied and interpreted for use, but this shouldn't deter anyone who wants to explore use of such programs.

Sources

Printed Sources

Articles and Books

Askevold, I. S. and C. W. O'Brien. 1994. DELTA, an invaluable computer program for generation of taxonomic monographs. *Annals of the Entomological Society of America* 87: 1–16.

Edwards, M. and D. R. Morse. 1995. The potential for computer-aided identification in biodiversity research. *Trends in Ecology and Evolution* 10: 152–158.

Fortuner, R., ed. 1993. *Advances in Computer Methods for Systematic Biology: Artificial Intelligence, Databases, Computer Vision.* Baltimore, The Johns Hopkins University Press.

Hall, A. V. 1970. A computer-based system for forming identification keys. *Taxon* 19: 12–18.

Leenhouts, P. W. 1966. Keys in biology: A survey and a proposal of a new kind. *Proceedings of the Koninklijke Nederlandsche Akademie van Wetenshappe Amsterdam* ser. C, 69: 571–576.

Metcalf, Z. P. 1954. The construction of keys. *Systematic Zoology* 3: 38–45.

Pankhurst, R. J. 1978. *Biological Identification.* London, Edward Arnold.

Pankhurst, R. J. 1991. *Practical Taxonomic Computing.* Cambridge, U.K., Cambridge University Press.

Tilling, S. M. 1984. Keys in biological identification: Their role and construction. *Journal of Biological Education* 18: 293–304.

Voss, E. G. 1952. The history of keys and phylogenetic trees in systematic biology. *Journal of the Science Laboratories, Denison University* 43: 1–25.

Westfall, R. H., H. F. Glen, and M. D. Panagos. 1986. A new identification aid combining features of a polyclave and an analytical key. *Botanical Journal of the Linnaean Society* 92: 65–73.

Wright, J. F., D. R. Morse, and G. M. Tardivel. 1995. An investigation into the use of hypertext as a user interface to taxonomic keys. *Computer Applications in the Biosciences* 11: 19–27.

Internet Sources

All addresses begin with **http://www.**

biodiversity.uno.edu/delta/
The DELTA System site includes print and online references to applications and documentation of the system.

rbge.org.uk/research/pankey.html
This site provides information on PANKEY.

Chapter 20

Description of Higher Taxa

To rise beyond the generic level in classification is to enter a world of much greater uncertainty. . . . Taxa at higher levels will be well-defined or ill-defined depending on the group in question. —STUESSY (1990:207)

The essential point . . . is to preserve a sense of proportion, for in dealing with higher categories, judgement must be exercised, and conservatism should be the rule. —MAYR ET AL. (1953:52)

The family, at least in the flowering plants, is widely regarded as having a greater correspondance with "reality" than any other supraspecific rank and also as an evolutionary phenomenon. They are often regarded as peaks or modes of evolutionary diversity and the so-called natural families have been known and recognized by man long before scientific taxonomy was invented. Yet the family, much more so than in the case of genera, is very much a perceptual category in that it cannot in any meaningful sense be handled, discussed or assessed. —HEYWOOD (1988:49)

This chapter covers family and higher group categories and the rules of nomenclature dealing with them and gives examples of descriptions of higher taxa. Descriptions of new orders, classes, or phyla are extremely rare; a new order is described in fewer than 1 in 1,000 taxonomic publications, a new class in 1 in 4,000, and a new phylum in about 1 in 10,000 (figure 6.1) (Winston and Metzger 1998). Changes to existing groups at these levels are also rare and are unlikely to be published, much less accepted by the biological community, unless they are done in conjunction with a thorough phylogenetic analysis, generally using many kinds of biological evidence. On the other hand, although descriptions of new families are not common (a new family is described in fewer than one in a hundred papers), it is certainly possible that study of an unknown habitat or region, or a new paleontological site, could result in your finding a new species so unusual and so distinct from known relatives that a new family as well as a new genus might be the best way to accommodate its unique combination of characters.

Family Concepts and Their Use in Taxonomy

The family level is the next step above the genus level in formal classification, and the highest level whose description is regulated by the codes of nomenclature. A family may consist of a single genus (with one or more species), but is more commonly made up of a group of closely related (monophyletic) genera. Like the other basic elements of biological classification and nomenclature, the idea of families developed over a long period of time. Some family-level groups of plants with easily recognizable members (e.g., those of the mint family, Labiatae) were recognized in ancient Greece, but the family concept didn't become part of taxonomic practice until after Linnaeus's time (Stuessy 1990).

Linnaeus did not use families as such, although the "natural orders" in the appendix to his sixth edition of *Genera Plantarum* (1764) are equivalent to families, some of them quite similar in composition to modern plant families. Many modern plant families derive from the work of two French botanists whose lives overlapped with that of Linnaeus: M. Adanson, who published *Familles des Plants* (1764), with 58 families of plants, and A.-L. de Jussieu, whose 1789 *Genera Plantarum* divided plants in 100 "natural orders," many of them equivalent to today's plant families. However, the term *family* did not come into general use for those units until after the mid-nineteenth century (Stace 1989; Stuessy 1990).

The familial concept in animals also began to develop at the very end of the eighteenth century, again with French biologists. Latreille (1796) divided insects into families, which he diagnosed but did not name. Duméril (1800) arranged the insects by orders and families, but used French (not Latin or Latinized) family names. Others began using vernacular family names, too, but started to base them on names of genera included in them. A British scientist, W. Kirby (1813), introduced the Latin ending **-idae,** meaning resembling, -like, from the Greek plural **eidos,** a resemblance (Mayr et al. 1953; Savory 1962). Taxonomists soon realized that many of Linnaeus's animal genera could be raised to the family level to organize and accommodate the influx of new species being found, and the family concept spread into general use as the need for an additional means of grouping the rapidly multiplying numbers of new species became urgent (Mayr et al. 1953).

Practical Significance in Biology

Family names have practical significance for several reasons. They can be an efficient way to teach students to recognize different groups of plants and animals. Take wildflowers, for example. *A Field Guide to Wildflowers, Northeastern and North-Central North America* by Roger Tory Peterson and Margaret McKenney (1968) covers 1,293 species. But *wildflower* is an artificial category that includes members of many different groups of flowering plants, from delicate grasses to large shrubs, with flower and leaf combinations that can seem almost infinite. The best way to master their identification is to become familiar with the plant families involved. Plant families are large enough categories to group plants with general similarities, but small enough categories not to have descriptions too general to be useful for identification purposes. Although there are still more than 100 families of wildflowers in North America, about 24 families contain most of the species found (Imes 1990). Not only that, but as Mayr et al. (1953) pointed out, families are commonly worldwide in distribution. Learning those 24 families would be a great help if a student later wanted to go on to identify wildflowers in Europe or Asia.

Because they provide suites of readily recognizable characters, families are the level on which keys are based for many groups of organisms. Therefore, they are the level to which many groups of organisms are sorted and identified in survey work. Because they appear in results and analyses in many faunal surveys and ecological studies, they are important to ecologists and applied biologists. In at least some groups of organisms, families appear to have a biological reality as well, sharing life history traits or occupying a particular niche. This is certainly true for some plant families (as pointed out by Heywood in the quotation at the beginning of this chapter). Ecologists define *guilds* as groups of species, usually but not always taxonomically related, that exploit a common resource (e.g., food) in a similar manner (Putman and Wratten 1984). Ecologists have sometimes tried to link ecological guilds to families, but although the connection has been useful in some cases, it does not hold up for all of them.

Because they are often used in applied biological work, stability of family names is a matter of practical concern (Reveal and Hoogland 1991). In some groups families are quite satisfactory, easily recognizable, monophyletic, and characteristic of a particular niche, or a portion of one. Yet in other groups of organisms, including many groups of invertebrates, families are poorly defined. Even their apparent ecological similarity may result from the fact that they have traditionally been grouped ecologically rather than phylogenetically on the basis of shared derived characters. In such groups of organisms, genera or even species may be the level on which keys and ecological generalizations must be based. For such groups especially, there is still need and scope for revision at the family level as well as opportunities for finding or creating entirely new families.

Describing Families

Family names, like those of other taxa above the genus level, are uninomial, plural Latin (or Latinized) nouns. They are written with the first letter capitalized, but are not italicized. Note, however, that when family names or other higher taxa are turned into English common names by dropping the Latin endings, they are no longer capitalized; for example, the squirrel family Sciuridae becomes sciurid (Council of Biology Editors, 1983).

Zoology (ICZN, Art. 35)

In zoology the family group includes the family, one category above it (the superfamily), and two categories below that (subfamily and tribe) in common use, although the ICZN allows for "any rank below superfamily and above genus that may be desired," such as subtribe. Family group names are formed from the stem of the name, to which the following endings (ICZN 4th draft, Art. 40) are added:

Family = **-idae**

Superfamily = **-oidea**

Subfamily = **-inae**

Tribe = **-ina**

Families are the core of the family group. You can have families without superfamilies, subfamilies, or tribes, but you cannot have superfamilies, and so on without families. In some groups of organisms, such as mammals and reptiles, these extra levels have described and studied extensively. In many other groups of organisms superfamilies and subfamilies have never been described at all.

The family group name is based on a type genus, chosen to represent the family. Of course, you'll have no problem selecting a type genus for a family if you are dealing with a family with only one genus. Otherwise, the type genus must be one of the genera included in the family. It does not have to be the oldest (first described) genus, but, if possible, it should be one that is well known, representative in its taxonomic characters, and common or with a large geographic distribution (Blackwelder 1967). If a family group taxon is subdivided, the subdivision containing the type genus (the nominotypical subdivision) keeps the original name; the other subdivisions must be given new names, based on one of the genera remaining in them.

When the genus name is a Latin or Latinized Greek word or ends in a Greek or Latin suffix, the stem of the family name is found by deleting the case ending from the genitive singular of the genus name, then adding **-idae**. When the name is from some other language, ends in a word from another language, or is made up of an arbitrary combination of letters, the author who establishes the name (you) chooses the stem to be used by adding **-idae** to it.

Family names in zoology are coordinate, so a name established for any taxon in the family group keeps the same author and date if it is raised to superfamily or lowered to subfamily or tribe. Only the suffix must be changed (e.g., from **-idae** to **-oidea**) to indicate its change in rank (ICZN, Art. 36).

Because standardization of family group names developed only gradually, the ICZN must address the problem of old family names that do not meet the criteria of today's nomenclature. Old family names that were based on an included genus are acceptable. Those not based on any particular genus, or based on a genus that is not included in the family, cannot be used. Those that were published with an incorrect Latinized suffix can be given the correct Latin ending by a later worker, but keep the date and authorship of the original description.

Otherwise, the same general rules for publication and priority that apply to the names of new species and genera apply to the description of new family group names. There is one exception (ICZN, Art. 40), a change that was made to attempt to increase stability at the family level. After 1960, if a valid family group name turns out to have been based on a genus name that is discovered to be a junior synonym, that family group name is *not* to be replaced unless the senior generic synonym is also the basis of a family group name or a reclassification also involves other family group names (then the principle of priority applies to all of them).

Another change expected to appear in the fourth version of the ICZN allows a person establishing a new family group name after 1999 to "adopt a stem of the name of the type genus which is not that derived from the genitive of the generic name according to strict grammar, and such a spelling is to be maintained by subsequent authors. For example, an author intending to establish a new family-group name which would be a homonym of an existing name, because the stems of the names of the respective type genera are identical, will be able (and is advised) to avoid homonymy by incorporating the entire name of the type genus into the spelling of the new family-group name" (International Trust for Zoological Nomenclature Web page 1998)

Botany (ICBN, Art. 18,19)

Family categories in botany consists of family, subfamily, tribe, and subtribe. Family names are plural adjectives used as nouns (substantives). All of the endings basically mean "-like" or "resembling" and come from Latin or Greek adjectival suffixes (Savory 1962; Stearn 1983). The name of a plant family (*Familia*) is formed by replacing the genitive singular ending with **-aceae,** (from **-aceus, -a, -um**) = -like:

Rosa, gen. singular **Rosae,** family name **Rosaceae,** "the rose-like ones"

The names of the other family categories are constructed in the same manner, using the appropriate endings: *Subfamilia,* subfamily, uses the ending **-oideae** (from **-oidius, -a, -um**) = indicates resemblance, -like. *Tribus,* tribe, uses the ending **-eae** (from **-eus, -a, -um**) = indicates material, color, or resemblance in quality; pertaining to. *Subtribus,* subtribe, uses the ending **-inae** (from **-inus, -a , -um**) = indicates possession or resemblance, of, belonging to.

The ICBN (Art. 10.6) states, "The type of a name of a family or of any subdivision of a family is the same as that of the generic name on which it is based. For purposes of designation or citation of a type, the generic name alone suffices. The type of a name of a family or subfamily not based on a generic name is the same as that of the corresponding alternative name." The last sentence is necessary because there are some exceptions to the rules. The ICBN makes exceptions for nine families whose names were not based on generic names, but have been traditionally used by botanists. For these families, either the standardized or the traditional name is acceptable (ICBN, 18.5, 18.6). They are the Palmae (standardized name, Arecaceae; type, *Areca* L.), Gramineae (Poaceae; type, *Poa* L.), Cruciferae (Brassicaceae; type, *Brassica* L.), Leguminosae (Fabaceae; type *Faba* Miller), Guttiferae (Clusiaceae; type *Clusia* L.), Umbelliferae (Apiaceae; type, *Apium* L.), Labiatae (Laminaceae; type, *Lamium* L.), Compositae (Asteraceae; type, *Aster* L.), and, when the Papilionaceae (Fabaceae; type, *Faba* Miller) are regarded as a family distinct from the remainder of the Leguminosae, the name Papilionaceae is conserved against Leguminosae.

Some of other family names, ending in **-aceae** and based on an included genus, have also been added to the list of conserved names because, although they turned out to be junior synonyms, they had become so firmly established by tradition that to change them would only cause botanists irritation and inconvenience, as well as information loss. These names are listed in Appendices IIA and IIB of the ICBN.

The biggest difference between dealing with families in botany and zoology is that (as at other levels) in botany priority is restricted at each rank (ICBN, Art. 11.2, 11.3). No name has priority outside the rank of the taxon to which it was applied. If a subfamily is raised to a family, the name does not go with it. Instead, the oldest name for the group as a family must be used (or, if there is none, a new name must be composed).

Family-Level Descriptions: Examples

Animals

A new family is described just like a new genus or a new species, covering the same topics as those descriptions cover: type, diagnosis or description, taxonomic remarks, and discussion of relationships.

If the new family has only a single genus, then that genus automatically become the type of the family. In the following example, Robert Bieri, had compiled a list of the hundred or so species in the phylum Chaetognatha. In the course of that project he discovered that one of the genera had never been assigned to a family. He decided to create a new family to include it:

Krohnittellidae, new family

Diagnosis.–The family lacks both anterior and posterior teeth. A single pair of lateral fins, completely rayed, partly on the trunk and partly on the tail do not reach the ventral ganglion. No ventral transverse musculature.

Discussion.–The family is monogeneric containing the single genus, *Krohnittella* Germain & Joubin, 1912. The authors based their new genus on two specimens taken in a 4000 m net tow that possibly struck bottom in the eastern North Atlantic. Because more than 60 years passed before the genus was reported again, considerable doubt existed about its validity. Tokioka in his 1965 revision retained it as valid, but did not assign it to any known family. Bieri (1974) rediscovered the genus in the eastern North Pacific where a single, nearly mature specimen in good condition was caught with an opening-closing sled trawl at 2000 m in the San Clemente Basin. The lack of ventral transverse musculature and teeth clearly differentiate the family from the Eukroniidae and Spadellidae. . . . (Bieri 1989:973)

In a 1989 paper, John Holsinger established two new families for two groups of cave-dwelling amphipods whose unique character state combinations had been recognized informally by previous authors, but which had not been given family status. One of these had two genera, making it necessary for him to decide upon and name a type genus in his description:

Pseudocrangonyctidae, new family

Type genus.–*Pseudocrangonyx* Akatsuka & Komai, 1922.

Diagnosis.–Typically without eyes (except on one species) and pigment, of stygobiont facies. Body generally smooth, except last seven body segments bearing dorsal setae and uronite 2 bearing few small dorsal spines. Sexually mature females larger than males. Interantennal lobe rounded anteriorly. . . .

Remarks.–At present this family is composed of two northeast Asian genera, *Pseudocrangonyx* and *Procrangonyx*. . . .

Relationship.–Although the Pseudocrangonyctidae are allied with the Holarctic family Crangonyctidae as indicated below, they differ from this group in a number of important characters and warrant recognition as a distinct family: (1) segment 3 of mandibular palp equal in length to segment 2; (2) molar of mandible weakly triturative (or perhaps not triturative in some species); (3) gnathopods and pereopods tending to be more setose, especially segment 2 (basis) of the gnathopods and pereopods 3 and 4. . . .

A relatively close phylogenetic relationship between the families Pseudocrangonyctidae and Crangonyctidae is indicated by similarity of the following characters, most of which are apparently synapomorphies: (1) 2-segmented accessory flagellum; (2) structure of mouthparts, except that mandibular palp segment 3 is proportionately a little longer and the molar is not as strongly developed in the Pseudocrangonyctidae . . . ; (3) similar shape of, and proportionately large, gnathopod propods (in combination with short carpi); (4) palms of gnathopod propods with double row of thick distally notched spine teeth (although possibly variable in Pseudocrangonyctidae); (5) rastellate setae on carpus of one or both gnathopods; (6) median sternal gills; (7) loss of inner ramus of uropod 3 (c.f. *Stygobromus* and *Synurella*); and (8) relatively short telson with shallow apical notch (variable). (Holsinger 1989:953–954)

Some descriptions of new families also include distributional information, as in the following description of a new ascidian family by Patricia Kott:

Family PYCNOCLAVELLIDAE new family

The family accommodates *Pycnoclavella* Garstang, 1891 and the new monotypic genus *Euclavella,* both containing species formerly in Clavelinidae.

The new family is characterised by its relatively small but thread-like zooids divided into thorax and abdomen, smooth apertures, no internal longitudinal branchial vessels, a long oesophageal neck, smooth stomach at the posterior end of the abdomen, and a posterior stomach in the pole of the gut loop rather than (as in Clavelinidae) in the descending limb. The anus opens at the base of the atrial cavity (unlike the Clavelinidae or Diazonidae where it opens some distance up the branchial sac). . . .

Though records are few, possibly because colonies are cryptic and usually not intertidal, *Pycnoclavella,* found in Atlantic and Pacific oceans, and in tropical as well as temperate waters is especially well represented in Australian waters. *Euclavella* is known by only one species recorded from the coast of New South Wales and the North Island of New Zealand. (Kott 1990:86–87)

The same procedure followed in describing a new family is followed in naming and describing a new superfamily, as is shown by this description of a new superfamily of stomatopod crustaceans. The only difference in a superfamily description is that the families included are mentioned.

Erythrosquilloidea, new superfamily

Diagnosis.–Propodi of third and fourth maxillipeds broad, ventrally beaded. Telson with distinct dorsal median carina. At most, submedian marginal teeth of telson with movable apices. No more than 2 intermediate denticles present on telson.

Type genus.–Erythrosquilla Manning & Bruce, 1984, herein designated.

Included families.–Erythrosquillidae Manning & Bruce, 1984.

Remarks.–The superfamily Erythrosquilloidea can be distinguished from Bathysquilloidea Manning, 1967, Gonodactyloidea Giesbrecht, 1910, and Squilloidea Latreille, 1803 by the propodi of the third and fourth maxillipeds being broad and ventrally beaded rather than being slender and not ventrally beaded; the superfamily can be distinguished from the Lysiosquilloidea Giesbrecht, 1910 by having a distinct dorsal median carina on the telson. . . . (Manning and Camp 1993:87–88)

A new subfamily is also described like a new family, as in this description by Michael Novacek of a new subfamily of hedgehoglike insectivores from the middle Eocene of California:

SYSTEMATICS

CLASS MAMMALIA LINNAEUS, 1758
ORDER INSECTIVORA CUVIER, 1817
SUBORDER ERINACEOMORPHA GREGORY, 1910
FAMILY DORMAALIDAE QUINET, 1964
SUBFAMILY SESPEDECTINAE, NEW

TYPE GENUS: *Sespedectes* Stock, 1935.

INCLUDED GENERA: *Proterixoides* Stock, 1935; *Crypholestes* Novacek, 1976.

DIAGNOSIS: Shares with the dormaaliids *Scenopagus, Ankylodon, Dormaalius,* and *Macrocranion* the following derived dental features: P_{1-2} reduced, single-rooted peglike, or procumbent. . . . Differs from the above genera in having a P_3 with a well-developed protoconid and a short heel. . . . Differs from (primitive) erinaceids in the small and simple structure of P_{1-2} and P_4. . . . Similar to the "amphilemurids" (*Amphilemur, Pholidocercus, Alsaticopithecus, Gesneropithex*) in having bunodont crowns of cheek teeth but differs from this group in having large, double-rooted P_3. . . .

DISTRIBUTION: Middle-later Eocene (Uintan), southern California.

DISCUSSION: The distinctive features shared by the southern California Eocene dormaaliids warrant recognition of a separate subfamily for these taxa. *Sespedectes,* by far the best represented form, is designated the nominal genus for the group. *Crypholestes* is the most conservative sespedectine; *Proterixoides,* in some aspects, the most derived.

The above diagnosis clarifies the inclusion of sespedectines within the Dormaaliidae (see also Novacek, 1982) but distinguishes them from more generalized dental taxa (*Scenopagus*) or divergently specialized forms. . . . (Novacek 1985:3–4)

A new tribe is described in a similar manner, with the names of included genera listed, as in this description of a new tribe of iguanian lizards (Frost 1992). Note that the name of this nominotypical tribe takes the date and author of the original family description:

TROPIDURINI BELL, 1843

Tropiduridae Bell, 1843: 1. Type genus: *Tropidurus* Wied-Neuwied, 1825.

Ptychosauri Fitzinger, 1843: 16. Type genus: *Ptychosaurus* Fitzinger, 1843 (= *Plica* Gray, 1831).

Steirolepides Fitzinger, 1843: 17. Type genus: *Steirolepis* Fitzinger, 1843 (= *Tropidurus* Wied-Neuwied, 1825).

DIAGNOSIS: (1) Superior fossa of quadrate enlarged (not penetrated by a quadrate process of the squamosal); (2) alveolar shelf of mandible somewhat eroded; (3) posterior process of the interclavicle anterior to contact with the sternum long; (4) elongate hemipenes. . . .

CONTENT: *Microlophus* (Duméril and Bibron, 1837); *Plesiomicrolophus,* new genus; *Tropidurus* Wied-Neuwied, 1825; *Uranoscodon* Kaup, 1826. (Frost 1992:44)

Plants

Those who study flowering plants agree that families are pretty well defined (Heywood 1988; Stuessy 1990), so changes at the family level should be rare. As pointed out at the beginning of this chapter, many of the more than 300 plant families have diagnostic characters with such useful recognition and predictive power that they are used in texts and field guides. However recent molecular work (e.g., sequencing of chloroplasts) seems to be challenging that stability to some extent (Zomlefer 1994).

The following is excerpted from the description of a new flowering plant family from Central America.

Ticodendraceae Gómez-Laurito & Gómez P. fam. nov. Magnoliophytarum
Typus: Ticodendron Gómez-Laurito & Gómez P.

Arbor dioicus vel raro polygamodioicus. Ramulis teretis, cicatricibus teretis stipulorum delapsorum notatis. Stipuli sabulatae, circundatae, caducae. . . .

Dioecious or less commonly polygamodioecious trees. Twigs terete with the scar of stipules evident. Stipules subulate, encircling the stem, caducous. (Gómez-Laurito and Gómez P. 1991:87)

Note that it is very similar in style to the description of a new genus. The description (diagnosis) is given in Latin and English; the new plant's taxonomic placement and type genus are given in the heading.

Redescriptions of Family-Level Taxa

Redescriptions of families are often carried out in connection with studies of a particular group of organisms in a particular region, especially when they include taxonomic revision of earlier surveys. For example, in the course of describing the ctenostome bryozoans for the series *Synopses of the British Fauna,* Peter Hayward (1985) redescribed families, genera, and species, as well as updating their nomenclature and distributional records within the region. This type of family redescription may be quite brief, giving salient diagnostic features only, because it will be followed by more detailed description of genera and species occurring in each family.

Family ARACHNIDIIDAE Hincks, 1877

> Colony encrusting; autozooids repent, budding in ramifying, linear series, forming an open network over the surface of the substratum. Autozooids usually finely tapered proximally; cross-connections may form between zooids through the linking of buds, and appear superficially to represent stolons. Kenozooids absent. Orifice distally situated, often papillate. A key to all described species currently classified within the Arachnidiidae has been published by D'Hondt and Geraci (1976). (Hayward 1985:77)

Even though such family descriptions are brief, they can be very valuable to others working on the group (even in other regions of the world), especially when, as in the preceding example, the author has reviewed the literature and incorporated new information and taxonomic changes into the work.

Revisions at the generic and family level, and the analysis of phylogenetic relationships in these groups, are among the most important projects systematists carry out. Old placements are reevaluated on the basis of new evidence and necessary changes are made. In the following example the recognition of homoplasy in the loss of the posterior tracheal system of certain spiders led to the combination of two previously separate spider families and a redescription of the combined family under the older of the two names.

MICROPHOLCOMMATIDAE HICKMAN

> Micropholcommatidae Hickman, 1944, p. 183 (type genus *Micropholcomma* Crosby and Bishop).
>
> Textricellidae Hickman, 1945, p. 136 (type genus *Textricella* Hickman). NEW SYNONYMY.
>
> DIAGNOSIS: Micopholcommatids can be distinguished from other spiders by the combined presence of an elevated cheliceral gland mound and anterior booklungs reduced to either a few elongated leaves or true tracheae; males typically have one or more apophyses on the palpal patella.

> DESCRIPTION: See Hickman (1944, 1945), Forster (1959), Forster and Platnick (1977, 1981, 1984), and Davies (1985).
>
> SYNONYMY: See Familial Placement, above; both families were previously synonymized with the Symphytognathidae by Forster (1959) but were removed from that group and considered valid by Forster and Platnick (1977, 1984), Brignoli (1983), and Davies (1985). (Platnick and Forster 1986:7)

Many phylogenetic analyses of genera, families, and superfamilies go beyond nomenclatural revision to present the author's conclusions regarding the relationships of the taxa involved graphically. These papers include a detailed description of the characters used, the methodology of the analysis, and the phylogenetic conclusions, as well as any taxonomic revisions necessitated by them, as two examples briefly illustrate. Bill Shear (1986) used 27 characters (table 20.1) to carry out a cladistic analysis of the opilionid (harvestman) superfamily Ischyropsalidoidea. His phylogenetic results were shown as a cladogram (figure 20.1) and, in the text of the paper, as a taxonomic rearrangement of the superfamily into three monophyletic families, including one new family, the Ceratolasmatidae, which is formally described in the paper. Darrel Frost (1992) analyzed the relationships of 44 species of the South American *Tropidurus* group of iguanian lizards using standard cladistic techniques (chapter 22) on transformation series for 77 characters. His conclusions are shown by a consensus tree and cladogram (figure 20.2). The analysis led him to elaborate the taxonomy of the subfamily Tropidurinae to include two new tribes: Stenocercini and Tropidurini.

Descriptions of Taxa Above the Family Level

The codes do not provide rules for the names of taxa above the level of family (ICBN) or superfamily (ICZN) (ICZN, 1(b)1; ICBN, Art.10.7), although in botany names of taxa above the rank of family are considered to be "automatically typified" if they are based on a generic name (i.e., they are typified by that name).

Descriptions of new higher taxa may be rare compared to descriptions below the family level, but we will certainly see more of them. Our understanding of the phylogenetic relationships of organisms has changed rapidly in the last 25 years and continues to change almost

Table 20.1 Characters for the Cladistic Analysis of Isochyropsalidoidea

Characters	Plesiomorphic	Apomorphic
Metapeltidial sensory cones	Absent	Present
Ovipositor	Segmented	Segmentation lost
Palpal claw	Large	Reduced or absent
Cheliceral glands of males	Absent	Present
Genital operculum	No suture	With suture
Palpus	Gracile	Tibia and tarsus enlarged
Front of carapace	Not indented	Deep slot in midline
Sclerotization	Heavy	Reduced
Legs	With setae and microtrichia	With raised scalelike tubercles
Abdominal scutum	Without spikes	With spikes
Chelicerae	Normal size	Enlarged and spiny
Metapeltidial sensory cones	2	More than 2
Plumose setae of palpus	Present	Lost
Seminal receptacles	4	More than 4
Eye tubercle	Rounded, low	With median spike
Cheliceral glands of males	Present	Lost
Metapeltidial sensory cones	Present	Reduced or lost
Microsculpture of scutum	Small acute denticles	Anvil-shaped tubercles
Microsculpture of scutum	Small acute denticles	Large, round warts
Troguloid facies	Absent	Present
Stridulatory mechanism	None	Cheliceral/papal
Spikes of abdominal scutum	Erect, acute	Recumbent, blunt
Macrosetae of legs	Erect	Recumbent, sinuous
Scutum	Parvum	Magnum
Palpal femora of males	Without glands	With glands
Muscles of penis	In truncus	Not in truncus
Metapeltidial sensory cones	2	

Source: Shear (1986), table 1.

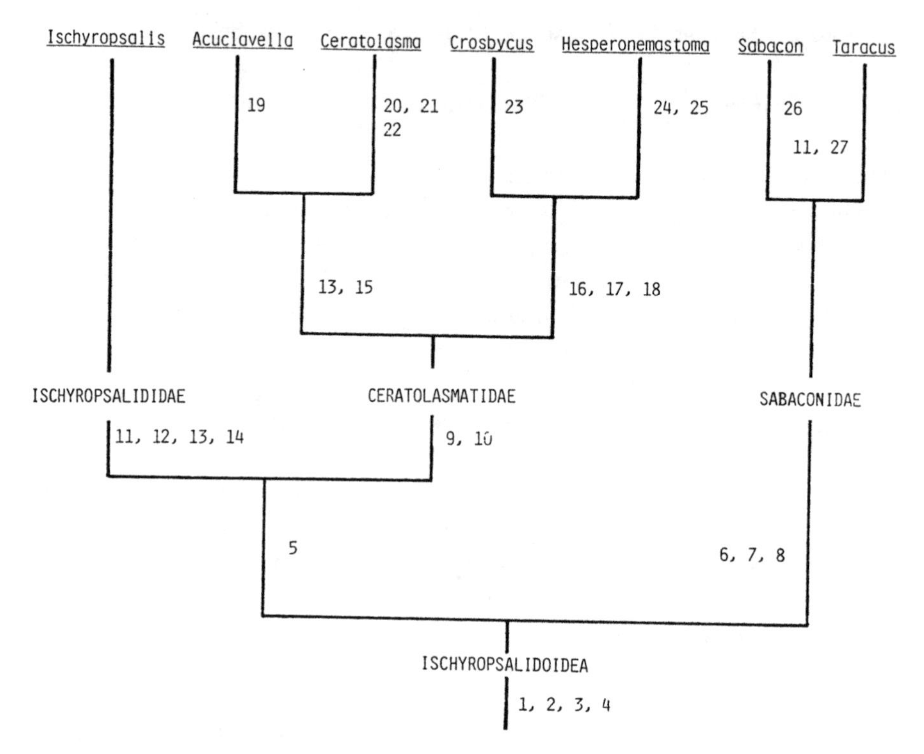

Figure 20.1 Phylogenetic analysis in the description of higher taxa. Cladogram of families and genera of the superfamily Isochyropsalidoidea. *–Source: From Shear (1986), courtesy of the American Museum of Natural History.*

daily as new evidence from molecular studies and cladistic analyses is published. This is true even at the highest levels. No longer do we think of organisms in the traditional way as "plants" or "animals." Now there are five kingdoms, possibly more. Many algal groups have been elevated to divisions; animal groups once thought to represent a single phylum (e.g., Platyhelminthes) have been fragmented, and bacteria are showing divergences greater than we had ever imagined. Between the level of family and that of phylum/division, many groups are in a state of flux, such as the orders of angiosperms (Stuessy 1990). Both molecular and cladistic studies are helping clarify relationships in angiosperms, yet the author of a recent book on plant families could still state, "This book does not stress higher ranks, such as subclass, superorder, order, and suborder, for they imply a level of knowledge that currently does not exist" (Zomlefer 1994:3).

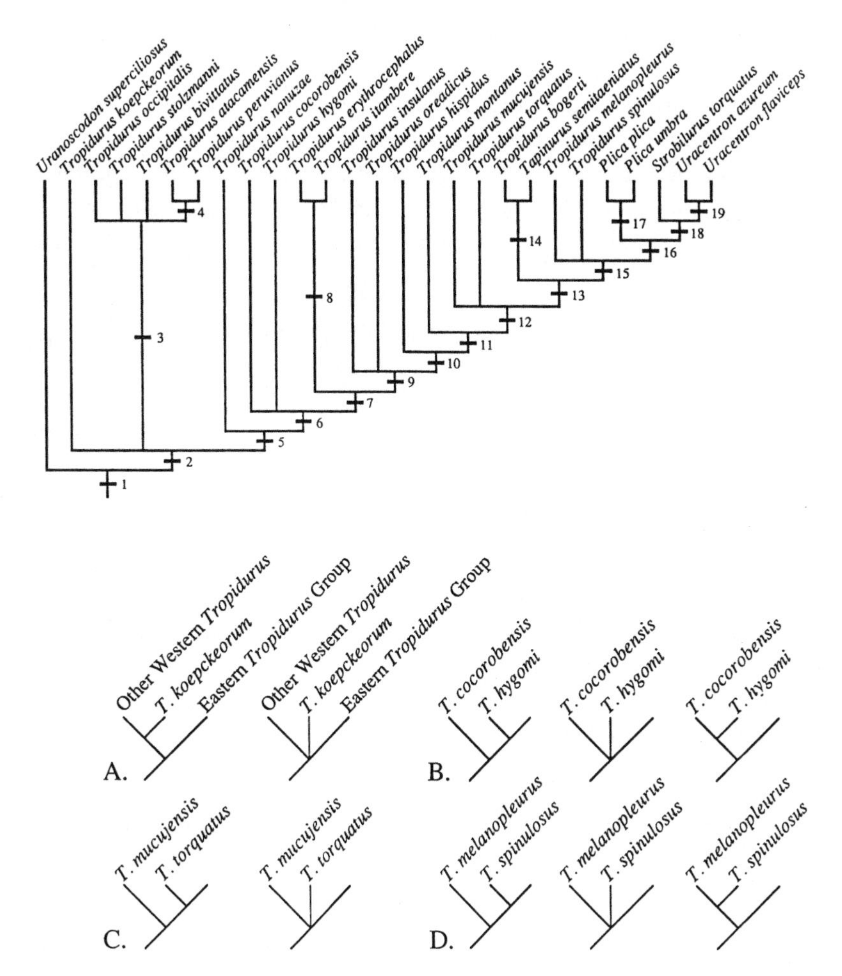

Figure 20.2 Phylogenetic analysis in the description of higher taxa. Strict consensus tree and cladograms from an analysis of the *Tropidurus* group of iguanian lizards. *–Source: From Frost (1992), courtesy of the American Museum of Natural History.*

Zoology

Although today there are no rules governing names and descriptions of groups above the family level, in zoology, at least, there once were. They were covered by versions of the code in place from 1901 (the first version of the *Règles Internationales de la Nomenclature Zoologique*) to 1961 (when a completely new version of the code was published). Thus, although any author has much more freedom with description of a higher-level taxon, a mixture of past practices and recommendations does exist. It generally parallels the rules for describing taxa in the lower-ranked categories (Savory 1962; Blackwelder 1967).

Taxa in these categories do not have a formal type. This is not a problem unless the group is later divided or combined and it is necessary to decide which group should keep the original name. Blackwelder (1967) made two recommendations to avoid confusion in these cases:

When a taxon is divided, the original name should be retained for the larger of the two subgroups.

When two taxa are combined, if one is significantly larger than the other, it should get the original name, but if they are both about the same size, a new name should be created for both of them.

Names of higher-category taxa are uninomial, made up of one word of classical or nonclassical origin, treated as if it were Latin, and in the Latin nominative plural (Blackwelder 1967).

Blackwelder (1967) made some common-sense recommendations for the description of higher taxa, adapting the rules used for descriptions of lower-ranking taxa. Paraphrased and updated (for more recent changes in the ICZN) they are as follows:

Names should consist of a single word with the initial letter capitalized, Latin or Latinized and spelled in the neo-Latin alphabet, with no italics used.

Names should be pronounceable and pleasant-sounding.

They should have no hyphens or diacritical marks.

They *may* be formed by changing the ending of some characteristic subordinate taxon, such as Arachnida (*Arachne*) or Hydrozoa (*Hydra*), but they do not have to be formed this way.

The names of higher-category taxa should be published (like other scientific names)

In an acceptable journal or other scientific publication, with a definite author and date

With the intent to establish a new name for a group of organisms clearly stated

With the diagnostic characteristics of the taxon clearly described

With the taxa to be included in the group clearly indicated

Examples of descriptions of some higher taxa follow.

A NEW ORDER The following, an excerpt from the description of a new order of Peracarid crustaceans by Thomas Bowman and colleagues (1985), shows the inclusion of *diagnosis, composition* (what species, genera, families, etc. make up the new taxon), and *etymology:*

Mictacea, new order

Body slender, cylindrical. Head fused to first thoracic somite (somite of maxilliped). Carapace not developed posteriorly but small lateral carapace folds covering bases of maxilla 1, maxilla 2, and maxilliped; elliptical area dorsal to carapace fold inflated, thin-walled, apparently functioning in respiratory exchange. Pereionites and pleonites all free. Telson, free entire. . . . Embryo flexed dorsally. Eggs hatching as manca lacking pereipod 7.

Composition.–Hirsutia bathyalis Sanders, Hessler and Garner, 1985, family Hirstuiidae; *Mictocaris halope* Bowman and Iliffe, 1985, family Microcarididae.

Etymology.–from the Greek "miktos" (mixed, blended) plus "acea," the most common ending for names of peracaridan orders. The name reflects the fact that very few of the characters of the Mictacea are unique to this order; most are found also in at least one other peracaridan order. (Bowman et al. 1985:74–75)

A NEW CLASS In paleontology, complete placement of a new higher taxon may not be possible, as in the following description of an extinct Paleozoic taxon whose relationships are still unclear:

Phylum uncertain

Class Multiplacophora classis n.

Diagnosis.–Animals with a single row of thick, calcareous anterior, posterior, and 10 intermediate left-and right-handed plates; distinct insertion plates on borders of plates; narrow, elongate spines bordering plates; plates and spines with numerous surface pores leading to complex internal canal systems.

Order Strobilepida ord. n.

Diagnosis.–as for the class.

Family Strobilepidae fam. n.
Diagnosis.–as for the class. (Hoare and Mapes 1995:118–119)

REDESCRIPTION OF A CLASS A new class (or an old one) may have to be redefined to include an important new discovery. The echinoderm class Concentricycloidea was described by Baker, Rowe, and Clark in 1986 to represent a new unique new asterozoan from New Zealand. Only one species, *Xyloplax medusiformis,* was known at the time of the description, so the diagnosis of the class was based on characters of that species. The discovery of another species of *Xyloplax* in the Atlantic Ocean necessitated a revised diagnosis to include characters of the new species as well.

Class CONCENTRICYLOIDEA Baker, Rowe and Clark (1986)

Definition (emended)

A free-living, pentamerous echinoderm, characterized by a weakly inflated disc-like body, without radiating arms. A shallow, sac-like stomach, with centrally placed mouth, may be present, but intestinal tract or anus absent in known forms. The oral surface of stomachless forms is covered by a complete velum. The water-vascular system, including the tube-feet, and supporting skeletal structures are arranged concentrically on the oral surface. The water-vascular canals form a double ring connected at interradial points.

Type-genus

*Xyloplax,*Baker, Rowe and Clark (1986).

Occurrence

New Zealand, South Pacific and Tongue of the Ocean, Bahamas, West Atlantic, 1000–2066 m.

Remarks

The class Concentricycloidea consists of an order (Peripodida) and family (Xyloplacidae), both defined as for the class, erected by Baker et al. (1986) to include their new species *Xyloplax medusiformis.* In light of new material from the Atlantic Ocean, which is described herein as a new species, the definition of the class has been amended. This has been necessary to accommodate the new form possessing a stomach which is lacking in *X. medusiformis.* (Rowe et al. 1988:454–455)

DESCRIPTION OF A NEW PHYLUM A description of a new phylum is really rare, but still necessary from time to time. The latest invertebrate phylum to be discovered was the lobster-lip animal or "ring-bearer," a creature that lives attached to the mouth appendages of the Norwegian lobster *Nephrops norvegicus.*

Cycliophora, new phylum

Diagnosis. Acoelomate, marine metazoans with bilateral symmetry and differentiated cuticle. Compound cilia engaged in filter feeding working as a downstream-collecting system. Sessile, solitary feeding stages with multiciliated cells in the gut, anus just behind a ciliated mouth ring, and internal budding with extensive regeneration of feeding apparatus. Feeding stages alternating with brief, non-feeding, free-swimming stages. Brooding of asexual larva (Pandora larva), male and female. Females brooding chordoid larva with a mesodermal, ventral rod of plate-like muscles cells (chordoid structure) and a pair of protonephridia with multiciliated terminal.

Etymology. Cyclion and phoros are Greek for "a small wheel" and "carrying" referring to the circular mouth ring. (Funch and Kristensen 1995:711)

The new organism has since been found on another species of lobster, but so far only one species (*Symbion pandora*) is known. In their description of the new phylum in *Nature,* Peter Funch and Reinhardt Kristensen also described the new species, new genus, new family, new order, and new class, but for the other higher categories the diagnoses could be very brief, only the names and etymologies differing:

Diagnosis. Same as the phylum. (Funch and Kristensen 1995:711)

Botany

In botany also, there are no regulations governing description of ranks above the family level. Most of them have become established over time and by tradition within a group and different authors may have used different names for the same groups. It is not necessary to cite an author's name when using them.

At least some botanists still recommend that a description of a new order be done in Latin and that a type family be indicated (Hawksworth 1974). For example, in this description of a new order of red algae, G. W. Saunders and G. T. Kraft used the latest molecular techniques (small-subunit rRNA gene sequences) to analyze phylogeny of a number of red algal families before deciding to establish a new order, but their formal description of the new order follows established tradition:

Halymeniales ord. nov.

Plantae multiaxiales; medulla filamentosa et cortex pseudo-parenchymatus foveis scundariis numerosis intra thallum inter cellulas medullosas et cellulas corticales interiores conjunctivis. . . .

Plants multiaxial; medulla filamentous, cortex pseudo-parenchymatous, secondary pit-connections numerous between cells in medullary and inter-cortical thallus layers. Pit plugs with bounding membrane on both surfaces; plug caps absent. Major galactans not exclusively carrageenans but apparent blends of [X]-type carrageenans and agars similar to those of the Rhodymeniales. Life histories triphasic, gametophytes and tetrasporophytes isomorphic. Carpogonial branches 2- to 4-celled and outwardly directed, on presumed fertilization issuing multiple septate, branched connecting filaments directly from carpogonium. Auxiliary cell on septate cortical branch system to supporting cells, diploidized by terminating or on-growing connecting filament; gonimoblast initial single, oriented thallus-outwardly; carposporophyte multilobed, most cells differentiating in carposporangia. Tetrasporangia regularly or decussately cruciate. Phylogenetic analysis of small-subunit rRNA and *rbcL* sequence data indicate closer affinity to representatives of the Rhodymeniales than to those of the Gigartinales.

TYPUS FAMILIAE: Halymeniacaeae Bory (1828, p. 158). (= Cryptonemiaceae Harvey 1849, p. 132; = Grateloupiceae Schmitz 1892, p. 18). Nomenclature and prioritization of family names discussed by Guiry (1978b, p. 192).

ADDITIONAL FAMILY: Sebdeniaceae Kylin (1932, p. 12).

The Halymeniaceae consists of 10 genera (Womersley and Lewis 1994) and around 160 species (Kraft 1981) that are dispersed worldwide from cold-temperate through tropical waters. The greatest abundance occurs in temperate Australasia and the western Pacific. . . . (Saunders and Kraft 1996:703)

Problems with Nomenclature of Higher Taxa

Problems encountered with the names and composition of higher-category taxa can be addressed with logic, but they are not necessarily resolved by logic. Tradition and even sociology play a large role here. If there are two names for a higher group taxon, they are indeed synonyms, but you are free to choose which one to use, regardless of priority, general acceptance, or any other factor. Some workers have stressed priority, rejecting junior synonyms, but this may turn out not to be as important to their colleagues as recognizability, tradition, or appropriateness. The group I work on, bryozoans, is a good example. Three names have been used for the phylum: Bryozoa, Ectoprocta, and Polyzoa. Tom Schopf (1968) made an excellent case for the use of Ectoprocta for the phylum as we now understand it, yet the British still favor Polyzoa, and the International Bryozoology Association keeps on meeting. Why? Perhaps we just don't like to think of ourselves as "ectoproctologists."

Sources

Balmford, A., M. J. B. Green, and M. G. Murray. 1996. Using higher-taxon richness as a surrogate for species richness: I. Regional tests. *Proceedings of the Royal Society of London* ser. B, 263: 1267–1274.

Balmford, A., A. H. M. Jayasuriya, and M. J. B. Green. 1996. Using higher-taxon richness as a surrogate for species richness: II. Local applications. *Proceedings of the Royal Society of London* ser. B, 263: 1571–1575.

Benton, M. J. 1995. Diversification and extinction in the history of life. *Science* 268: 52–58.

Burlando, B. 1990. The fractal dimension of taxonomic systems. *Journal of Theoretical Biology* 146: 99–114.

Chitwood, B. G. 1958. The designation of official names for higher taxa of invertebrates. *Bulletin of Zoological Nomenclature* 15: 860–895.

Ferraro, S. P. and F. A. Cole. 1990. Taxonomic level and sample size sufficient for assessing pollution impacts on the Southern California Bight macrobenthos. *Marine Ecology Progress Series* 67: 251–262.

Gaston, K. J. and P. H. Williams. 1994. Mapping the world's species: The higher taxon approach. *Biodiversity Letters* 1: 2–8.

Lee, M. S. Y. 1996. The phylogenetic approach to biological taxonomy: Practical aspects. *Zoologica Scripta* 25: 187–190.

Lee, M. S. Y. 1997. Documenting present and past biodiversity: Conservation biology meets palaeontology. *Trends in Ecology and Evolution* 12: 132–133.

Schaefer, C. W. 1976. The reality of the higher taxonomic categories. *Zeitschrift für Zoologische Systematik und Evolutions-forschung* 14: 1–10.

Simonetta, A. M. 1990. On the significance of ecotones in evolution and on the significance of morphologic differences in primitive taxa: Reflections and hypotheses. *Bolletino di Zoologia* 57: 325–330.

Smith, S. D. A. and R. D. Simpson. 1993. Effects of pollution on holdfast macrofauna of the kelp *Ecklonia radiata:* Discrimination at different taxonomic levels. *Marine Ecology Progress Series* 96: 199–203.

Starobogatov, Y. I. 1991. Problems in the nomenclature of higher taxonomic categories. *Bulletin of Zoological Nomenclature* 48: 6–18.

Van Valen, L. 1971. Adaptive zones and the orders of mammals. *Evolution* 25: 420–428.

Van Valen, L. 1973. Are categories in different phyla comparable? *Taxon* 22: 333–373.

Warwick, R. M. 1988. The level of taxonomic discrimination required to detect pollution effects on marine benthic communities. *Marine Pollution Bulletin* 19: 259–268.

Warwick, R. M. 1993. Environmental impact studies on marine communities: Pragmatical considerations. *Australian Journal of Ecology* 18: 63–80.

Williams, P. H. and K. J. Gaston. 1994. Measuring more of biodiversity: Can higher-taxon richness predict wholesale species richness? *Biological Conservation* 67: 211–217.

Williams, P. H., C. J. Humphries, and K. Gaston. 1994. Centres of seed-plant diversity: The family way. *Proceedings of the Royal Society of London* ser. B 256: 67–70.

Chapter 21

Common Problems

With the vast increase in numbers of known forms of animals and with the change in concepts of classification brought about by acceptance of the theory of evolution, the mechanics of modern taxonomy have become so complex as to discourage the beginning student. –Schenk and McMasters (1936:1)

The bringing to light of overlooked names in the old literature is perhaps nearing completion . . . it is hoped that this will lead to name-changes for nomenclatural reasons becoming ever fewer and fewer until eventually they cease to trouble us. Unfortunately, the same cannot be said of name-changes that become necessary for taxonomic reasons. These arise from taxonomic research itself and are inevitable accompaniments of our systems of classification. –Jeffrey (1989:31)

Describing a new species is often a simple process, taking only straightforward background research, plus a little motivation, to accomplish. But some cases can be more difficult, either because of problems discovered while carrying out the background research or because of nomenclatural difficulties encountered in the course of writing up the description for publication. This chapter covers a few of the most common problems that may arise.

Missing Types

One of the problems you might have to deal with during background research for your project is what to do when you cannot locate the type material for a species you need to study or compare. Perhaps no type material was ever deposited in an institutional collection, or, although deposited, it has since been lost or destroyed.

Sometimes type material has been lost or damaged because a taxonomist did not make arrangements for the preservation of his or her private collection after death. C. McLean Fraser, a hydroid taxonomist who worked at the University of British Columbia, described more than 200 new hydroid species during his career. Although he designated holotypes for some of those species and they were duly deposited in Allan Hancock Foundation, USNM, and MCZ collections, he kept a much larger number in his private collection at the university. For a long time after his death in 1946, no one was able to use these specimens because they had been stored and forgotten. Finally, in the 1970s the collection was rescued from storage by Mary Arai, who salvaged it as well as possible (unfortunately, the preservative in some of the jars had evaporated by then) and deposited it in the Royal British Columbia Museum (Arai 1977; D. Calder, personal communication 1996).

Types have even been destroyed deliberately. Sig Thor, a Norwegian mite taxonomist who also described many new species during his career, became so angry at certain of his colleagues that he told his wife to destroy his type-containing personal collection when he died. Fortunately, he had also informed some other colleagues of his decision, and they were determined to save the types from destruction. Unfortunately, when he died World War II was in full swing. Travel was extremely difficult for civilians, and his friends could not get to the distant city where he had lived in time to save the specimens (K. Fauchald, personal communication 1996).

Sadly, whole museums and their collections have perished in floods, fires, and especially in times of war, when museums were bombed and all their contents destroyed or when portions of hurriedly hidden collections were damaged or lost.

There are several ways to solve the problem of missing types. In some cases descriptions and original illustrations are so indicative that there is no doubt about a species's identity and no need to search for type material. For example, many Linnaean species of birds and mammals are common or distinctive and have also been studied by biologists for years, so there is no question about their identity.

For many species described in the eighteenth and early nineteenth century, no type material was preserved. Nevertheless, even some of those for which the name was based on an illustration alone, or an illustration plus an inadequate written description, may be identifiable from that original illustration. In other cases (sometimes even with a species depicted in another illustration from the same work), the intended identity is completely obscure because the old illustration does not show the characters now required to distinguish species in that genus or family. For example, in 1826 J. V. Audouin listed 67 bryozoan species from the Mediterranean and the Red Sea (Dumont 1981). But Audouin's "descriptions" consisted only of his captions to figures in plates drawn by J. C. Savigny and published in Savigny's book *Descriptions de l'Égypt Histoire Naturelle* (Audouin 1826). No type material for these species has ever been found. Even so, some of the species are clearly recognizable today and their names are still in use, such as *Lepralia cecilii,* now called *Arthropoma cecilii* (figure 21.1a and b). Other illustrations do not show the characters needed to determine to which of numerous species of the genus inhabiting the area it belonged. This is the case with *Lepralia jacotini* (figure 21.1 c). This illustration shows a species that clearly belongs in the family Smittinidae, but the characters of orifices and ovicells now used to distinguish species in this group cannot be seen clearly; the magnification of the drawing is too low.

Anyone working on a species that resembles one in a questionable illustration or an unclear old description would do better to describe the species as new, making a note of the early name as a possibility. Should the original specimens someday be discovered, restudy might result in your name becoming a junior synonym, but at least the history of your name and your decision would be clear. This was the course John and Dorothy Soule took when, in their revision of Hawaiian and eastern Pacific smittinid bryozoans, they had to deal with a Hawaiian species that seemed quite similar to Audouin's *Cellepora parsevalii* (Soule and Soule 1973). They published both Savigny's illustration and their own SEM photographs of Hawaiian material, pointing out the apparent distinctions between their material and the illustration, to substantiate their decision to name and describe the Hawaiian material as a new species: *Parasmittina parsevaliformis.*

> The Hawaiian material might be considered the same species as the *Cellepora parsevalii* of Audouin (1826, p. 238), which Harmer classed as *Smittina parsevalii.* In the present paper (fig. 8G), our copy of the Savigny illustration . . . shows that small lateral oral avicularia were present, presumably placed in an ascending position, but there is no suggestion of

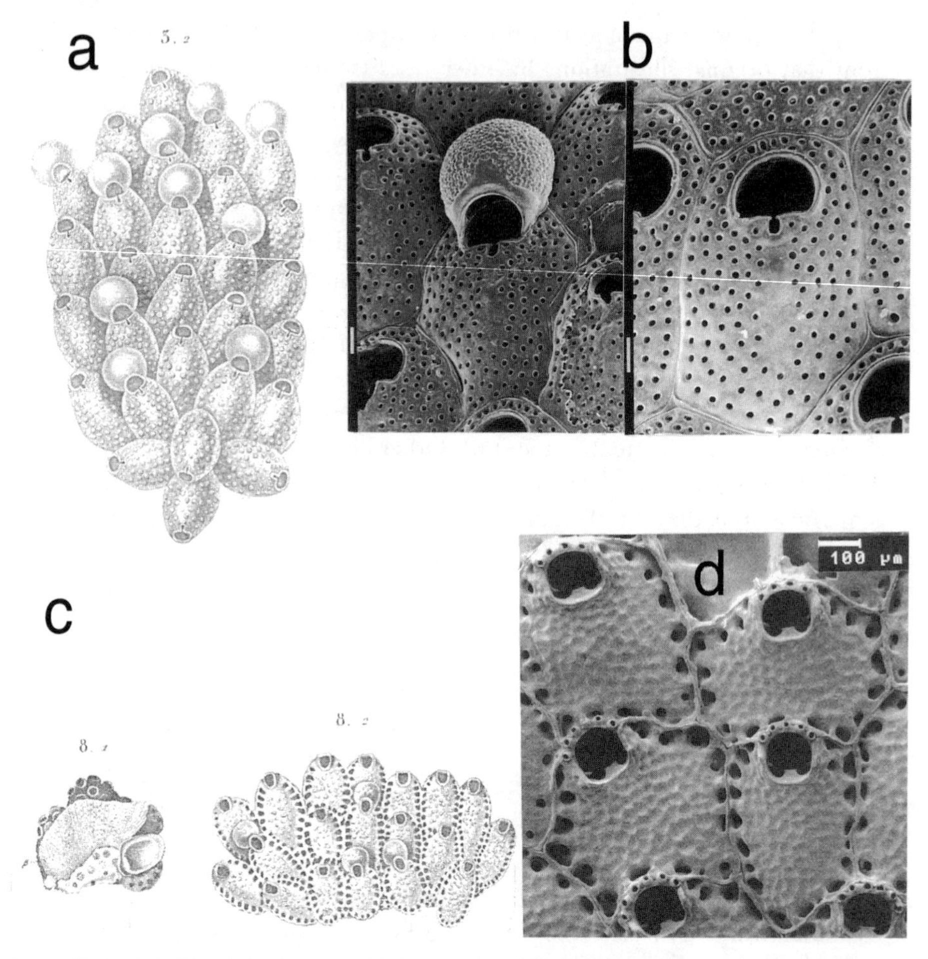

Figure 21.1 Early "descriptions" of some Mediterranean and Red Sea bryozoans were just captions by one author for figures in plates in a book by another author. Some species can be recognized from the figures; others cannot. **A,** *Lepralia cecilii.* Savigny (?1817), plate 8, figure 3. **B,** *Athropoma cecilii,* Scanning electron micrographs courtesy of P. J. Hayward. Scale: Bars = 100 μm. This species can be recognized from the old figure. **C,** *Lepralia jacotini.* Savigny, plate 7, figure 8. This species cannot be recognized from the figure. Although it can be placed in a family (the Smittinidae), the illustration lacks the details of orifice, oral spine, and ovicell characters used to distinguish species in that family, as shown in this modern scanning electron micrograph **(d)** of a related species.

the large spoon-shaped avicularia which are so characteristic of the Red Sea, Atlantic and Pacific material ascribed to that species. . . .

In the Savigny illustration . . . , the ovicells can be seen to have few pores, and to have the triangular avicularia encroaching on the ovicells from adjacent frontals. In the Hawaiian material, the colonies that have ovicells with few pores and encroaching avicularia also have the giant spoon-shaped avicularia. (Soule and Soule 1973:413)

Another good approach to the problem of missing type material is to examine *topotypic* material, specimens from the locality where the species was originally collected. You may be able to find such material in a museum collection. You may have to visit the type locality and make your own collection. Comparing that material to your specimens should permit you either to feel reasonably sure about your identification or to decide that your material does not belong to the previously described species and should be described as new. Topotypic material can be extremely useful when a species named in the last century is restudied. For example, the description of the inarticulate brachiopod *Lingula reevii,* given by Davidson in 1880, described the color and shape of the valves but did not contain the information necessary to identify the species according to current criteria for species identification in *Lingula.* So Christian Emig used topotypic material, specimens from Hawaii (the type locality for the species) to redescribe it using modern diagnostic characters such as arrangement of musculature (Emig 1978).

Generic types may be lost, and topotypic material may be valuable there, too. The type specimen (and only specimen known to have been collected) of *Segestrioides bicolor,* the type species of the spider genus *Segestrioides,* was lost many years ago from the collection where it had been deposited, and the taxonomic position of the genus was considered uncertain by later workers. Norman Platnick was finally able to redescribe this species when a colleague, Frederick Coyle, doing field work in Peru, was able to make a collection at the type locality of *S. bicolor.* By turning rocks at this locality, a eucalyptus forest on the western edge of San Mateo, Peru, Coyle and his field crew of students collected a series of juveniles and females of the species, which allowed the genus to be redescribed and its provisional placement in the family Diguetidae confirmed (Platnick 1989).

Lectotypes

In the past an author of a new species sometimes deposited the material on which its description was based into a museum collection without picking a particular specimen to serve as the holotype. Sometimes a series of syntypes was deposited. In other cases (particularly when the entire personal collection of a deceased taxonomist was donated to a museum) there might be no indication as to which specimen or specimens of a particular species formed the basis for the original description. A later taxonomist studying such a species may choose a specimen from the original material and designate it as the *lectotype* for that species by writing and publishing a paper to that effect (ICZN, Art. 73, 74; ICBN, Art. 9.9). This designation does not have to be lengthy. It can consist of a single brief paragraph, as in the following example for the crab spider, *Xysticus emertoni.* After a short discussion of the available material collected by the original author of the species, Richard Hoffman wrote,

> In the belief that (with other considerations being equal) a lectotype should be selected from material deposited with the original owner of a syntype series, and from an unimpeachable locality, I hereby designate the female numbered 2.415 as lectotype of *Xysticus emertoni* Keyserling, and the two females MHNP 2.416 and that under BMNH 1890.7.1 3386 as lectoparatypes. The White Mountains of New Hampshire (specifically Mount Washington) are thus fixed as the restricted type locality. (Hoffman 1996:48)

and so designated a lectotype for the species. Note that the other syntype specimens thereby become paralectotypes (lectoparatypes) (ICZN, Art. 73b (ii)). In a much earlier paper on *Xysticus,* W. J. Gertsch (1939) had stated that the type locality of *emertoni* was "Georgia" and that the "type" specimen so labeled was in the Paris Museum. However, Gertsch did *not* stipulate that he was selecting a lectotype from among the several syntype possibilities, as required by the code. Hoffman was therefore free to alter Gertsch's informal type locality designation to one consistent with the species's known boreal range by citing all of the syntype material and selecting a specimen from Mount Washington as lectotype of *emertoni.* The type locality in such cases is determined by the provenance of the lectotype, not the reverse. Circumstantial evidence indicated that the indication "Georgia" was almost certainly incorrect. During the process of tracing locations of the original syntypes of *emertoni,* Hoffman found that the "Georgia" specimen actually had been retained by Keyserling and not returned to the Paris Museum, as Gertsch had assumed. In fact, it had later gone from Keyserling to the British Museum instead. Because of its uncertain origin, Hoffman did not designate it as a lectoparatype. The brevity of the actual designation doesn't lessen the importance of such a search and description. In this case, in the process of designating of a lectotype by choosing a specimen from a verifiable type locality, Hoffman was able to resolve a long-standing biogeographic puzzle and correct a mistake made in 1880, which had been passed down by subsequent workers for over a hundred years.

For well-studied groups such as birds and mammals, it may desirable to designate lectotypes for subspecies. For example, in the course of revising the "Catalog of Type Specimens of Recent Mammals in the American Museum of Natural History" Marie Lawrence and Guy Musser came across a problem with a subspecies of a spotted or leopard cat, *Prionailurus bengalensis.* The subspecies name, *alleni,* had been assigned to a series of specimens from the Chinese island of Hainan by H. J. V. Sody (1949), but Sody did not select a holotype. Therefore, they chose one specimen from the original series and designated it as a lectotype. Their paper included an introduction describing the history of the specimens and literature about them, a materials and methods section listing the available specimens and measurements made on them, and a section on the selection of the lectotype, with tables of measurements documenting the cranial and dental evidence that, along with tooth wear patterns, had enabled them to pick an adult specimen as the lectotype. This was followed by a detailed description of the chosen specimen, illustrated with photographs of study skin and skull:

DESIGNATION AND DESCRIPTION

> We designate AMNH 59957, the skin and skull of an adult female, as the lectotype of *Prionailurus bengalensis alleni* Sody (1949). It was collected by Clifford Pope at Nodoa (Nada or Danxian in modern orthography) on Hainan Dao in the Guandong Province of The People's Republic of China. The original field number is 220. Its traits exemplify the distinctions between samples from mainland and island that Sody interpreted as significant in identifying morphological boundaries of *P. bengalensis alleni.* . . .
>
> The tanned study skin is illustrated in figures 3 and 4. . . .
>
> The cranium of the lectotype (figs. 1 and 2) appears gracile and elongate compared with crania of other small-bodied felid species such as *F. catus, F. sylvestris, F. geoffroyi,* or *F. weidii,* for example. . . .
>
> The upper and lower teeth are moderately worn and demonstrate deterioration. . . .

> In its general conformation, the mandible resembles that in other species of small felids. The following characters are distinctive: the body of the ramus is robust for the overall size of the dentary; each coronoid process is round; and there is a bony flange on the medial side of each angular process (fig. 2). (Lawrence and Musser 1990:5–7)

and concluded with a discussion of the controversy still existing in the literature concerning taxonomic placement of both species and subspecies.

After 1999, the fourth edition of the ICZN requires that lectotype designation use the term *lectotype* (or its translation in the language used in the paper) and carry a statement declaring that the lectotype designation is being made to clarify the application of a name to a taxon.

Lectotypification is important in botany also. As in zoology, older names that applied to common or distinctive species have been used in a fairly consistent way over the years, but for less familiar plants, the fact that for a long period of time botanists diverged greatly in their methods of naming plants led to many later problems in deciphering an earlier worker's intentions (Fosberg 1992; Stevens 1994; chapter 2). Thus, there is still a need to designate lectotypes to clarify the use of many older names. For example, the genus *Sargassum* is one of the largest and most morphologically complex groups of brown algae. It also shows considerable variability in the morphological taxonomic characters that have been used to distinguish species and subspecies. Dennis Hanisak and John Kilar were working with *Sargassum* in Florida. Their studies included populations that fit the generally accepted description of *Sargassum filipendula,* described by Carl Agardh in 1824. When they requested type specimens from the Agardh Herbarium in Lund, Sweden (abbreviated LD), they found that only a series of syntypes existed. Looking at the borrowed material, they realized that Carl Agardh's son, Jacob, had later used the same material to describe two varieties, *S. filipendula* var. *contractum* and *S. filipendula* var. *laxum.* They decided to designate lectotype specimens for the species and those varieties:

> *Lectotypification of* Sargassum filipendula *C. Agardh*
>
> *Sargassum filipendula* C. Agardh, Sys. alg. 300. 1824. LT: "India Occidentalis, Aspegren" [cited as "In sinu mexicano"] Agardh Herbarium no. *3253* (LD), here first designated.
>
> *Sargassum filipendula* var. *contractum* J. Agardh, Spec. gen. ord. alg. 315. 1848, "contracta" LT: "India Occidentalis, Aspegren" [cited as "in sinu Mexicano"] Agardh Herbarium no. *3244* (LD), here first designated.

> *Sargassum filipendula* var. *laxum* J. Agardh, Spec. gen. ord. alg. 315. 1848, "laxa." LT: "India Occidentalis, Aspegren" [cited as "in sinu Mexicano"] Agardh Herbarium no. *3252* (LD), here first designated. . . .
> . . . given the fragmentary habit of these specimens we designate the most complete specimen, Agardh Herbarium specimen (LD) number *3253* (fig. 1), as the lectotype of *Sargassum filipendula* C. Agardh (1824). This selection is consistent with the protologue of this taxon and agrees with J. Agardh's (1848: 314; 1889: 120) concept of this species, as he annotated this specimen as *S. filipendula*. . . . J. Agardh differentiated the two varieties primarily by blade features (length, degree of serration) and degree of branching. . . . Consistent with the protologues of the two varieties, we designate Agardh Herbarium (LD) specimen number *3244* as the lectotype of *S. filipendula* var. *contractum* J. Agardh, and Agardh Herbarium (LD) specimen number *3252* as the lectotype of *S. filipendula* var. *laxum* J. Agardh. This lectotypification is consistent with J. Agardh's concept of the varieties; both specimens are annotated in his handwriting to their respective varieties and cite p. 315 of Agardh (1848). . . . (Hanisak and Kilar 1990:96–97)

The term *protologue* used in the preceding excerpt means the original description, including illustrations or any other supporting material. It is more commonly used by botanists than zoologists. Note that designating a lectotype solves only the taxonomic problem relating to clarification of the original use of the name. It does not solve problems created by the use of variable taxonomic characters. Hanisak and Kilar (1990) concluded their paper with the recommendation that in future studies of the group, greater taxonomic emphasis should be placed on reproductive and developmental characters.

Although anyone who designates a lectotype should have thoroughly studied the original material, once in a while a mistake is made. The botanical code allows a lectotype to be replaced by a new one if the first one seriously conflicts with the original description and its material *and* there is additional material available that does not conflict (ICBN Art. 9.13). This can lead to name changes as well, as J. P. Roux found in studying a fern from the Cape of Good Hope. The species had originally been described in 1768 as *Asplenium lucidum.* Two specimens from the original material still existed in the Burman herbarium at the Conservatoire et Jardin Botanique de Genève. The following excerpts from his paper describe the problem and its resolution:

The first specimen, belonging to a *Polystichum* species, has two labels attached to it. The first reads "Planta Capensis" and the second "*Asplenium lucidum* Burm." Becherer (1937) regarded this sheet as the "échantillon-type" (type specimen) and, based on his lectotypification, made the combination *Polystichum lucidum* (Burm. f.) Bech. However, this specimen is in serious conflict with the protologue in (1) not being shiny, (2) the pinnae being pinnate (not decompound), and (3) not being glabrous. There is no evidence that it was part of the Burman herbarium prior to the publication of the *Flora indica.*

The second specimen, representing *Asplenium adiantum-nigrum* L., has two labels. The first reads "*Filix lucida pinnis Apii in morem laciniatis*" and "Hermann Cat pl. Afr p 9." The second label reads "*lucidum Asplenium fronde decomposita, pinnis obsolete cuneiformibus, serratis alternis*". This sheet presumably formed part of the original Houttuyn herbarium partly acquired by Burman . . . as both labels are in Houttuyn's hand. The inscription . . . refers to the catalog of Cape plants observed by Paul Hermann, published as an appendix to J. Burman (1737). . . . No publication of the phrase name on the second label could be traced. On the (reasonable) assumption that the latter specimen was already known to the elder Burman, it may be considered a possible lectotype for *A. lucidum*. . . .

Burman's second specimen is here formally designated as the superseding lectotype of *Asplenium lucidum* Burm. f. . . .

Asplenium lucidum Burm. f. therefore falls into synonymy under *A. adiantum-nigrum* L.

Asplenium adiantum-nigrum L., Sp. PL.: 1081. 1753. = *Asplenium lucidum* Burm. f., Fl. Indica, Prodr. Fl. Cap.: 28. 1768–Lectotype (here designated): Cape of Good Hope, *Hermann s.n.* ex Herb. Burman (G). (Roux 1994:641–642)

In zoology a mistake like this must be handled by an appeal to the ICZN, as with the second example in the next section.

Neotypes

If no type material exists, designation of a *neotype* is sometimes necessary to maintain nomenclatural stability and to solve problems of doubtful and confusing identities. A neotype is a specimen chosen to fix the identity of a previously described species whose original type material has been lost or destroyed. Like a holotype specimen, it should be chosen carefully. If possible, it should come from the same locality as the original (Quicke 1993).

Designation of a neotype should not be undertaken lightly. The zoological rules state that neotypes should be created "only in exceptional circumstances when a neotype is necessary in the interest of stability of nomenclature" (ICZN, Art. 75 (b), and see also for botany ICBN Art. 9.6–9.13). They cannot be created merely to have a type for each species (even though its identity is not in doubt), to create more types for a particular collection, or because the original type is looking shabby, as actually occurred in some museums in the early nineteenth century according to Mayr (1969).

However, the exceptional circumstances in which designation of a neotype is appropriate are probably not so exceptional for many invertebrates, especially those described early on that, like the Audouin example given earlier, have only illustrations to serve as their description.

Even when a fairly good description was written originally, designation of a neotype may later become necessary. For example, *Ophiostigma isocanthum* (Say, 1825) was one of the first brittle stars to be described by an American scientist. The description was based on a single specimen from the Florida Keys, but no type material has ever been found, either at the Philadelphia Academy of Natural Sciences (where other species described at the same time were deposited) or in any other museum collection. The species seemed recognizable from Say's description, however, and as long as no similar species were known from the area, the lack of a type created no problem. However, when Gordon Hendler described a new species, *Ophiostigma siva,* occurring in the same region and morphologically very similar to *O. isocanthum,* it became necessary to select a neotype specimen to represent *O. isocanthum* (Hendler 1995). He chose a specimen from the Florida Keys because that was the original collecting locality. In this case, designation of the neotype was not done as a separate paper, but as part of his redescription of Say's species via special sections under material examined and in the description, plus a separate plate illustrating the neotype specimen (figure 21.2).

> **MATERIAL EXAMINED:** Neotype. FLORIDA, ATLANTIC: (LACM 85.265.3), alc. LK 45, 15 May 1985, 24°33.7'N 81°25.7'W, Looe Key National Marine Sanctuary, Florida Keys, 11 m, coll. G. Hendler et al. . . .

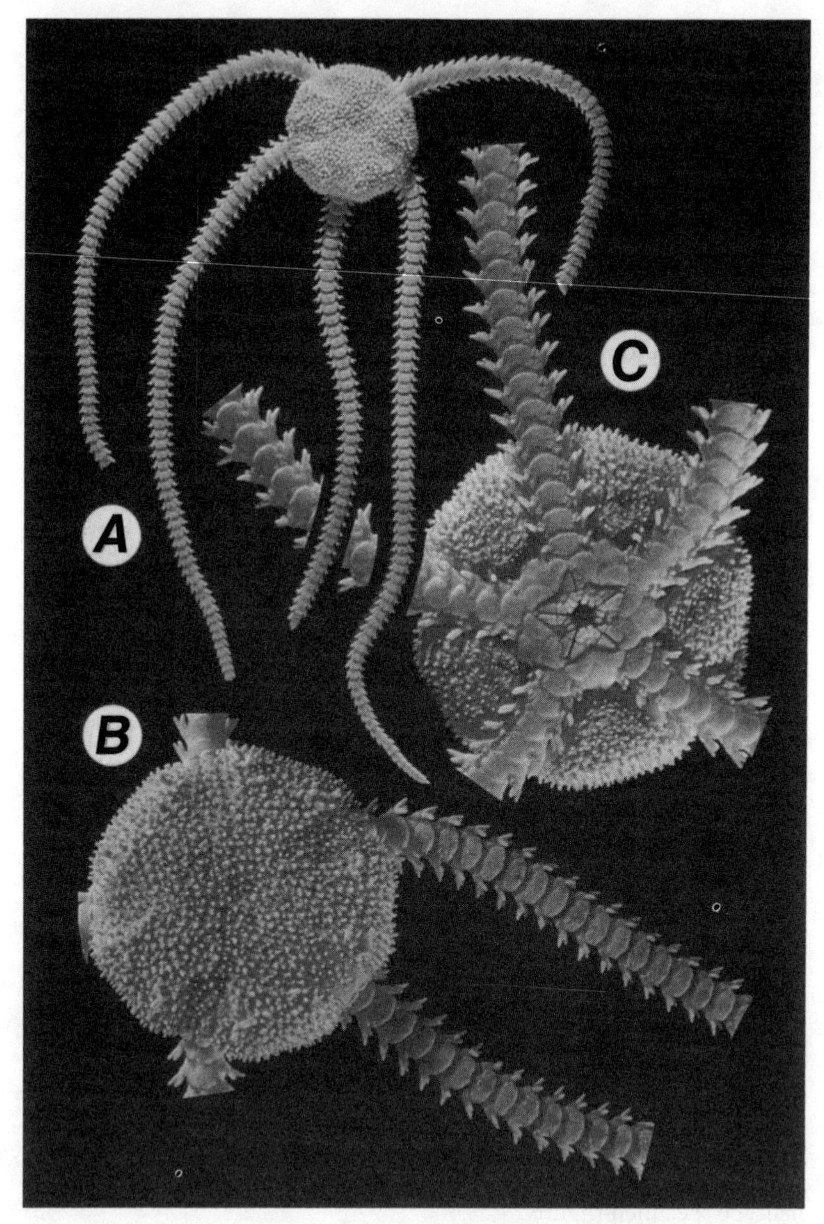

Figure 21.2 Neotype of *Ophiostigma isocanthum* (Say). Neotype LACM 85–265.3. **A,** Entire, dorsal view; **B,** Disk, dorsal view; **C,** Disk, ventral view. Disk diameter = 6.0 mm. *–Source: Figure 2 of Hendler (1995), courtesy of Gordon Hendler.*

DESCRIPTION OF NEOTYPE: disk diameter 6.0 mm; length of longest arm 28 mm. Disk round in outline, dorsal surface flattened, inflated interradially, bearing short, blunt tubercles above and below. Arm joints below disk narrower than those beyond edge of disk. Arm gradually tapers form edge of disk to tip, dorsal surface rounded, ventral surface relatively flattened, arm tip dorsoventrally flattened; joints set off by constrictions of lateral arm plates. Major ossicles of disk and arms thick, somewhat swollen, opaque. (Hendler 1995:5, 7)

Caution is still necessary in describing neotypes, however, for they do have the potential to cause more problems than they solve. Even if they appeared to have been properly designated at the time they were published, additional knowledge or new techniques of study can later show that material once believed to belong to one species really contains a mixture of species. In zoology a replacement neotype must be proposed to the International Commission on Zoological Nomenclature, the proposal must be published in the *Bulletin of Zoological Nomenclature,* and, after a period of time for comment by other biologists, its adoption must be voted on by the members of the commission. The commission's ruling is then published in the *Bulletin* and also summarized in several other journals (e.g., the *Proceedings of the Biological Society of Washington,* the *Journal of Paleontology*). This was the course of action John Bishop had to undertake in trying to fix the identity of *Cribrilina punctata,* the type species of the genus *Cribrilina,* itself the type genus of the cosmopolitan Cretaceous to Recent family Cribrilinidae. Knowing the true identity of this species is important to workers on both living and fossil cheilostomes, but it had long been confused with similar species. Nineteenth-century workers had traditionally accepted a good deal of morphological variation in bryozoan species. With cribrilinids, which are among the smallest cheilostomes, they really didn't have much choice. The zooecial characters now used to distinguish species could not readily be studied before the development of scanning electron microscopy (SEM). In 1972, a proposal was adopted by the ICZN to establish a neotype of *Cribrilina punctata.* Unfortunately, the specimens suggested as possible neotypes included what later SEM studies by Bishop would show were two different species. Worse, the one actually chosen as neotype belonged not only to another species, but to another genus (*Collarina* Jullien, 1886). So Bishop (1986) had to publish a proposal in which he requested that the original neotype be set aside and replaced by a specimen belonging to the other species present in the material suggested in the first neotype proposal. This material did represent the species *Cribrilina punctata,* and so would keep the original meaning of all the generic and family group names intact. He illustrated his proposal with SEM photographs that clarified the morphology of the four British and European species that had previously been confused (figure 21.3). His proposal was voted on by the commission and approved in Opinion 1499, published in the *Bulletin of Zoological Nomenclature* in June 1988:

PROPOSAL

25. The replacement neotype proposed below, BMNH 1985.11.20.1, which belongs to sp. **D**, encrusts a bivalve shell fragment collected in *c.* 45 m of water at 58° 06.8'N 03° 05.2'W, in the Moray Firth *c.* 20 km from the Scottish coast. The specimen is illustrated as Fig. **D**. . . .
26. Under the following proposal, *Collarina* would be a valid and useful genus, and sp. **A** could be described as a new species within it. The genus *Cribrilina* would be closer to its original concept, as would the family CRIBRILINIDAE. The names *Cribrilina punctata* and *C. cryptooecium* would be retained in the accustomed usage. . . .
28. To restore the original meaning of the genera *Cribrilina* and *Collarina* and of the family CRIBRILINIDAE, the Commission is therefore requested:

(1) (a) to use its plenary powers to set aside the neotype designated in Opinion 1016 for *Lepralia punctata* Hassall, 1841;
(b) to designate as replacement neotype for *Lepralia punctata* Hassall, 1841, specimen number BMNH 1985.11.20.1, whose details are given in paragraph 25;
(2) to amend the entry on the Official List of Specific Names in Zoology (Name Number 2523) arising from Opinion 1016 to read: *punctata* Hassall, 1841, as published in the binomen *Lepralia punctata,* as defined by the neotype designated in (1) (b) above. (Bishop 1986:293–295)

The corresponding ruling from the *Bulletin of Zoological Nomenclature* 45(2), p. 177 (1988):

OPINION 1499:

Lepralia punctata **Hassall, 1841 (currently** ***Cribrilina punctata;*** **Bryozoa, Cheilostomata): replacement neotype designated**

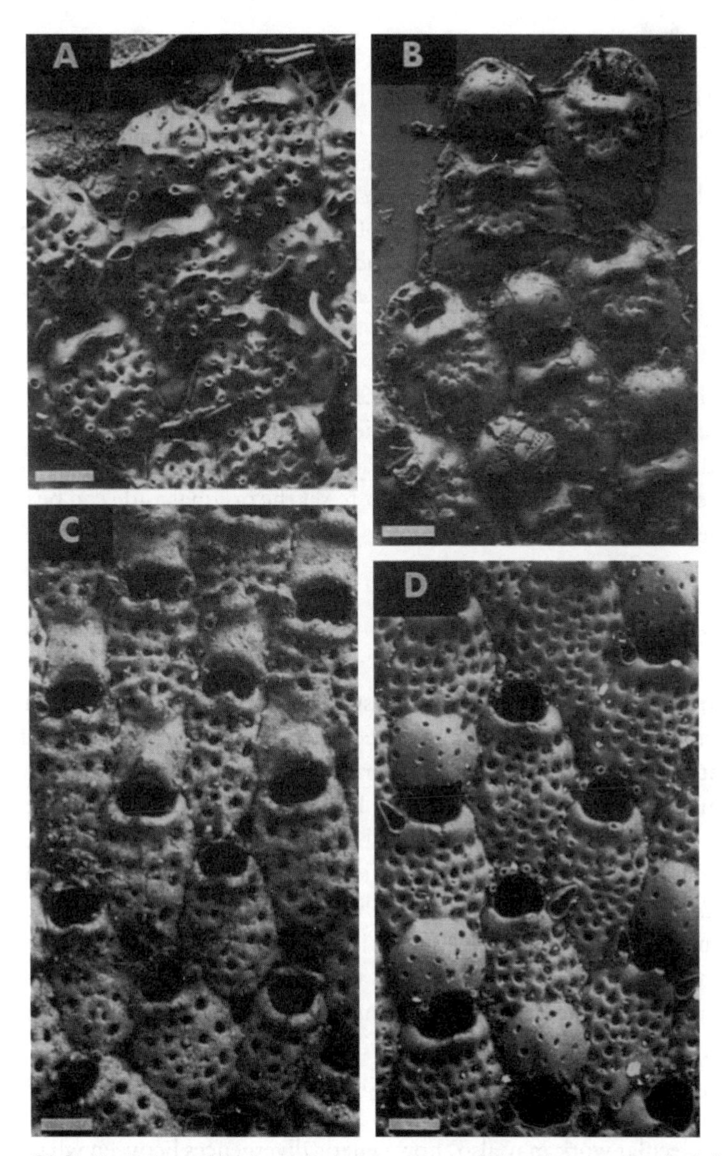

Figure 21.3 Illustrations of the four principal species of cribrimorph bryozoans. **A,** *Collarina* n. sp. BMNH 1973.4.6.1 (sp. A in text); **B,** *Collarina cribrosa* Manchester Museum 1060 (sp. B); **C,** *Cribrilina cryptooecium* BMNH 1847.9.16.118 (sp. C); **D,** *Cribrilina punctata,* BMNH 1985.11.20.1. (sp. D). Scale bars = 0.15 mm. –*Source: Bishop (1986).*

Ruling:

(1) (a) Under the plenary powers the neotype designated in Opinion 1016 for *Lepralia punctata* Hassall, 1841, is hereby set aside;

(b) A replacement neotype, specimen number BMNH 1985.11.20.1, is hereby designated for *Lepralia punctata* Hassall, 1841.

(2) The entry on the Official List of Specific Names in Zoology arising from Opinion 1016 is hereby amended to read: *punctata* Hassall, 1841, as published in the binomen *Lepralia punctata,* as defined by the neotype designated in (1) (b) above.

The fourth edition of the ICZN makes automatic allowance for rediscovery of a lost type of a species or subspecies. The original type, if rediscovered, automatically displaces the neotype. However, in cases where this causes confusion a request to set the original aside can be made to the commission.

Necessary Name Changes

Although the codes are intended to stabilize the system of nomenclature, occasionally names must be changed. Some name changes result from scientific progress, increases in our knowledge of the biology of a group as a result of more research. Increasing biological familiarity may show that what had once been considered to be a single species is actually several morphologically similar sibling species. For example, the jellyfish *Cyanea capillata,* widely distributed along the east coast of North America, was originally divided by Louis Agassiz into three species, but later workers had merged them into varieties of a single species. However, a decade of studying morphological variations and reproductive isolation in two Long Island Sound populations convinced Robert Brewer that even within a single bay and estuary system there might be two separate species (Brewer 1991).

Molecular work may also show genetic divergences between what were previously considered to be morphotypes of a single species. For example, in the sea off the Isle of Man, three morphotypes of a common sponge, *Suberites ficus,* could be distinguished by habitat and coloration. A pale orange morph grew on snail shells occupied by hermit crabs, whereas red-orange and pale yellow forms encrusted scallop shells. But color is not usually a very reliable taxonomic character in sponges, and spicule morphology, the character that is usually used, showed no differentiation.

However, molecular studies by A. M. Solé-Cava and J. P. Thorpe (1986) showed that the three morphotypes could be clearly differentiated on the basis of isozyme patterns, evidence that together with their sympatric distribution, argued that they represent different species.

At the opposite extreme, further research may show that what were described as separate species may turn out to represent intraspecific variation between populations of the same species. In the sponge genus *Cliona,* for example, three growth stages, papillate, boring, and massive, were considered separate species by some taxonomists and lumped as one species by others. Sponge biologists knew that this species or species complex varied in coloration, growth form, and spicule size under different environmental conditions, but this only made the situation more confusing. Finally, detailed study of the pattern of biological variation by Dolors Rosell and Maria-Jesus Uriz (1991) showed that color and growth forms varied in abundance along a vertical bathymetric gradient, whereas spicule morphology followed an allometric growth curve. The authors concluded that only one species, *Cliona viridis,* was present.

Recently (as of the Tokyo Code, 1994, Art. 9.7) botanists developed a new kind of type, the *epitype,* to deal with situations in which previously selected type material is ambiguous (as could happen if the plant selected as a holotype was so heavily parasitized that taxonomic characters were altered or destroyed) (J. McNeill, TAXACOM Listserv, 16 Mar. 1995). Article 9.7 of the ICBN states, "An epitype is a specimen or illustration selected to serve as an interpretive type when the holotype, lectotype or previously selected neotype, or all original material associated with a validly published name, is demonstrably ambiguous and cannot be critically identified for purposes of the precise application of the name of a taxon. When an epitype is designated, the holotype, lectotype or neotypes that the epitype supports must be explicitly stated," as in the following example:

> From among the herbarium collections prepared from the cultivated plants that were collected by the original author when the species was described, the only fertile specimen (*N. E. Brown s. n.,* 3/19/1884, K) is here designated as the epitype of the name *Godwinia gigas* Seemann and consequently of *Dracontium gigas* (Seemann) Engler (Fig. 2).
>
> **Dracontium gigas** (Seemann) Engler in DC., Monogr. Phan. 2: 284, 1889. *Godwinia gigas* Seemann, J. Bot. 7: 313–315. t. 96 & 97, 1869. LECTOTYPE: Figure 1 in J. Bot. 7: t. 96 and 97. 1869. EPITYPE: Cultivated plant in Royal Botanic Gardens, Kew, originally collected by

Seemann between the Javali Gold Mine and the Quebrada de las Lajas, Chontales, Nicaragua, *N. E. Brown s. n.* 3/19/1884 (K). (Zhu 1994:407)

Replacement Names: Homonymy

The second type of necessary nomenclatural change results from additional library or museum research on a group. In the course of background research you may find that the species you are studying has been described by a number of authors working separately. This occurred more commonly in the days when communication was slower and more difficult. However, it can still happen that a worker has described a new species without realizing that the same name has already been used for another species in that genus. In such cases the name must belong to the species that has priority. The second (or any later) use of the name becomes a *junior homonym* and it must be replaced with a new name. In a straightforward case this is a simple matter that may take the form of a brief note, a paragraph or two in length. It should include the following information: a history of the mix-up, with references for both senior and junior homonyms, the new name, and its etymology. For example, in describing a new species of psocopterid insect from Moorea, Ian Thornton inadvertently used a name that was already in use for another species. When he realized he had done so, he published a replacement name for the species he had described.

Aaroniella andrei nom. nov.

Replacement for *Aaroniella badonelli* Thornton, 1989, *Bull. Mus. Natl. Hist. Nat.,* Paris. 4ᵉ sér., 11, 1989, section A, no. 4: 810–812. Preoccupied by *Aaroniella badonelli* (Danks): DANKS, L. (1950), State Museum of Natural History, Riga, Latvian Soviet Republic, Information Sheet No. 1, 2 pp. Transferred to *Aaroniella* by THORNTON, I. W. B. (1981). *Systematic Entomology,* 6: 425, 432.

The new name retains the purpose of the preoccupied one: to acknowledge the great debt that I, along with other students of the Psocoptera, owe to the late André BADONNEL in the study of this order of insects. (Thornton 1991:483)

The best way to get avoid having to publish a replacement name for a species you have described is to do a through job of background research, including, if possible, corresponding or working with a specialist. A specialist will know what projects other people in the field have going and may be able to save you from using a name that someone has used in a manuscript already in press or published. Even so, mistakes will happen; no one has access to all the literature as soon as it comes out, and the volume of scientific literature produced means that indexing and abstracting of names still lag months or years behind publication.

The problem of describing a replacement name can get more complicated. For example, in 1894, a Black Sea population of copepods was described by V. Karavaev as *Calanus finnmarchicus* var. *pontica.* Soon after that, however, he reexamined his specimens in the light of new information published on Mediterranean populations, decided that the differences on which he had based his variety did not hold up, and in a second paper withdrew his var. *pontica* (Karavaev 1895). But later workers continued to notice distinctive characteristics in the Black Sea populations, and in 1987 Fleminger and Hulsemann proposed that it be recognized as a species and given the name *Calanus ponticus* Karavaev. However, as they later learned, another Russian worker, Krichagin, had used the name *Calanus ponticus* for a completely different Black Sea copepod some 20 years before Karavaev's work was published (Krichagin 1873). Because Krichagin's name had priority, the *C. ponticus* of Karavaev had to be given another name. After recounting the entire confusing history of the species, this was done by Hulsemann (1991):

> I hereby propose *Calanus euxinus* as a new replacement name for *C. ponticus* Karavaev, 1894. The species-group name is taken from the Greek *euxeinos* meaning hospitable, a classical epithet of the Black Sea. (Hulsemann 1991:62)

Conservation of a Name

Normally the principle of priority applies to binomens: The oldest name takes precedence, the rest become synonyms. However, in rare instances, as when a commonly used species turns out to be going under a later name, sticking to the rule would cause a great deal of inconvenience and loss of information. For example, all the references in the literature under the old name would be impossible to find if you did not happen to know what the species had previously been called. For such reasons, exemptions to the rules allow certain names to be conserved.

In zoology this is done via an application to the Commission on Zoological Nomenclature, the 18-member board covering all major groups of living and fossil animals, part of whose job is to approve or disapprove such changes. For example, in 1956 the ICZN ruled that generic names published by Lorenz Oken in his *Lehrbuch der Naturgeschichte, Vol. 3* (1815–1816) were to be rejected for nomenclatural purposes, but zoologists were allowed to propose conservation of any names dating from this work, where they felt changes would lead to confusion in the taxonomy of a particular group. Bryozoan worker John Ryland decided that two bryozoan genera named in Oken, *Bugula* and *Scruparia,* had been so commonly used (*Bugula* is the usual textbook example of a bryozoan) and the genera themselves were so widespread and familiar that replacing them with new or unfamiliar, long synonymized names would definitely be a mistake. So he wrote a proposal to the commission, giving all the information on the case and concluding with this request:

> 6. The International Commission on Zoological Nomenclature is requested to take the following action:
>
> (1) to use its plenary powers:
>
> (a) to validate the generic name *Bugula* Oken, 1815, as allowed in Opinion 417, and to designate *Sertularia neritina* Linnaeus, 1758, as the type-species;
>
> (b) to validate the generic name *Scruparia* Oken, 1815, as allowed by Opinion 417, and to designate *Sertularia chelata* Linnaeus, 1758, as the type-species;
>
> (2) to place the following generic names on the Official List of Generic Names in Zoology:
>
> (a) *Bugula* Oken, 1815 (gender: feminine), type-species by designation under the plenary powers in (1) (a) above, *Sertularia neritina,* Linnaeus, 1758;
>
> (b) *Scruparia* Oken, 1815 (gender: feminine) type-species by designation under the plenary powers in (1) (b) above, *Sertularia chelata,* Linnaeus, 1758. (Ryland 1967:25)

He also had to request that the junior synonyms that would have otherwise replace these names be invalidated by being placed on the *Official Index of Rejected and Invalid Generic Names in Zoology.* The changes requested by Ryland were approved and biologists still use the old familiar names.

In botany, proposals to amend the International Code of Botanical Nomenclature are published in a special section of *Taxon* (the Journal of the International Association for Plant Taxonomy). If you encounter a problem in botany you would need to write a statement of the cases for and against its rejection (pretty similar to the point-by-point case you would make in a proposal to the ICZN) and submit it to the attention of the General Committee. All the proposals they gather are discussed and voted on at a meeting of the Nomenclature Section of the International Botanical Congress and finally authorized by the plenary session at the next International Botanical Congress (ICBN, Div. III, McNeill and Greuter 1987). The ICBN itself includes the conserved names in special appendices. Conserved generic names are listed in boldface type in the left-hand column of Appendix IIIA, with synonym and earlier homonyms in the right-hand column. List of conserved and rejected specific names (a much smaller list because originally only the names of economically important species could be conserved) are listed in Appendix IIIB. A list of names rejected because each had been "widely and persistently used for a taxon or taxa other than its type" makes up Appendix IV. Although the appendixes take up more room in the ICBN volume than the code itself, this system ensures that the information is gathered conveniently in one book.

Emendations

The other name changes are nomenclatural. Just saying *emendation* brings to mind the kind of excessive pickiness that once gave taxonomy a bad name. It is true that, just as they used to "correct" types if the originals didn't seem good enough, some taxonomists in former days were fond of correcting names if the Latin was incorrect or the Greek improperly transliterated, and keeping up with their changes caused headaches for other biologists (Mayr 1969). But now the codes strictly limit the changes that can be made. In zoology ICZN Articles 32–34 cover these cases. The original spelling is not to be altered just because it was improperly Latinized or transliterated. Emendations are allowed only when it is clear from the original publication that a mistake in copying or typesetting the name has occurred. For example, if you stated in your etymology section that you were naming your new barnacle species *Xxxxus darwini* in honor of Charles Darwin, but a typographic error turned it into *Xxxxus barwini,* it is the kind of error that should be corrected by the first person to spot it; to do so is considered a justified emendation. If an original description contains more than one spelling, the first person to revise the group gets to choose which spelling should be used thereafter (ICZN, Art. 32 b (i)).

On the other hand, when, in the research described in the neotype example mentioned earlier, Gordon Hendler noted that the Latin of Say's original spelling of the species name (which should have been *Ophiura isacantha,* rather than *Ophiura isocantha*) was incorrect, he did not change it. He realized that according to the ICZN this kind of minor error in Latinization does not need correction and he left the species name spelling as it had been (Hendler 1995).

The zoological code also states that you cannot change the **-ii** of an old honorific personal species name to **-i**. However, changes made by removing diacritical marks and hyphens and by compounding separate word species terms are required.

Article 60 of the ICBN covers original spelling of names. As in zoology, the original name is kept unless there have been typographic or writing errors, or if it is to delete hyphens or diacritical marks or to standardize compound names. In botany, corrections do need to be made when personal names do not follow the recommendations laid down (see chapter 8).

New Combinations

In zoology, the new combinations of genus and species that occur when a genus is renamed or the taxonomic placement of a species changes are often carried out without any special attention being called to them. The author of the paper usually gives the evidence for moving the species to a new genus in the discussion section of the description, but may or may not note in the heading for the species description that it is a new combination (new comb., comb. nov., etc.). However, some journals require authors, and some authors prefer, to note these changes directly, either in the title of the paper (especially if it concerns only one species) or in the heading of the species description for that species:

Namikawa, H., S. Kubota, and S. F. Mawatari. 1990. Redescription of *Stylactaria uchidai* (YAMADA, 1947), comb. nov. (Hydrozoa: Hydractiniidae) in Hokkaido, Japan. *Proceedings of the Japanese Society of Systematic Zoology* 42: 2–9.

in which *Stylactis uchidai* Yamada, 1947 becomes *Stylactaria uchidai* (Yamada, 1947).

Kubota, S. 1991. Second finding of *Stylactaria piscicola* (Komai, 1932) comb. nov. (Hydrozoa: Hydractiniidae) from off Atsumi Peninsula, Japan. *Publications of the Seto Marine Biological Laboratory* 35: 11–15.

in which *Stylactis piscicola* Komai, 1922 becomes *Stylactaria piscicola* (Komai, 1932). Both of these changes were necessary because a previous worker (Calder 1988) had changed the name of the genus to *Stylactaria.*

Alvarez, F. and J. L. Villalobos. 1994. Two new species and one new combination of freshwater crabs from Mexico (Crustacea: Brachyura: Pseudothelphusidae). *Proceedings of the Biological Society of Washington* 107: 729–737.

in which *Pseudothelphusa guerreroensis* Rathbun, 1933 becomes *Tehuana guerreroensis* (Rathbun, 1933). In this case the species was reassigned to the genus *Tehuana* for taxonomic reasons: "We propose the assignment of *Pseudothelphusa guerreroensis* to the genus *Tehuana,* based on the presence of a distinct frontal border of the carapace and on the partially fused marginal and mesial processes in the male gonopod" (p. 732).

New combinations can be made more casually in zoology because the changes do not change the author and date of the name. In zoology each combination of a species name with a different generic name is *not* considered to be a different scientific name, as it is in botany. Because of the noncoordinate status of genus and species in botanical nomenclature, a change in generic placement is a little more complicated because it affects the dates of establishment and authorship as well as the operation of priority (Jeffrey 1986). For example, Harold Robinson (1988) described the new genus *Chrysolaena* for seven Neotropical members of the Asteraceae (Aster or Composite family):

Chrysolaena flexuosa (Sims)
H. Robinson, comb. nov.

Vernonia flexuosa Sims, Bot. Mag. 51, pl. 2477, 1824. *Vernonia montevidensis* Nees ex Otto & Dietr., Allg. Gartenszeitung 1: 229. 1833. *Cacalia flexuosa* (Sims) Kuntze, Rev. Gen. Pl. 2: 970. 1891. Argentina, Brazil (Rio Grande do Sul, Santa Catarina, São Paulo), Paraguay, Uruguay. (Robinson 1988:956–957)

A change in rank (e.g., from variety to species) also results in changes that have to be formally recognized, as when Norman Robson elevated this tropical tree from variety to species:

Vismia tenuinervia (E.v.d. Berg) N. Robson, comb. et stat. nov. *Vismia cayennensis* var. *tenuinervia* E. v.d. Berg, Acta Amazonica 4: 16, f. 19. 1974. (Robson 1990:411)

Sometimes further study may show that a plant variety not only deserves recognition as a species, but also must have a new name because the variety name has already been used as a species name in that genus.

Globba corneri **A. WEBER, stat. et nom. nov.**
≡ *Globba unifolia* RIDL. var. *sessiliflora* HOLTTUM, *Gard. Bull. Sing.* **13:** 34 (1950); LIM, *Not. Roy. Bot. Gard. Edinburgh* **31:** 252 (1972).– Non. *G. sessiliflora* SIMS., *Bot. Mag.* tab. 1428 (1811). (Weber 1991:1)

In the preceding example, Anton Weber had photographed and collected living specimens of this tropical garland-lily, a plant of the ginger family, in the Malay Peninsula. He also collected the specimens of *Globba unifolia,* the species of which it was thought to be a variety. He found several differences between the two, some of which had not been apparent to the author of the varietal name, who had worked from dry material collected many years before. So he decided that *Globba unifolia* var. *sessiliflora* should be raised to species rank. But as a species it could not be called *Globba sessilifora;* another *G. sessiliflora* already existed. He chose to rename it *Globba corneri* instead, in honor of Professor E. J. H. Corner, the University of Cambridge botanist who, in 1935, had first collected it.

Lack of Information

A final problem is lack of information. Sometimes you just don't have the information you need to place your new species taxonomically, or even to describe it completely. This happens most often in paleontology. It can also happen in studies of living groups of plants and animals when there just isn't enough material (e.g., there are no adult specimens or no female or male specimens) or the material is in fragments or otherwise damaged by collecting techniques.

In some cases you can hold off description and publication until you can obtain more material. In other cases (e.g., when you are dealing with writing up an entire fauna or flora), it is probably best to publish whatever information you have. Even a limited amount of information may be more useful to other researchers than none.

In paleontological surveys in particular, it common to have enough material of some members of the assemblage to describe them and identify them as new or known species, whereas other taxa may be represented by a single tooth or bone. For example, when Jason Lillegraven and Malcolm McKenna described the Late Cretaceous fossil mammal fauna from the Bighorn and Wind River Basins of Wyoming (1986), they described to species level the taxa for which there was sufficient material, but for others could only state

> In addition to the specimens discussed above, a number of isolated teeth and jaw fragments remain unidentified. . . . Although most are fragmentary or heavily worn, some are excellently preserved and should be identifiable once larger samples and better comparative series become available. (Lillegraven and McKenna 1986:39)

and

> They are (1) "Multituberculata, new genus and species unnamed," representing a presently unidentifiable species of extraordinary large body size; (2) "Dryolestidae, genus and species unidentified," a phylogenetic relict represented by materials inadequate for diagnosis. (Lillegraven and McKenna 1986:39)

A nonfossil example of dealing with inadequate information is given in a paper by Øresland and Pleijel (1991). These authors studied and described a polychaete worm ectoparasitic on the Antarctic chaetognath, *Eukrohnia hamata.* Although members of the holopelagic polychaete family Typhoscolecidae had been reported to prey on chaetognaths by decapitating them, no one had actually studied the feeding ecology of the worms. They discovered that the polychaete parasitized the chaetognath by attaching behind its head and feeding (probably on coelomic fluid), eventually causing the chaetognath to lose its head. Because no mature specimens of the parasite were found in their study, its species identity could not be determined. Based on body shape and morphology of cirri, the worms were tentatively assigned to a genus (*Typhoscolex*). The authors also provided measurements and SEM photographs that might make it possible to identify or describe the species once adults have been found.

Chapter 22

Further Studies in Systematics

> *Thus a paleontologist unearthing skeletons in an Asian desert and a molecular biologist sequencing a strand of deoxyribonucleic acid (DNA) can both claim to be systematists if they share an interest in how species are related and how they arose over time. All these issues depend on theories of patterns of descent, of organisms branching off from each other in a way that accurately reflects their histories. When such theories continue to successfully explain new observations, they form the basis for many statements about the biological world.*
>
> –NOVACEK (1992:103)

> *It has been said that for every problem concerning living things there is an organism ideal for its solution. It is probable that there are still undiscovered species living that hold the answers to problems that face us now or will in the future.*
>
> –EVANS (1993:267)

Although the words *taxonomy* and *systematics* are sometimes used interchangeably, the definition followed in this book (chapter 1) considers *taxonomy* to be the part of systematics dealing with the description, naming, and classification of organisms. *Systematics,* by this definition, is the entire scientific field that "deals with the organization, history, and evolution of life. It ultimately asks, how did life forms originate? How did they diversify and how are they distributed both in space and time?" (Novacek 1992:103). Carrying out the process of describing a new species through to publication is a very satisfying endeavor, if for no other reason than that of finally having a name for an organism you've been studying for months or years. But many who begin this process discover that species description is just a prelude to the real fun of systematics: trying to answer some of the most difficult and universal questions of biology.

If you want tackle some of these problems, you will need to know more about the methods and the theoretical aspects of systematics than can be covered in this book. This final chapter briefly describes the main philosophies, methods, and topics and gives citations for some of the many excellent references from which you can learn more.

Before starting that discussion, it is important to review two important points from the beginning of this book. First, phylogeny and classification are not necessarily one and the same; in fact, complete concordance may never be possible. Second, a "natural" or phylogenetic classification may not be the easiest or best to use in terms of practical identification needs (chapter 2; Stace 1989). The classification systems of biological nomenclature consist of hierarchical groupings of taxonomic categories, ordered in an ascending series for our convenience, whereas a phylogeny consists of a branching pattern of descent. Even when such patterns can be reconstructed for organisms, it is difficult to show them in a single linear sequence, as you would need to do for a taxonomic monograph, for example (Stace 1989; Stuessy 1990). This compromise is still unresolved; suggested solutions have ranged from ignoring phylogeny to abolishing the Linnaean hierarchy, but a really practical solution has not yet appeared and most systematists have managed to work within the present system's limitations without too much difficulty.

Evolutionary Systematics

Many early classifications (among them Linnaeus's mistaken but heavily self-promoted sexual system of plant classification) were *artificial,* based on the most useful diagnostic characters, whether or not they reflected relationships. They were often *monothetic,* relying on single or a very few characters chosen in advance by the taxonomist. They resulted in some very strange groupings that had little predictive value. Even before Linnaeus's time, taxonomists had begun to attempt to develop *natural classifications,* classifications that did reflect genetic and phylogenetic relationships. These were *polythetic,* relying on as many shared characters as possible, characters selected by the taxonomist because they were correlated with other characters. These classifications did have some predictive value (Stuessy 1990).

Obviously, before the publication of the evolutionary theories of Darwin and Wallace in 1859, classifications could not intentionally reflect the evolutionary history of the group involved, but it took many years after that date for the idea that classifications should be phylogenetic to percolate through the system. In the last part of the nineteenth century, paleontologists and biologists began seeking *homologous* characters (those inherited from a common ancestor), using evidence from anatomical comparison and embryology (e.g., Owen and Haeckel). But it was not until the early 1900s, when the newly developed field of genetics had begun provide explanations for observed patterns, that scientists consciously began to develop phylogenetic classifications (Stace 1989; Minelli 1993). Classifications sought by taxonomists became "natural" not just in the sense of overall similarity, but in the sense of seeking primitive and derived characters and lines of descent and branching, often depicted as dendrograms called phyletic or phylogenetic trees (Stace 1989; Stuessy 1990). Figures 22.1 and 22.2 show examples of such dendrograms.

The incorporation of genetics and population biology, as well as information from paleontology, into systematics led to what was called "The New Systematics," after a book edited by Julian Huxley (1940) that documented its approaches and development. This era of systematics

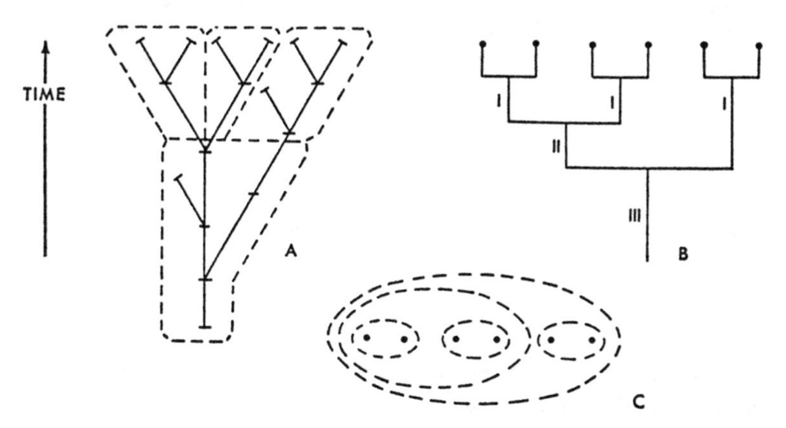

Figure 22.1 Three ways of presenting tree diagrams. **A,** A hypothetical phylogeny. The solid lines equal species, which have a duration (length in time) along the line (shown by the separating cross bars). The dotted lines represent one of the ways in which the species might be assigned to genera. **B,** A dendrogram with no time dimension. In this diagram the terminal circles represent extant species. The lines represent the relationships among the species. The numbered stems could be used to group them into phylogenetically valid higher taxa. **C,** A group-in-group Venn diagram arrangement of the terminal species of A or the species of B. *–Source: Figure 3 of Simpson (1961).*

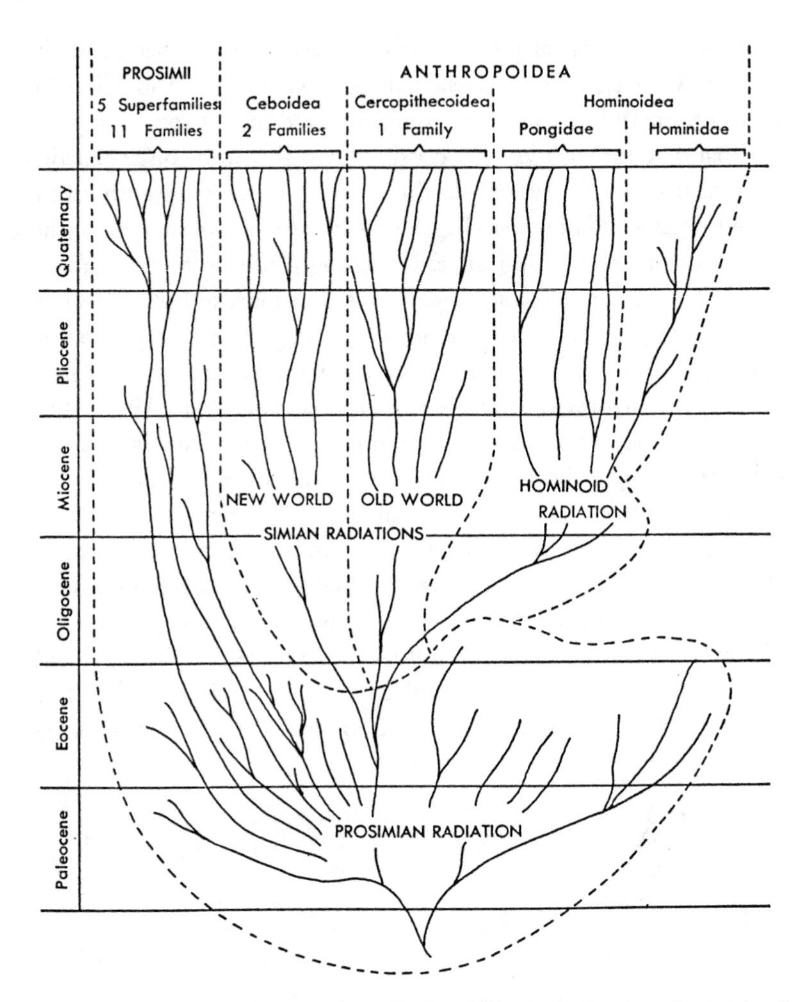

Figure 22.2 Diagram of phylogeny and classification of the primates (a schematic phylogenetic tree). *–Source: Figure 22 of Simpson (1961).*

(1940s–1960s) focused extensively on experimental methods, the use of nonmorphological characters, the development of the biological species definition, and on *microtaxonomy,* studies carried out at the species and subspecies level. But it also took a stand on the issues of *macrotaxonomy,* the classification of higher taxa (Mayr and Ashlock 1991; Minelli 1993). This new approach made systematics exciting and attracted many biology and paleontology students to the field. Important contributions included the clear distinction between *grades,* or levels of complexity, and *clades,* monophyletic groups of taxa (Huxley 1958; Quicke 1993), and detailed studies of patterns of speciation: *allopatric* (during and after geographic isolation), *parapatric* (by divergence of two contiguous but nonoverlapping populations), and *sympatric* (without geographic isolation, via isolating mechanisms that develop within a population) (Mayr and Ashlock 1991).

The principles of classical evolutionary systematics are presented in books and articles by Simpson (1945, 1961), Bock (1977), Mayr (1969, 1982), and Mayr and Ashlock (1991) (see this chapter's *Sources*). Evolutionary systematists believe that classifications must reflect the genealogical, ancestor–descendant relationships of taxa. But they also follow Darwin's example in believing that the amount of differentiation taxa have undergone since branching should be incorporated in into classifications (Mayr 1988). Evolutionary systematic classifications should show homogeneity and similarity within taxa as well as reflecting the group's evolutionary origin. This means, for instance, that a group such as birds, which has diverged greatly from its ancestral and sister groups, would be classified separately (e.g., as a separate vertebrate class) rather than being classified as a sister order to the Crocodilia (with whom the birds are considered to have shared most recent common ancestry) (Marcus 1993). Evolutionary systematists agree with cladistic systematists that the branching points in a phylogeny can be determined only by shared derived characters or *synapomorphies,* but they do not think that *plesiomorphic,* primitive or ancestral characters should be ignored in phylogenetic schemes because they may include some of a group's most outstanding characteristics (Mayr 1988; Mayr and Ashlock 1991).

What evolutionary taxonomy lost in its attempts to deal with both phylogeny and the amount of differentiation shown by descendant taxa was the ability to show phylogenetic relationships clearly. In evolutionary classifications the phylogenetic interpretation had to be discussed separately, with additional diagrams and discussion (Marcus 1993). Relationships in many groups of organisms still could not be untangled, and some taxonomists began to feel that their approach had been too intuitive, with classifications relying too much on a taxonomist's judgment as to what characters to use and how much weight to give them. They thought there should be some way to make classifications more objective, to frame their questions clearly as testable hypotheses, and to incorporate rigorous analysis of characters.

This dissatisfaction was occurring at about the time that computers were becoming standard features of universities, although they were not yet in every office and lab. The conjunction of increasing accessibility of computers with this disillusion with the attempts at phylogenetic classifications led to the development in the late 1950s and early 1960s of a new methodology called *phenetics,* or classification by similarity.

Phenetics

Similar procedures were developed at about the same time by P. H. A. Sneath, who worked on bacteria, and by C. D. Michener and R. R. Sokal, who worked on bees. Sokal and Sneath produced two books on the subject: *Principles of Numerical Taxonomy* (1963) and *Numerical Taxonomy* (1973). As summarized by Stace (1989), phenetic classifications are based on the following principles. First, the more information content and the greater the number of characters used, the more predictive the resulting classification will be. Second, phenetics is empirical, rather than deductive in method; every character is given equal weight. Third, overall similarity between two entities is a function of their individual similarities. Fourth, taxa can be distinguished by differing correlations of characters. Finally, classifications are based on phenetic similarity, but by making certain assumptions about evolutionary pathways and mechanisms, phylogenetic interpretations can be made.

Phenetics groups from the bottom up. Similar species are grouped into similar genera, similar genera into families, etc. Its basic unit is the operational taxonomic unit (OTU), the lowest taxonomic entity being used in a particular study. These might be species or higher taxa, or they might be samples of populations or individuals. Phenetic taxonomists seek to identify and sample as many characters as possible in each study (with 60 being a minimum number and 100 or more being desirable). They draw the characters from as many data sets as possible–morphological, behavioral, chemical, ecological, and so on–and organize them into a data matrix of OTUs times characters. The character states are usually coded into binary (positive/negative) or multistate rankings. The matrix is then processed in a computerized cluster analysis that produces a similarity matrix or distance coefficients, giving the similarity or dissimilarity of all possible pairwise combinations of OTUs. The results can be grouped by computer into hierarchical branching diagrams called *phenograms,* tree diagrams with OTUs combined into more and more inclusive groups (figure 22.3), as well as cluster diagrams (figure 22.4) of various sorts (Stace 1989).

The original hope of the pheneticists was that as the number of characters used increased and more groups were analyzed, the results would be repeatable and would eventually result in a natural classification. Unfortunately, although results of a particular analysis on a particular matrix are repeatable, there are a number of different cluster programs available (e.g., UPGMA, WPGMA, centroid clustering, nearest neighbor clustering, furthest neighbor clustering) and different analyses carried out on the same matrix can lead to very different clusters and thus to very different classifications (Stace 1989; Marcus 1993; Quicke 1993).

Principal component analysis, canonical variate analysis, and principal coordinate analysis are also used in phenetics. They are used to show the gaps between taxa or to find the best ways to tell similar taxa apart (Quicke 1993).

The chief contribution of phenetics to modern taxonomy was that it forced taxonomists to think carefully about their goals and methods, their characters and character states, and to define their terms clearly (as was necessary for computer analysis). It also made taxonomists familiar with (some expert in) computer analysis, which became a factor in the next development in systematics: cladistics. Phenetics is still valuable in microbiology and in the taxonomy of groups of plants in which reticulate patterns of evolution predominate. It is most useful where other methods have failed or are just too difficult to use. It is also used by botanists to check their classifications in order to modify and improve them (Stace 1989). The taxonomic literature shows that the number of papers using phenetic methods (figure 22.5) is still increasing yearly, although not at the same rate as papers using cladistic methods (Winston and Metzger 1998).

Cladistics

In 1950, when a computer still filled an entire room, German entomologist Willi Hennig published a book called *Grundzüge einer Theorie der phylogenetischen Systematik.* This book laid the groundwork for *phylogenetic systematics* or *cladistics.* It was revised by Hennig and translated into English in 1966. Despite its cumbersome terminology (e.g., "character-bearing semaphoront" for organism), the logical clarity of its method of

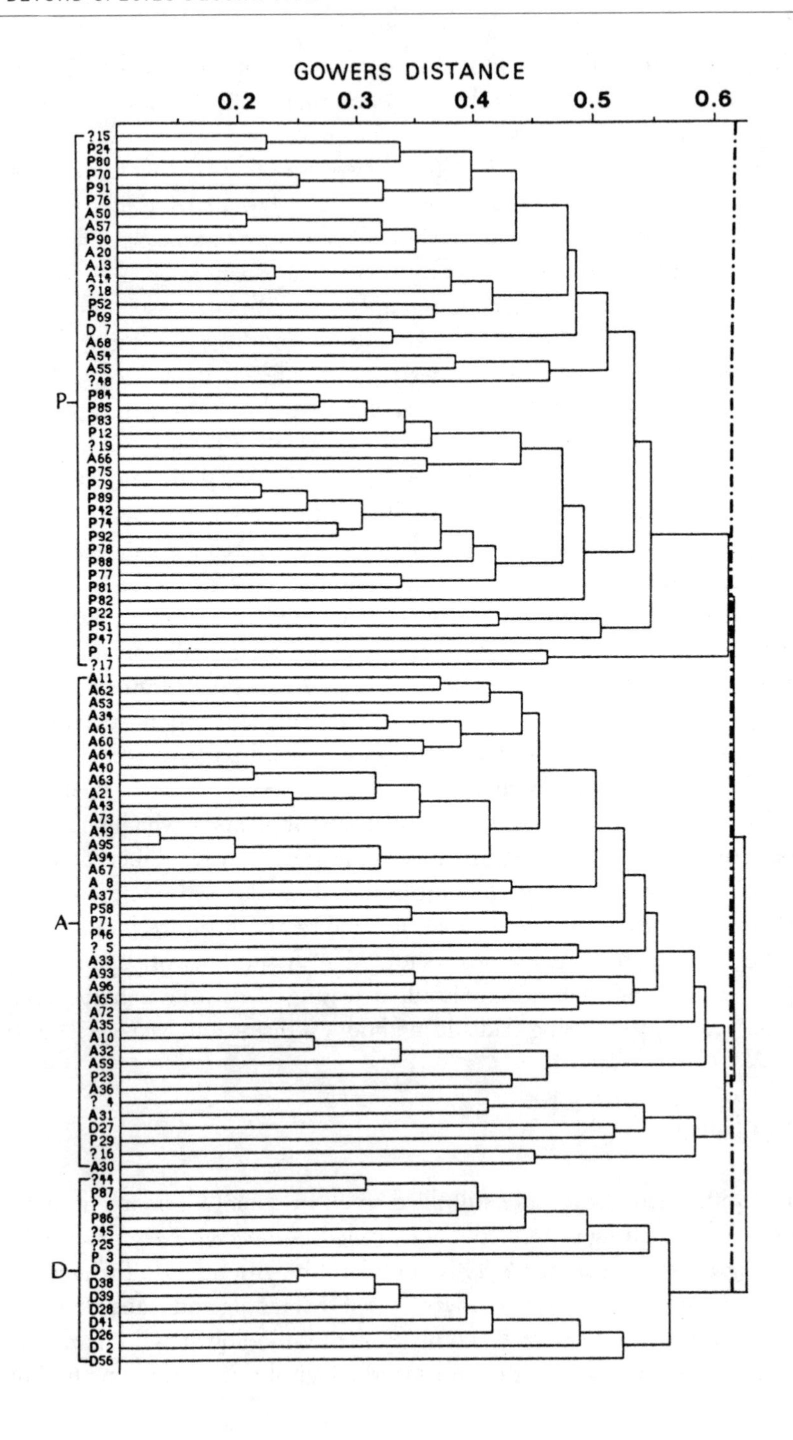
GOWERS DISTANCE
0.2
0.3
0.4
0.5
0.6
P
A
D

Figure 22.3 (opposite) Phenogram showing cluster analysis of populations in the *Chenopodium atrovirens* (A), *Chenopodium dessicatum* (D), and *Chenopodium pratericola* (P) complex (Goosefoot family). Specific identity of the populations with question marks was uncertain before analysis. Dashed line represents the level of similarity at which the three series of populations are believed to be specifically distinct. *–Source: Reproduced from figure 7.5 of Stuessy (1990), from Reynolds and Crawford (1980:1358).*

constructing phylogenies had a great impact on systematics. Cladistics was especially appealing to students and systematists already familiar with statistical computer techniques. And although simple cladistic analyses can be carried out without a computer, the more complicated analyses

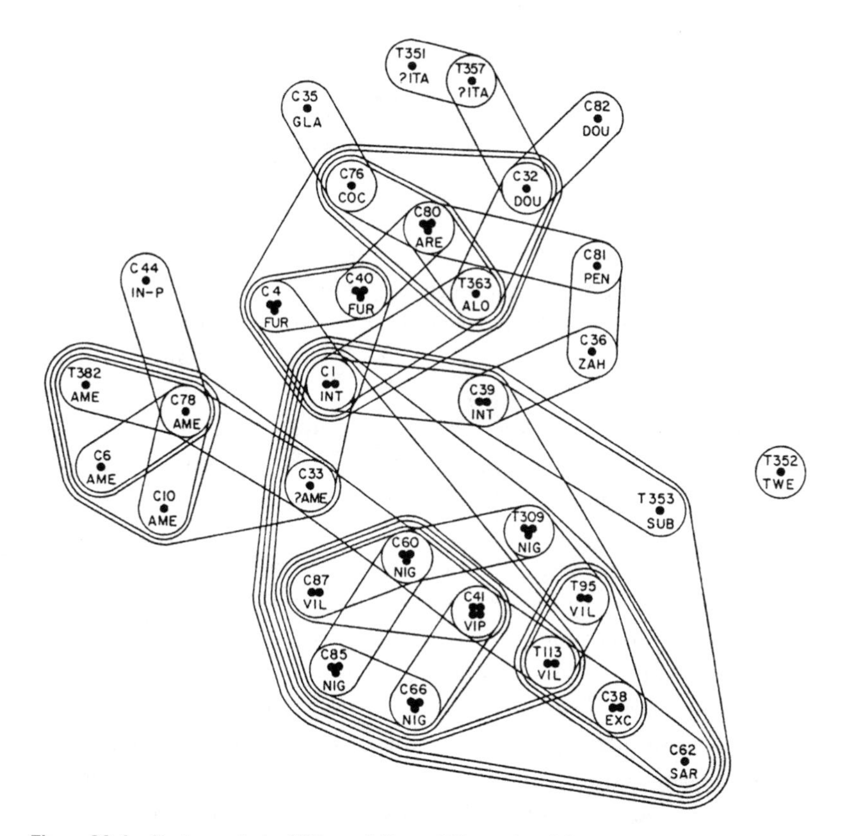

Figure 22.4 Cluster analysis of 32 populations of 19 species of *Solanum* sect. *Solanum* (Solanaceae) based on a dissimilarity matrix. OTUs positioned to facilitate drawing of clusters. Lines enclose levels of dissimilarity. Populations coded by letter and number code and species by acronyms with dots representing their ploidy level: one dot = diploid, two dots = tetraploid, etc. *–Source: Reproduced from figure 7.8 of Stuessy (1990), from Edmonds (1978:43).*

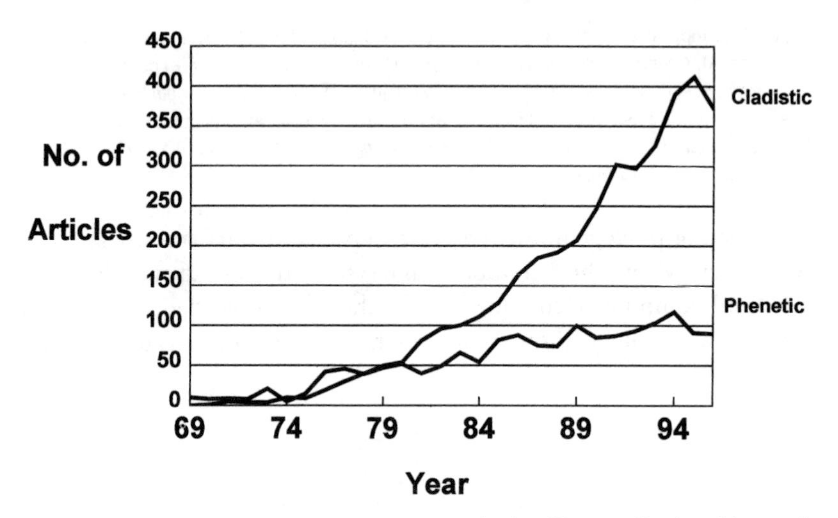

Figure 22.5 Trends in taxonomic methodology over the last 28 years. Number of taxonomic articles using cladistic and phenetic methods by year. *–Source: Figure 4a of Winston and Metzger (1998).*

do require computers. In fact, most systematic innovation in the last 30 years has revolved around the development of more and better computer methods of phylogeny construction and better understanding of the mathematics and theory underlying them.

The three basic principles of cladistics are *apomorphy, monophyly,* and *parsimony* (Funk and Brooks 1990). An apomorphy is a derived character state, as opposed to the primitive or *plesiomorphic* state, or as Wiley (1981:9) defines it, "an evolutionary novelty and its preexisting homologue." For example, if a crustacean with a small claw on a particular limb segment evolves into a species with a large claw on the same limb segment, the claws of the two are homologous, and the large claw is the derived, or apomorphic state. Monophyly or *holophyly* is the establishment of groups of taxa in which all members have a common ancestor and that include all descendants of that ancestor. Its opposite is *paraphyly,* having a group that includes a common ancestors and shared derived character states but does not include all the descendants of that ancestor (Quicke 1993). Parsimony is the principle that the evolutionary pathway or its phylogenetic explanation is most likely to be the one requiring the fewest steps (Quicke 1993).

Cladistic methods directly address the phylogenetic relationships of organisms by the use of branching diagrams called *cladograms.* Each node or branching point on a cladogram represents a speciation event. It shows the shared derived characters that define *sister taxa,* those related most closely by common ancestry (figure 22.6). According to this method, only *synapomorphies,* shared derived characters, are useful in constructing phylogenies. Ancestral states (*plesiomorphies*) present in all members of a group are not useful (Mayr and Ashlock 1991; Marcus 1993).

First a set of homologous characters must be selected. The next step in a classical cladistic analysis is to determine whether each character state is ancestral or derived, referred to as *polarization* of character states. This is often done by outgroup comparison. A more distantly related group is included in the analysis and its character state for the character is taken as being primitive (on the assumption that an ancestral state is distributed throughout a larger group of taxa). A character state is clearly derived if it is found in only one member of a taxon. However, when it is found in some but not all members, it must be placed in a transformation series showing the states that could have been sequentially derived from the more ancestral condition during the evolution of the group (Mayr and Ashlock 1991 Marcus 1993).

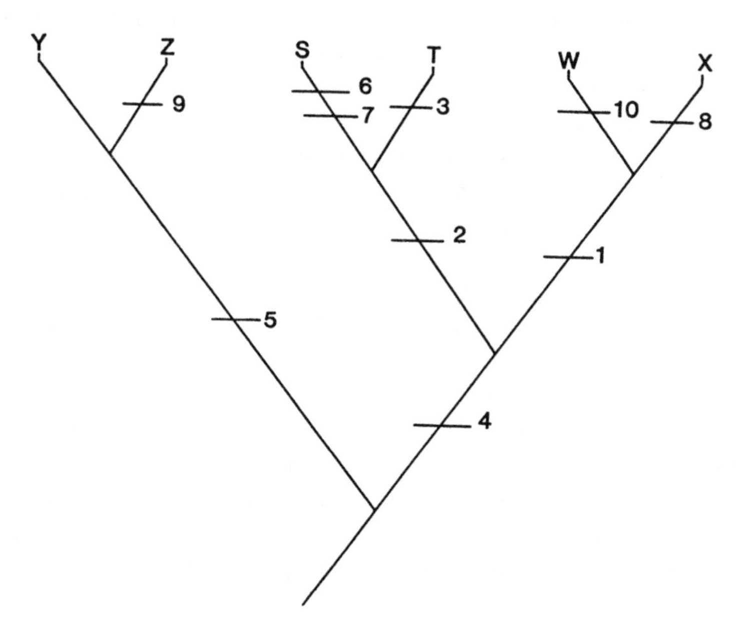

Figure 22.6 Hypothetical cladogram showing relationships among six taxa: S, T, W, X, Y, and Z. Synapomorphies (shared derived characters) are numbered and the point of change from the primitive to the derived state is indicated by bars. *–Source: Figure 8.1 of Stuessy (1990).*

Character polarization is sometimes determined by studying ontogeny in the group, on the assumption that the characters mostly widely distributed are the most likely to be ancestral (but, as developmental biologists know, this doesn't always hold true, especially for larval characters). Once polarity has been determined for the character states and a data set has been organized into plesiomorphic and apomorphic states, the analysis can be carried out. When a group has a fossil record, stratigraphic evidence can also be used to determine which modification arose earliest and whether any reversals have taken place. The addition of further derived characters that support the same polarity provides another check of the hypothesis (Stace 1989; Mayr and Ashlock 1991; Marcus 1993).

The pairs of character states in the matrix are then scored according to the number of derived (apomorphic) characters and connected by the number of shared derived (*synapomorphic*) characters. Various computer programs are used to do this, including HENNIG 86, PHYLLIP, PAUP, and MacClade. They work by finding the *most parsimonious solution,* the one that takes the fewest connecting steps to construct a cladogram (figure 22.7 shows a cladogram produced by one of these programs). The computer problems involved in finding the most parsimonious cladograms are called nondeterministic polynomial problems, and they are computationally difficult. Even though the ones that can run on microcomputers look only for subsets of solutions, they require lots of memory and time to run (Marcus 1993).

Some taxonomists also use character compatibility analysis and maximum likelihood analysis. Character compatibility analysis is a procedure for finding the largest number of mutually compatible characters. It has not received as much use as other methods, partly because the final tree produced usually contains less than half of the characters that were put into the original matrix. Maximum likelihood methods are probability methods that assume evolutionary change to be a random process. They are used mostly in the analysis of molecular sequence data (Quicke 1993).

One of the chief benefits of cladistics is the fact that it has forced taxonomists to clarify their thinking even more rigorously than was necessary with the phenetic approach. They must examine their methods and they must know exactly what questions they are asking. They must also be aware of the assumptions and consequences of the different types of computer analyses used.

Cladistic analysis has led to some remarkable progress in creating phylogenies. As shown in figure 22.5, the yearly percentage of papers

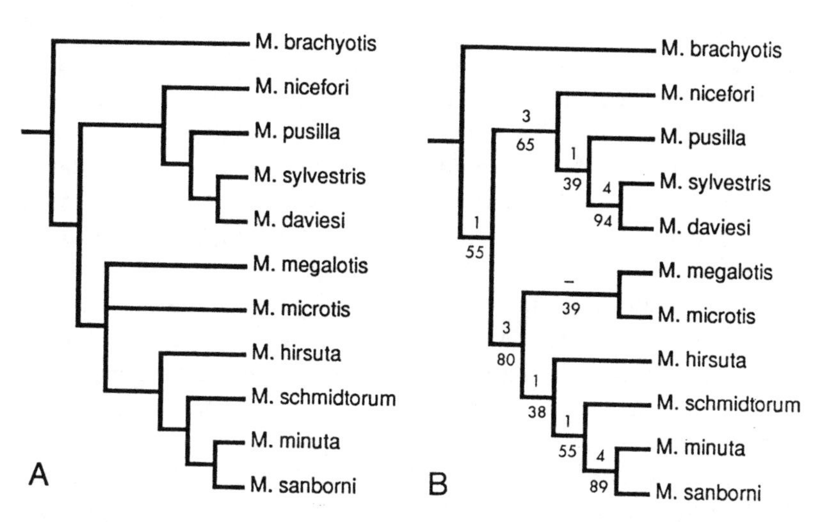

Figure 22.7 Real-life cladograms showing relationships of species of *Micronycteris* (a phyllostomid bat genus) based on a data set that included morphological characters, allozyme data, and karyotype data. **A,** Strict consensus tree (based on three equally parsimonious trees). The analysis was carried out using PAUP. The data were then further manipulated using decay analysis and branch-and-bound bootstrap analysis with 1,000 replicates, resulting in the final cladogram of **B.** *–Source: Reproduced from figure 9 of Simmons (1996), courtesy of the American Museum of Natural History.*

using cladistic methods has grown rapidly in the last 28 years. However, cladistic methods also have some weaknesses. The first concerns what to do when evolution in a group has not occurred in a neat stepwise branching pattern. Sometimes character reversal occurs. Derived character states revert to ancestral ones in descendant taxa or evolve in a parallel or convergent manner. Convergence is usually not too difficult to recognize because a convergent trait usually won't be correlated with other characters and so stands out as odd. *Homoplasy,* false synapomorphy resulting from convergence, parallelism, or character reversals causes problems with parsimony methods. In such cases the most parsimonious cladogram might not be the one that shows the true pattern (Stace 1989).

Another criticism of cladistic methodology relates to character polarization. A regular cladogram is a *rooted tree;* its evolutionary polarity is already determined. In many groups this is just not possible to do. So methods for working with *unrooted trees* have been added. These result in a number of possible alternatives; the taxonomist decides which end of the tree is ancestral after the analyses have been carried out (Stace 1989).

Another problem some systematists have with cladistic methods is the necessity for taxa to be *holophyletic* to be acceptable. To satisfy strict Hennigian criteria, they must be monophyletic, not just in the sense of all being descended from a common ancestor, but in containing all the descendants of that common ancestor (Marcus 1993). This may be achievable in groups in which all species are known, but those who work on invertebrates must cope with the fact that for many if not most of the groups they study, new species will continue to be found for years to come. Should they wait until that time even attempt some understanding of relationships in the group? Most would say no. Yet by this strict definition of monophyly, any classifications they produce would be *paraphyletic,* and not acceptable. The truth is that even in well-known groups such as mammals, there are still a few surprises every year, as the examples in chapter 1 showed. If you have the hubris to state in print that your classification provides the final answer, you will surely be proven wrong.

So far, cladistic methods have worked best in groups in which relationships were already somewhat known (e.g., families of insects) and where polarization of characters is fairly clear. They have been less useful for groups or cases in which these conditions could not be met. However, newer methods for working with unrooted trees have made the methodology more widely applicable. Plugging the same old set of characters into the most sophisticated cladistics program will not result in a better classification, but it can help you see what was wrong with the old one. From that point, however, you may need to find new characters or underused old ones if your classification is to add to scientific knowledge.

So far, cladistic methods have been used more often for animals than for plants. The reticulated patterns of relationship found in many higher plants means that phenetic approaches may still be the most useful for those groups (Stace 1989). However, they have been used to work out relationships in some plant groups also (Barnes 1989), and the increasing number of botanical cladistics papers produced in the last few years seems to indicate that cladistic methods can be productive for plant as well as animal systematics.

Molecular Systematics

Because all the inherited aspects of organisms are coded into their nucleic acid sequences, it seems that if we could only decode them directly we would have no more need for systematics. In fact, in the 1980s plenty of people were saying just that. But it hasn't turned out to be that simple. For one thing, it has turned out that a lot of garbage or nonsense information is incorporated into those sequences as well. Furthermore, preparing and analyzing DNA sequences is still a time-consuming and expensive process. That is changing rapidly, but molecular analysis is still at the stage computer analysis was at the time phenetic and cladistic analyses started, the stage in which a university might be able to afford one shared molecular lab, but not one for every biologist.

Electrophoresis

As figure 22.8 shows, studies using electrophoresis have been in use longest. This technique relies on the physical properties of proteins (enzymes) to work out genetic information. It depends on the fact that proteins have net electrical charges and soluble proteins can be made to migrate along an electrical gradient. Slightly different forms vary in charge and migrate different distances. A range of procedures are used (e.g., Avise 1994; Hillis et al. 1996), but basically the protein sample is placed at one end of a suitable medium such as a starch or acrylamide gel and an electric field is applied. The resulting patterns can be stained, photographed, and analyzed. Most proteins studied by electrophoresis are stainable enzymes. Allozymes are enzyme variants at single gene locus. Isozymes are genes at different loci. Each slightly different allele moves differently along the gradient. Enzymes involved in basic cellular processes (e.g., glycolysis or the Krebs cycle) are most useful in these studies for two reasons: They are found in almost all living organisms and, because of their necessary function, they are present in sufficient amounts in almost any tissue sample. In the simplest case the results show a single band for each homozygote and a pair of bands for heterozygotes. Of course, things are not usually that simple, and it can take considerable knowledge to interpret real-life results. Electrophoresis does not have the glamor of the newer approaches, but it is still one of the best and least expensive ways to study intraspecific variation and population genetics (Quicke 1993). For example, Nancy Moncrief (1993) used allozyme electrophoresis along with morphometric analysis of standard cranial and mandibular characters to study geographical variation in fox squirrels (*Sciurus niger*) and gray squirrels (*Sciurus carolinensis*) in the lower Mississippi River valley. She found that morphological and allozyme variation were not congruent.

Morphological analysis showed that within each species populations from the delta and floodplain region were smaller than other populations. However, for both species, allozyme analysis showed variation between populations east and west of the Mississippi, indicating that the river has served as a barrier to gene flow in these species.

Because it can be used to separate genetic and ecological factors, the technique has become a valuable tool in conservation biology and management. Allozyme variation has been used to assess genetic variation in animals as diverse as gray wolves, northern elephant seals, Florida tree snails, and Sonoma topminnows (Avise 1994).

DNA Hybridization

DNA hybridization was the next molecular technique to become popular. It is based on the fact that when double-stranded DNA is heated to 100°C in an aqueous solution, the two strands break apart. On cooling, the complementary base pairs realign and the strands reassociate. If strands from different individuals or taxa are mixed and melted together and then cooled, the rate at which they reassociate can be used as an estimate of genetic distance. The analysis of the results, carried out by constructing melting curves, is much more complicated. DNA hybridization has been criticized because of findings that reciprocal distances between taxa were

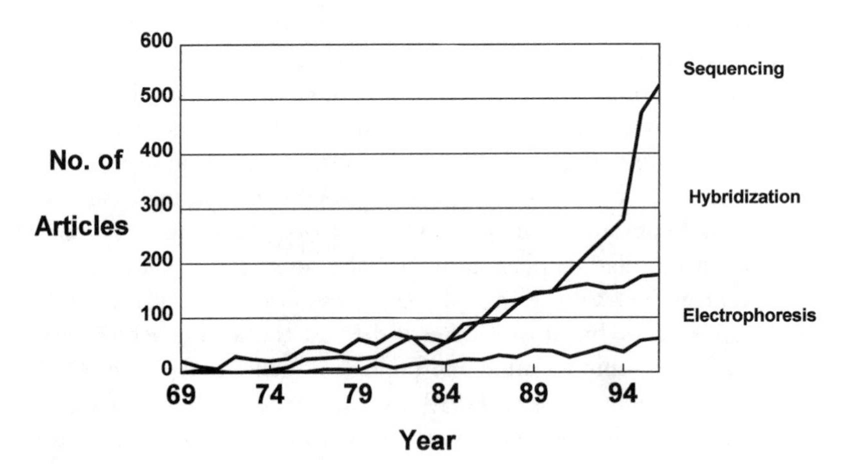

Figure 22.8 Trends in use of molecular methods in taxonomy over the last 28 years. Number of articles using different molecular methods by year. *–Source: Figure 4b of Winston and Metzger (1998).*

not always equal. Whether the measurements they produce are suitable for phylogenetic reconstruction has also been questioned (Quicke 1993).

Methods for the analysis of mitochondrial DNA were developed beginning in the late 1970s. DNA fingerprinting developed in the mid- to late 1980s, and polymerase chain reaction (PCR)-mediated DNA sequencing in the 1990s (Avise 1994). Thanks to the development of PCR methods, DNA sequencing techniques that only a few labs in the country could carry out 20 years ago are now within the reach of many biologists and their students.

Fragments and partial sequencing of the smaller mitochondrial DNA molecules have been used to look at differences in organisms that result entirely from mutation. Because mtDNA is transmitted maternally, the changes caused by reassortment or recombination during meiosis are factored out. Plant chloroplasts also have circular DNA molecules called cpDNA, which also can be sequenced and used taxonomically. The method is especially valuable in determining the origins of hybrid species (Quicke 1993).

DNA fingerprinting techniques are based on restriction fragment analysis of portions of microsatellite DNA sequences. They are affected by recombination and are highly polymorphic. Therefore, unrelated individuals are extremely unlikely to have identical restriction fragment patterns and very small samples of tissue can be used to determine individual identity and degree of relationship (Quicke 1993). We hear most often about its use in forensics, but in systematics it has been used to study parentage (e.g., of migrant organisms or larvae) and to map gene flow in populations (Avise 1994; Strassman et al. 1996).

Sequencing

The development of the PCR technique has made it possible to take the DNA in a very small sample of tissue and copy it until there is enough for restriction fragment analysis or sequencing. Usually, double-stranded DNA is split into single-stranded and then a primer DNA segment complementary to a sequence on the target DNA. This is divided into four subsamples and then processed with the primer extension reaction, which separates them into fragments of varying lengths. The fragments are then separated electrophoretically and made visible by fluorescence or autoradiography and the DNA sequences are read directly (Quicke 1993; Avise 1994).

Like other approaches to systematics, molecular methods have their advantages and disadvantages. Their two greatest advantages are direct access to the genotype and a rich supply of new characters (Mayr 1988). They also have obvious disadvantages: high cost and intensive training time (Avise 1994). Aside from that, there is a major philosophical difference between distance methods such as DNA hybridization, which is analyzed by phenetic methods, and sequencing techniques, which are analyzed by cladistic methods. Finally, as Avise (1994) points out, most molecular analyses support earlier phylogenies based on nonmolecular characters. Therefore, their best use is to analyze the problems in systematics, evolution, and ecology that have proven to be beyond the capabilities of traditional methods.

Biogeography

Another aspect of systematics is the study of biogeography, the distributions of plants and animals, of patterns of biodiversity, and the ecological and historical (geological and genealogical) reasons for them. Most new species seem to arise when a parent species is divided into two or more isolated populations by geographic barriers (Quicke 1993). Geographic distributions of a group of species or other taxa may have arisen in various ways, depending in part on the organisms' dispersal capabilities. For example, terrestrial mammals can be isolated by a seaway whose currents create dispersal avenues for marine invertebrate larvae. Meanwhile, tiny windborne seeds or insects might be carried high above, avoiding the barrier entirely.

Older biogeographic studies relied on dispersal as a mechanism, attempting to identify the center of origin of the group and connecting the distributions of its members to identify a dispersal route (e.g., Mayr 1942, 1963). Recent advances in geological knowledge, plate tectonics, and continental drift have led to the recognition that the geography of the earth itself has changed over time. This has led to new *vicariance* explanations of biogeographic patterns. These explain the occurrence of close relatives in widely separated geographic regions by the formation of natural barriers through these geological processes (Funk and Brooks 1990; Quicke 1993). These allow many more explanations of distributions and add possibilities for the use of cladistic analyses in explaining them. Three methods have been used to answer biogeographic questions. Strict consensus analysis is used to look for the portions of different cladograms that show identical distribution patterns. Adams consensus analysis is used to show what parts of phylogenies for different groups show the same distribution patterns. Mapping methods are used to summarize all the data and to attempt to place taxa that don't fit a general pattern. Finally, a combination of phylogenetic and biogeographic analysis can help determine modes of speciation (sympatric, allopatric, and parapatric) in different groups (Funk and Brooks 1990).

Comparative Biology

In addition to biogeography, phylogenetic techniques can be applied to studies of ecology, life histories, and behavior in an exciting new form of comparative biology. Systematists can use this approach to study the evolutionary origins of a character. Ecologists may need to use it to discover the phylogeny of the group they are working on in order to evaluate origins and associations of ecological characters through evolutionary time (e.g., a study of adaptation in some group is useless if the group is polyphyletic). Comparative methods can be used to study speciation, adaptation, coevolution, and community ecology, among other topics (Brooks and McLennan 1991). Their results promise a more complete evolutionary biology and bring us back to the premise of the first chapter of this book: that systematics is the core of evolutionary biology, the core that will determine the future health and biodiversity of this planet.

Sources

Books and Textbooks on Systematics

Benson, L. 1962. *Plant Taxonomy, Methods and Principles.* New York, Ronald Press.
Benson, L. 1979. *Plant Classification,* 2d ed. Lexington, Mass., Heath.
Blackwelder, R. I. 1967. *Taxonomy, A Text and Reference Book.* New York: Wiley.
Davis, P. H. and V. H. Heywood. 1965. *Principles of Angiosperm Taxonomy.* Princeton, N.J., Van Nostrand.
Eldredge, N., ed. 1992. *Systematics, Ecology and the Biodiversity Crisis.* New York, Columbia University Press.
Goto, H. E. 1982. *Animal Taxonomy.* London: Edward Arnold.

Grande, L. and O. Rieppel, eds. 1994. *Interpreting the Hierarchy of Nature: From Systematic Patterns to Evolutionary Process Theories.* San Diego, Academic Press.
Grant, V. 1991. *The Evolutionary Process: A Critical Study of Evolutionary Theory,* 2d ed. New York, Columbia University Press.
Hall, B. K., ed. 1994. *Homology: The Hierarchical Basis of Comparative Biology.* San Diego, Academic Press.
Hawksworth, D. L. 1974. *Mycologist's Handbook. An Introduction to the Principles of Taxonomy and Nomenclature in the Fungi and Lichens.* Kew, U.K., Commonwealth Mycological Institute.
Hillis, D. M., C. Moritz, and B. K. Mable. 1996. *Molecular Systematics,* 2d ed. Sunderland, Mass., Sinauer.
Jeffrey, C. 1982. *An Introduction to Plant Taxonomy,* 2d ed. Cambridge, U.K., Cambridge University Press.
Mayr, E. 1969. *Principles of Systematic Zoology.* New York, McGraw-Hill.
Mayr, E. and P. D. Ashlock. 1991. *Principles of Systematic Zoology,* 2d ed. New York, McGraw-Hill.
Mayr, E., E. G. Linsley, and R. L. Usinger. 1953. *Methods and Principles of Systematic Zoology.* New York: McGraw-Hill.
Minelli, A. 1993. *Biological Systematics: The State of the Art.* London, Chapman & Hall.
Quicke, D. L. J. 1993. *Principles and Techniques of Contemporary Taxonomy.* London, Blackie Academic and Professional.
Radford, A. E. et al. 1986. *Fundamentals of Plant Systematics.* New York, Harper & Row.
Ross, H. H. 1974. *Biological Systematics.* Reading, Mass., Addison-Wesley.
Sivarajan, V. V. 1991. *Introduction to the Principles of Plant Taxonomy,* 2d ed. Cambridge, U.K., Cambridge University Press.
Smith, A. B. 1994. *Systematics and the Fossil Record: Documenting Evolutionary Patterns.* Oxford, U.K., Blackwell Scientific.
Solbrig, O. T. 1970. *Principles and Methods of Plant Biosystematics.* New York, Macmillan.
Stace, C. 1989. *Plant Taxonomy and Biosystematics,* 2d ed. London: Edward Arnold.
Stuessy, T. F. 1990. *Plant Taxonomy: The Systematic Evaluation of Comparative Data.* New York: Columbia University Press.
Talbot, P. H. B. 1971. *Principles of Fungal Taxonomy.* London, Macmillan.
Wiley, E. O. 1981. *Phylogenetics. The Theory and Practice of Phylogenetic Systematics.* New York, Wiley.

Phenetics

Heywood, V. H. and J. McNeill, eds. 1964. *Phenetic and Phylogenetic Classification: Symposium.* London, Systematics Association.

Sneath, P. H. A. 1995. Thirty years of numerical taxonomy. *Systematic Biology* 44: 281–298.

Sneath, P. H. A. and R. R. Sokal. 1973. *Numerical Taxonomy.* San Francisco, W.H. Freeman.

Sokal, R. R. 1986. Phenetic taxonomy: Theory and methods. *Annual Review of Ecology and Systematics* 17: 423–442.

Sokal, R. R. 1988. Unsolved problems in numerical taxonomy, pp. 45–56 in H. H. Bock, ed. *Classification and Related Methods of Data Analysis.* Amsterdam, Elsevier.

Sokal, R. R. and P. H. Sneath. 1963. *Principles of Numerical Taxonomy.* San Francisco, W.H. Freeman.

Cladistics

Brooks, D. R. and D. H. McLennan. 1991. *Phylogeny, Ecology, and Behavior: A Research Program in Comparative Biology.* Chicago, University of Chicago Press.

Cracraft, J. and N. Eldredge, eds. 1979. *Phylogenetic Analysis and Paleontology.* New York, Columbia University Press.

Eldredge, N. and J. Cracraft. 1980. *Phylogenetic Patterns and the Evolutionary Process.* New York, Columbia University Press.

Forey, P. L., C. J. Humphries, I. L. Kitching, R. W. Scotland, D. J. Sierbert, and D. M. Williams. 1992. *Cladistics: A Practical Course in Systematics.* Oxford, U.K., Clarendon Press.

Funk, V. A. and D. R. Brooks. 1990. Phylogenetic systematics as the basis of comparative biology. *Smithsonian Contributions in Botany* 73: 1–45.

Harvey, P. H. and M. D. Pagel. 1991. *The Comparative Method in Evolutionary Biology.* Oxford, U.K., Oxford University Press.

Hennig, W. 1966. *Phylogenetic Systematics.* Urbana, University of Illinois Press.

Novacek, M. J. and Q. D. Wheeler, eds. 1992. *Extinction and Phylogeny.* New York, Columbia University Press.

Wiley, E. O., D. Siegel-Causey, D. R. Brooks, and V. A. Funk. 1991. *The Complete Cladist: A Primer of Phylogenetic Procedure.* Lawrence, University of Kansas Museum of Natural History.

Molecular Systematics

Avise, J. C. 1994. *Molecular Markers, Natural History and Evolution.* New York, Chapman & Hall.

Ferraris, J. D. and S. R. Palumbi. 1996. *Molecular Zoology: Advances, Strategies, and Protocols.* New York, Wiley.

Gillespie, J. H. 1991. *The Causes of Molecular Evolution.* Oxford, U.K., Oxford University Press.

Hewitt, G. M., A. W. B. Johnston, and P. R. W. Young, eds. 1991. *Molecular Techniques in Taxonomy.* Berlin, Springer-Verlag.

Hillis, D. M. and M. T. Dixon. 1991. Ribosomal DNA: Molecular evolution and phylogenetic inference. *Quarterly Review of Biology* 66: 411–453.

Hillis, D. M., C. Moritz, and B. K. Mable. 1996. *Molecular Systematics,* 2d ed. Sunderland, Mass., Sinauer.

Sobral, B. W. S. 1996. *The Impact of Plant Molecular Genetics.* Cambridge, Mass., Birkhauser.

Soltis, P. S., D. E. Soltis, and J. J. Doyle, eds. 1992. *Molecular Systematics of Plants.* New York, Chapman & Hall.

Phylogeny Software

Adachi, J. 1996. *MOLPHY Version 2.3: Programs for Molecular Genetics Based on Maximum Likelihood.* Tokyo, Institute of Statistical Mathematics.

Kumar, S., K. Tamura, and M. Nei. 1993. *MEGA: Molecular Evolutionary Genetic Analysis,* version 1.0. University Park, Pennsylvania State University.

Swofford, D. L. 1990. *PAUP: Phylogenetic Analysis Using Parsimony,* version 3. Champaign, Illinois Natural History Survey.

Swofford, D. L. 1998. *Phylogenetic Analysis Using Parsimony (PAUP),* version 4.0. New York, Sinauer.

Literature Cited

Abbott, R. J. 1992. Plant invasions, interspecific hybridization and the evolution of new plant taxa. *Trends in Ecology & Evolution* 7: 401–405.

Abramenko, M. I. 1990. A study of gynogenetic progeny of cyprinid hybrids as a model for determining the potential mechanism for the formation of clones of hybrid fishes. *Journal of Ichthyology* 30: 80–88.

Abrantes, I. M. de O. and M. S. N. de A. Santos. 1991. *Meloidogyne lusitanica* n. sp. (Nematoda: Meloidogynidae), a root-knot nematode parasitizing Olive Tree (*Olea europaea* L.). *Journal of Nematology* 23: 210–224.

Adanson, M. 1763–64. *Familles des plantes,* 2 vols. Paris, Vincent.

Ahnelt, H. 1995. Two new species of *Knipowitschia* Iljin, 1927 (Teleostei: Gobiidae) from Western Anatolia. *Mitteilungen aus dem Hamburgischen Zoologischen Museum und Institute* 92: 155–168.

Allen, G. A. and R. A. Cannings. 1985. Museum collections and life history studies. *Occasional Papers of the British Columbia Provincial Museum* 25: 169–194.

Allen, J. A. and F. M. Chapman. 1893. On a collection of mammals from the island of Trinidad, with descriptions of new species. *Bulletin of the American Museum of Natural History* 5: 203–234.

Allen, W. H. 1993. The rise of the botanical database. *BioScience* 43: 274–279.

Allmon, W. D. 1992. Genera in paleontology: Definition and significance. *Historical Biology* 6: 149–158.

Anderson, I. 1996. World's wetlands sucked dry. *New Scientist* 149: 9.

Anderson, M. E. 1995. The eelpout genera *Lycenchelys* Gill and *Taranetzella* Andriashev (Teleostei: Zoarcidae) in the Eastern Pacific, with descriptions of nine new species. *Proceedings of the California Academy of Sciences* 49: 55–113.

Andersson, L. 1990. The driving force: Species concepts and ecology. *Taxon* 39: 375–382.

Anonymous. 1990. Primate find surprises biologists. *Science* 249: 20–21.

Anonymous. 1991. Funding unsexy science. *Science* 251: 371.

Anonymous. 1995. Old World tree found in Colombia. *Science* 269: 1049.

Arai, M. N. 1977. Specimens of new hydroid species described by C. M. Fraser in the collections of the British Columbia Provincial Museum. *Syesis* 10: 25–30.

Arnett, R. H., N. M. Downie, and H. E. Jacques. 1980. *How to Know the Beetles.* Dubuque, Iowa, Wm. C. Brown.

Arnold, N. 1991. Biological messages in a bottle. *New Scientist* 131(1783): 25–27.

Arthur, J. C., J. H. Barnhart, N. L. Britton, F. Clements, O. F. Cood, F. V. Coville, F. S. Earle, A. W. Evans, T. E. Hazen, A. Hoolick, M. A. Howe, F. H. Knowlton, G. T. Moore, H. H. Rusby, C. L. Shear, L. M. Underwood, D. White, and W. F. Wight. 1907. American code of botanical nomenclature. *Bulletin of the Torrey Botanical Club* 34: 167–178.

Asker, S. E. and L. Jerling. 1992. *Apomixis in Plants.* Boca Raton, Fla., CRC Press.

Askew, T. M. 1988. A new species of pleurotomariid gastropod from the western Atlantic. *Nautilus* 102: 89–91.

Attems, C. 1928. Myriopoda of South Africa. *Annals of the South African Museum* 26: 1–431.

Attems, C. 1929. Diplopoden des Belgischen. I. Polydesmoidea. *Revue de Zoologie et de Botanique Africaines* 17: 253–278.

Auduoin, J. V. 1826. Explication sommaire des Planches de Polypes de l'Egypte et de la Syrie, in *Description de l'Egypte, Hist. Nat., I., 4.* Savigny, J. C. ?date, pls. i–xv, Paris.

Avise, J. C. 1994. *Molecular Markers, Natural History and Evolution.* New York, Chapman & Hall.

Avise, J. C. and W. S. Nelson. 1984. Molecular genetic relationships of the extinct dusky seaside sparrow. *Science* 243: 646–648.

Ayre, D. J. and S. Duffy. 1994. Evidence for restricted gene flow in the viviparous coral *Seriatopora hystrix* on Australia's Great Barrier Reef. *Evolution* 48: 1183–1201.

Baba, K. and S.-C. Oh. 1990. *Galathea coralliophilus,* a new decapod crustacean (Anomura: Galatheidae) from Singapore, Gulf of Thailand, and West Irian. *Proceedings of the Biological Society of Washington* 103: 358–363.

Baeyer, H. C. von. 1984. Rainbows, whirlpools, and clouds. *The Sciences* July/August: 24–37.

Ballantine, D. L. and N. E. Aponte. 1995. *Laurencia coelenterata* (Rhodomelaceae, Rhodophyta), a new diminutive species from the Dry Tortugas, Florida. *Botanica Marina* 38: 417–421.

Bamber, R. N. 1993. A new species of *Kalliapseudes* (Crustacea: Tanaidacea: Kalliapseudidae) from Trinidad. *Proceedings of the Biological Society of Washington* 106: 122–130.

Bänziger, H. 1989. Lardizabalaceae: New plant family for Thailand predicted by rare moth on Doi Suthep. *Natural History Bulletin of the Siam Society* 37: 187–208.

Barnes, R. D. 1989. Diversity of organisms: How much do we know? *American Zoologist* 9:1075–1084.

Barnes, R. S. K., ed. 1984. *A Synoptic Classification of Living Organisms.* Sunderland, Mass., Sinauer.

Barnes, R. S. K. and R. N. Hughes. 1982. *An Introduction to Marine Ecology.* London: Blackwell.

Barry, J. P., C. H. Baxter, R. D. Sagarin, and S. E. Gilman. 1995. Climate-related, long-term faunal changes in a California rocky intertidal community. *Science* 267: 672–675.

Bayer, F. M. 1990. The identity of *Fannyella rossii* J. E. Gray (Coelenterata: Octocorallia). *Proceedings of the Biological Society of Washington* 103: 773–783.

Beerli, P., H. Hotz, H. G. Tunner, S. Heppich, and T. Uzzell. 1994. Two new water frog species from the Aegean islands Crete and Karpathos (Amphibia, Salientia, Ranidae). *Notulae Naturae* 470: 1–9.

Bengtsson, J., H. Jones, and H. Setälä. 1997. The value of biodiversity. *Trends in Ecology and Evolution* 12: 334–336.

Benson, L. 1962. *Plant Taxonomy, Methods and Principles.* New York, Ronald Press.

Bentham G. 1874. On the recent progress and present state of knowledge of systematic botany. *Reports of the British Association for the Advancement of Science for 1874* 27–54.

Bergquist, P. R., R. C. Cambie, and M. R. Kernan. 1990. Scalarane Sesterterpenes from *Collospongia auris,* a new thorectid sponge. *Biochemical Systematics and Ecology* 18: 349–357.

Bergquist, P. R. and M. Kelly-Borges. 1991. An evaluation of the genus *Tethya* (Porifera: Demospongiae: Hadromerida) with descriptions of new species from the southwest Pacific. *The Beagle, Records of the Northern Territory Museum of Arts and Sciences* 8: 37–72.

Bieri, R. 1989. Krohnittellidae and Bathybelidae, new families in the phylum Chaetognatha; the rejection of the family Tokiokaispadellidae and the genera *Tokiokaispadella* Zahonya, and *Aberrospadella. Proceedings of the Biological Society of Washington* 102: 973–976.

Bieri, R. 1991. Six new genera in the chaetognath family Sagittidae. *Gulf Research Reports* 8: 221–225.

Bishop, J. D. D. 1986. *Lepralia punctata* Hassall, 1841 (Bryozoa: Cheilostomata). Proposed designation of a replacement neotype. Z.N.(S.) 2562. *Bulletin of Zoological Nomenclature* 43: 288–296.

Blackburn, T. M. and K. A. Gaston. 1995. What determines the probability of discovering a species? A study of South American oscine passerine birds. *Journal of Biogeography* 22: 7–14.

Blackwelder, R. I. 1967. *Taxonomy: A Text and Reference Book.* New York, Wiley.

Blake, J. A. 1972. Two new species of polychaetous annelid worms from Baffin Bay and the Davis Strait. *Southern California Academy of Sciences Bulletin* 71: 127–132.

Blatchley, W. S. 1928. Quit-claim specialists vs. the making of manuals. *Bulletin of the Brooklyn Entomological Society* 23: 10–18.

Bock, W. J. 1977. Foundations and methods of evolutionary classification, pp. 851–895 in M. Hecht, P. C. Good, and B. M. Hecht, eds. *Major Patterns in Vertebrate Evolution.* New York, Plenum.

Bock, W. J. 1985. Adaptive inference and museological research. *Occasional Papers of the British Columbia Provincial Museum* 25: 123–138.

Bock, W. J. 1994. History and nomenclature of avian family group names. *Bulletin of the American Museum of Natural History* 222: 1–281.

Bogart, J. P. 1980. Evolutionary implications of polyploidy in amphibians and reptiles. In W. H. Lewis, ed. *Polyploidy: Biological Relevance.* New York, Plenum.

Böttger-Schnack, R. and G. A. Boxshall. 1990. Two new *Oncaea* species (Copepoda: Poecilostomatoida) from the Red Sea. *Journal of Plankton Research* 12: 861–871.

Bowman, T. E., S. P. Garner, R. R. Hessler, T. M. Illiff, and H. L. Sanders. 1985. Mictacea, a new order of Crustacea Peracarida. *Journal of Crustacean Biology* 5: 74–78.

Brake, D. 1997. Lost in cyberspace. *New Scientist* 154 (2088): 1213.

Brewer, R. H. 1991. Morphological differences between, and reproductive isolation of, two populations of the jellyfish *Cyanea* in Long Island Sound, USA. *Hydrobiologia* 216/217: 471–477.

Bridson, G. D. R. 1968. *The Zoological Record:* A centenary appraisal. *Journal of the Society for Bibliography of Natural History* 5: 23–34.

Brooks, D. R. and D. A. McLennan. 1991. *Phylogeny, Ecology, and Behavior.* Chicago, University of Chicago Press.

Brown, R. W. 1956. *The Composition of Scientific Words,* rev. ed. Washington, D.C., Smithsonian Institution Press.

Brown, W. C. and E. L. Alcala. 1995. A new species of *Brachymeles* (Reptilia: Scincidae) from Catanduanes Island, Philippines. *Proceedings of the Biological Society of Washington* 108: 392–394.

Bulmer R. 1970. Which came first, the chicken or the egg-head? pp. 1069–1091 in J. Poullon and P. Marada, eds. *Echanges et Communication, Mélanges offerts á Claude Lévi-Strausse á l'occasion de son 60ème Anniversaire.* The Hague, Mouton.

Bulmer, R. N. H. and M. J. Tyler. 1968. Karam classification of frogs. *Journal of the Polynesian Society* 77: 333–387.

Buss, L. W. 1987. *The Evolution of Individuality.* Princeton, N.J., Princeton University Press.

Buss, L. W. and E. W. Iverson. 1981. A new genus and species of Sphaeromatidae (Crustacea: Isopoda) with experiments and observations on its reproductive biology, interspecific interactions and color polymorphisms. *Postilla* 184: 1–23.

Buss, L. W. and P. O. Yund. 1989. A sibling species group of *Hydractinia* in the north-eastern United States. *Journal of the Marine Biological Association of the United Kingdom* 69: 857–874.

Byrne, M. and M. J. Anderson. 1994. Hybridization of sympatric *Patiriella* species (Echinodermata: Asteroidea) in New South Wales. *Evolution* 48: 564–576.

Cadman, P. S. and A. Nelson-Smith. 1990. Genetic evidence for two species of lugworm (*Arenicola*) in South Wales. *Marine Ecology Progress Series* 64: 107–112.

Cain, A. J. 1958. Logic and memory in Linnaeus' system of taxonomy. *Proceedings of the Linnaean Society of London* 169: 144–163.

Cain, A. J. 1962. The evolution of taxonomic principles., pp. 1–13 in G. C. Ainsworth and P. H. A. Sneath, eds. *Microbial Classification.* London, Cambridge University Press.

Cairns, S. D. 1987. *Conopora adeta,* new species from Australia, the first known unattached stylasterid. *Proceedings of the Biological Society of Washington* 100: 141–146.

Calder, D. 1988. Shallow-water hydroids of Bermuda: The Athecatae. *Royal Ontario Museum, Life Sciences Contributions* 148: 1–107.

Caldwell, J. P. and C. W. Myers. 1990. A new poison frog from Amazonian Brazil, with further revision of the *quinquevittatus* group of *Dendrobates. American Museum Novitates* 2988: 1–21.

Campos, N. and R. Lemaitre. 1994. A new *Calcinus* (Decapoda: Anomura: Diogenidae) from the tropical western Atlantic, and a comparison with other species of the genus from the region. *Proceedings of the Biological Society of Washington* 107: 137–150.

Candolle, A. L. de. 1867. *Lois de la Nomenclature botanique adopteés par le Congrès International de Botanique tenu à Paris en Août, 1867, suivies a une deuxième édition de l'introduction historique et du commentaire qui accompagnaient la rédaction préparatoire présentée au Congrès.* Genève and Bale, H. Georg; Paris, J.-B. Baillière.

Carlton, J. T., J. K. Thompson, L. E. Schemel, and F. H. Nichols. 1990. Remarkable invasion of San Francisco Bay (California, USA) by the Asian clam *Potamocorbula amurensis.* I. Introduction and dispersal. *Marine Ecology Progress Series* 66: 81–94.

Chamberlain, Y. M. 1965. Marine algae of Gough Island. *Bulletin of the British Museum (Natural History). Botany Series* 3: 175–232.

Choate, H. A. 1912. The origin and development of the binomial system of nomenclature. *Plant World* 15: 257–263.

Chisman, J. K. 1989. Zoological Record, Biological Abstracts and Biological Abstracts/RRM: A comparison of overlap. *RQ* Winter 1989: 242–247.

Claridge, M. and F. M. Ingrouille. 1992. Systematic biology and higher education in the U.K., pp. 23–31 in D. L. Hawksworth, ed. *Improving the Stability of Names: Needs and Options.* Taunus, Germany, Koeltz Scientific Books.

Clark, K. B. and A. Goetzfried. 1976. *Lomanotus stauberi,* a new Dendronotacean nudibranch from central Florida (Mollusca: Opisthobranchia). *Bulletin of Marine Science* 26: 474–478.

Clayton, D. H. and N. D. Wolfe. 1993. The adaptive significance of self-medication. *Trends in Ecology & Evolution* 8: 60–63.

Cole, C. J. 1979. Chromosome inheritance in parthenogenetic lizards and evolution of allopolyploidy in reptiles. *Journal of Heredity* 70: 95–102.

Committee on Biological Diversity in Marine Systems. 1995. *Understanding Marine Biological Diversity.* Washington, D.C., National Academy Press.

Conard, H. S. 1956. *How to Know the Mosses and Liverworts,* rev. ed. Dubuque, Iowa: Wm. C. Brown.

Costello, M. J., C. S. Emblow, and B. E. Picton. 1996. Long term trends in the discovery of marine species new to science which occur in Britain and Ireland. *Journal of the Marine Biological Association of the United Kingdom* 76: 255–257.

Cottarelli, V. and A. C. Puccetti. 1988. *Indolaophonta gemmarum* n. sp. from interstitial littoral waters of Sri-Lanka (Crustacea, Copepoda, Harparcticoida). *Fragmenta Entomologica* 20: 129–136.

Coues, E. 1874. *Field Ornithology.* Salem, Mass., Naturalists' Agency.

Coues, E. et al. 1892. *The Code of Nomenclature Adopted by the American Ornithologists' Union,* 2d ed. New York, American Ornithologists' Union.

Council of Biology Editors Scientific Illustration Committee. 1988. *Illustrating Science: Standards for Publication.* Bethesda, Md., Council of Biology Editors.

Council of Biology Editors Style Manual Committee. 1983. *CBE Style Manual,* 5th ed. Bethesda, Md., Council of Biology Editors.

Coyle, F. A. 1986. *Chilehexops,* a new funnelweb mygalomorph spider genus from Chile (Araneae, Dipluridae). *American Museum Novitates* 2860: 1–10.

Crane, J. 1975. *Fiddler Crabs of the World.* Princeton, N.J., Princeton University Press.

Critchley, A. T. 1983. *Sargassum muticum:* A taxonomic history including world-wide and western Pacific distributions. *Journal of the Marine Biological Association of the United Kingdom* 63: 617–625.

Cronquist, A. 1978. Once again, what is a species? *Beltsville Symposium on Agricultural Research* 2: 3–20.

Culotta, E. 1990. Scientists protest museum cuts. *Science* 249: 219–220.

Culotta, E. 1992. Museums cut research in hard times. *Science* 256: 1268–1271.

Culotta, E. 1993. Mass job extinctions at L.A. Museum. *Science* 260: 1584.

Culotta, E. 1996. Exploring biodiversity's benefits. *Science* 273: 1045–1046.

Dall, W. H. 1877. Nomenclature in zoology and botany. *Proceedings of the American Association for the Advancement of Science* 1877: 7–56.

Dallwitz, M. J. 1980. A general system for coding taxonomic descriptions. *Taxon* 29: 41–46.

Daugherty, C. H., A. Cree, J. M. Hay, and M. B. Thompson. 1990. Neglected taxonomy and continuing extinctions of tuatara (*Sphenodon*). *Nature* 347: 177–179.

Davis, P. H. and V. H. Heywood. 1963. *Principles of Angiosperm Taxonomy.* Edinburgh, Oliver & Boyd.

Davis, P. H. and V. H. Heywood. 1965. *Principles of Angiosperm Taxonomy.* Princeton, N.J., D. Van Nostrand.

Day, R. A. 1988. *How to Write & Publish a Scientific Paper,* 3d ed. Phoenix, Az., Oryx Press.

de Queiroz, K. and J. Gauthier. 1990. Phylogeny as a central principle in taxonomy: Phylogenetic definitions of taxon names. *Systematic Zoology* 39: 307–322.

Diamond, J. M. 1966. Zoological classification system of a primitive people. *Science* 151: 1102–1104.

Dobshansky, T. F., J. Ayala, G. L. Stebbins, and J. W. Valentine. 1977. *Evolution.* San Francisco, W.H. Freeman.

Douvillé, H. 1882. *Régles proposées par le Comité de la Nomenclature paleontologique.* Congrès Géol. International. Compt. Rend. 2me Session, Boulougne, 1881; 594–595.

Downey, J. C. and J. L. Kelly. 1982. *Biological Illustration: Techniques and Exercises.* Ames, Iowa State University Press.

Doyen, J. T. and C. N. Slobodchikoff. 1974. An operational approach to species classification. *Systematic Zoology* 23: 239–247.

Dubois, A. 1988. The genus in zoology: A contribution to the theory of evolutionary systematics. *Mémoires du Muséum National d'Histoire Naturelle, Zoologie* 140: 1–122.

Duméril, A. M. C. 1800. *Leçons d'anatomie comparée de M. G. Cuvier, recueillies et publiées par Duméril et Duvernoy.* Paris, Baudoin.

Dumont, J. P. 1981. A report on the cheilostome Bryozoa of the Sudanese Red Sea. *Journal of Natural History* 15: 623–637.

Dung, V. V., P. M. Giao, N. N. Chinh, D. Tuoc, P. Arctander, and J. MacKinnon. 1993. A new species of living bovid from Vietnam. *Nature* 363: 443–445.

Dunn, R. A. and R. A. Davidson. 1968. Pattern recognition in biologic classification. *Pattern Recognition* 1: 75–93.

Du Rietz, G. E. 1930. The fundamental units of biological taxonomy. *Svensk Botanisk Tidskrift* 24: 333–428.

Earle, S. 1991. Sharks, squids and horseshoe crabs: The significance of marine biodiversity. *BioScience* 41: 506–509.

Edmonds, J. M. 1978. Numerical taxonomic studies on *Solanum* L. section *Solanum* (*Maurella*). *Botanical Journal of the Linnaean Society* 76: 27–51.

Edmunds, M. 1963. *Berthelinia caribbea* n. sp., a bivalved gastropod from the West Atlantic. *Journal of the Linnean Society of London. Zoology* 44: 731–739.

Edwards, M. and D. R. Morse. 1995. The potential for computer-aided identification in biodiversity research. *Trends in Ecology and Evolution* 10: 152–158.

Edwards, R. Y. 1985. Research: A museum cornerstone. *Occasional Papers of the British Columbia Provincial Museum* 25: 1–12.

Edwards, S. R., G. M. Davis, and L. I. Nevling. 1987. *The Systematics Community.* Lawrence, Kans., Association of Systematics Collections.

Ehrlich, P. R. and R. W. Holm. 1962. Patterns and populations. *Science* 137: 652–657.

Emerson, W. K. and M. K. Jacobson. 1976. *The American Museum of Natural History Guide to Shells, Land, Freshwater, and Marine, from Nova Scotia to Florida.* New York, Knopf.

Emig, C. C. 1978. A redescription of the inarticulate brachiopod *Lingula reevii* Davidson. *Pacific Science* 32: 31–31–34.

Endler, J. A. 1990. On the measurement and classification of colour in studies of animal colour patterns. *Biological Journal of the Linnaean Society* 41: 315–352.

Eriksson, G. 1983. Linnaeus the botanist, pp. 63–109 in T. Frängsmyr, ed. *Linnaeus, the Man and His Work.* Berkeley, University of California Press.

Erwin, T. L. 1982. Tropical forests: Their richness in Coleoptera and other Arthropod species. *Coleopterists' Bulletin* 36: 74–75.

Erwin, T. L. 1988. The tropical forest canopy: The heart of biotic diversity, pp. 123–129 in E. O. Wilson, ed. *Biodiversity.* Washington, D.C., National Academy Press.

Evans, H. E. 1993. *Pioneer Naturalists: The Discovery and Naming of North American Plants and Animals.* New York, Holt.

Evans, W. A. 1992. Five new species of marine Gastrotricha from the Atlantic coast of Florida. *Bulletin of Marine Science* 51: 315–328.

Fauchald, K. 1992. A review of the genus *Eunice* (Polychaeta: Eunicidae) based upon type material. *Smithsonian Contributions to Zoology* 523, 422 pp.

Fautin, D. G. 1984. More Antarctic and Subantarctic sea anemones (Coelenterata: Corallimorpharia and Actinaria). *Biology of the Antarctic Seas XVI. Antarctic Research Series* 41: 1–42.

Fautin, D. G. 1987. *Stylobates loisettae,* a new species of shell-forming anemone (Coelenterata: Actinidae) from Western Australia. *Proceedings of the California Academy of Sciences* 45: 1–7.

Feldmann, R. S. and R. B. Manning. 1992. Crisis in systematic biology in the "Age of Biodiversity." *Journal of Paleontology* 66: 157–158.

Flint, O. S. 1992. Studies of neotropical caddisflies, XLIX: The taxonomy and relationship of the genus *Eosericostoma,* with descriptions of the immature stages (Tricoptera: Helicophidae). *Proceedings of the Biological Society of Washington* 105: 494–511.

Flynn, L. J. 1986. Late Cretaceous mammal horizons from the San Juan Basin, New Mexico. *American Museum Novitates* 2845: 1–30.

Forster, R. B., N. I. Platnick, and J. Coddington. 1990. A proposal and review of the spider family Synotaxidae (Araneae, Araneoidea), with notes on theridiid interrelationships. *Bulletin of the American Museum of Natural History* 193, 116 pp.

Fortuner, R., ed. 1993. *Advances in Computer Methods for Systematic Biology: Artificial Intelligence, Databases, Computer Vision.* Baltimore, Johns Hopkins University Press.

Fosberg, F. R. 1992. An essay on lectotypification. *Taxon* 41: 321–323.

Fosberg, F. R. and M.-H. Sachet. 1977. Flora of Micronesia, 3: Convolvulaceae. *Smithsonian Contributions to Botany* 36: 1–34.

Foster, A. B. 1984. The species concept in fossil hermatypic corals: A statistical approach. *Palaeontographica Americana* 54: 58–69.

Francki, R. I. B., C. M. Fauquet, D. L. Knudson, and F. Brown. 1991. *Classification and Nomenclature of Viruses.* New York, Springer-Verlag.

Fraser, C. M. 1912. Some hydroids from Beaufort, North Carolina. *Bulletin of the Bureau of Fisheries* 30: 337–389.

Frost, D. R. 1992. Phylogenetic analysis and taxonomy of the *Tropidurus* group of lizards (Iguania: Tropiduridae). *American Museum Novitates* 3033: 1–68.

Fugate, M. 1993. *Branchinecta sandiegoensis,* a new species of fairy shrimp (Crustacea: Anostraca) from western North America. *Proceedings of the Biological Society of Washington* 106: 296–304.

Funch, P. and R. M. Kristensen. 1995. Cycliophora is new phylum with affinities to Entoprocta and Ectoprocta. *Nature* 378: 711–714.

Funk, V. A. and D. R. Brooks. 1990. Phylogenetic systematics as the basis of comparative biology. *Smithsonian Contributions in Botany* 73:1–45.

Gaffney, E. S. 1992. *Ninjemys,* a new name for *"Meiolania" oweni* (Woodward), a horned turtle from the Pleistocene of Queensland. *American Museum Novitates* 3049: 1–10.

Gaffney, E. S. and P. A. Meylan. 1992. The Transylvanian turtle, *Kallokibotion,* a primitive cryptodire of Cretaceous age. *American Museum Novitates* 3040, 37 pp.

Gaffney, E. S. and P. A. Meylan. 1988. A phylogeny of turtles. pp. 157–219 in M. J. Benton, ed. *The Phylogeny and Classification of the Tetrapods.* Oxford, U.K., Clarendon.

Gahan, A. B. 1923. The role of the taxonomist in present day entomology. *Proceedings, Entomological Society of Washington* 25: 68–78.

Gali , F., A. Schwartz, and A. Suarez. 1988. A new subspecies of *Leiocephalus personatus* (Sauria: Iguanidae) from Haiti. *Proceedings of the Biological Society of Washington* 101: 1–3.

Gardner, H. 1983. *Frames of Mind.* New York, Basic Books.

Gaston, K. and R. M. May. 1992. Taxonomy of taxonomists. *Nature* 356: 281–282.

Gertsch, W. J. 1939. A revision of the typical crab-spiders (Misumeninae) of America north of Mexico. *Bulletin of the American Museum of Natural History* 76: 277–442.

Ghosh, H. C. and R. B. Manning. 1993. A new deep-sea crab of the genus *Chacedon* from India (Crustacea: Decapoda: Geryonidae). *Proceedings of the Biological Society of Washington* 106: 714–718.

Gibbons, A. 1992. Plants of the apes. *Science* 255: 921.

Gibbons, M. J. and J. S. Ryland. 1989. Intertidal and shallow water hydroids from Fiji. I. Athecata to Sertulariidae. *Memoirs of the Queensland Museum* 27: 377–432.

Gómez-Laurito, J. and L. D. Gómez. 1991. Ticodendraceae: A new family of flowering plants. *Annals of the Missouri Botanical Garden* 78: 87–88.

Gonzales, P. C. and R. S. Kennedy. 1990. A new species of *Stachyris* babbler (Aves: Timaliidae) from the Island of Panay, Philippines. *Wilson Bulletin* 102: 367–379.

Gonzales, P. C. and R. S. Kennedy. 1996. A new species of *Crateromys* (Rodentia: Muridae) from Panay, Philippines. *Journal of Mammalogy* 77: 25–40.

Goodbody, I. 1996. *Pycnoclavella belizeana,* a new species of ascidian from the Caribbean. *Bulletin of Marine Science* 58: 590–597.

Goodbody, I. and L. Cole. 1987. A new species of *Perophora* (Ascidiacea) from the western Atlantic, including observations on muscle action in related species. *Bulletin of Marine Science* 40: 246–254.

Goodman S. M. and F. Ravoavy. 1993. Identification of bird subfossils from cave surface deposits at Anjohibe, Madagascar, with a description of a new giant *Coua* (Cuculidae: Couinae). *Proceedings of the Biological Society of Washington* 106: 24–33.

Gopen, G. D. and J. A. Swan 1990. The science of scientific writing. *American Scientist* 78: 550–558.

Gosliner, T. M. 1995. The genus *Thuridilla* (Opisthobranchia: Elysiidae) from the tropical Indo-Pacific, with a revision of the phylogeny and systematics of the Elysiidae. *Proceedings of the California Academy of Sciences* 49: 1–54.

Gosner, K. L. 1978. *A Field Guide to the Atlantic Seashore.* Boston, Houghton Mifflin.

Gosse, P. H. 1855. Notes on some new or little-known marine animals. *Annals and Magazine of Natural History,* ser. 2, 16: 27–36.

Goto, H. E. 1982. *Animal Taxonomy.* London: Edward Arnold.

Gould, S. J. 1989. *Wonderful Life: the Burgess Shale and the Nature of History.* New York, W.W. Norton.

Gradstein, S. R., R. Klein, L. Kraut, R. Mues, J. Spörle, and H. Becker. 1992. Phytochemical and morphological support for the existence of two species in *Monoclea* (*Hepaticae*). *Plant Systematics and Evolution* 180: 115–135.

Grant, B. R. and P. R. Grant. 1993. Evolution of Darwin's finches caused by a rare climatic event. *Proceedings of the Royal Society of London B* 251: 111–117.

Grant, P. R. 1986. *Ecology and Evolution of Darwin's Finches.* Princeton, N.J., Princeton University Press.

Grant, V. 1981. *Plant Speciation,* 2d ed. New York, Columbia University Press.

Grant, V. 1991. *The Evolutionary Process: A Critical Study of Evolutionary Theory,* 2d ed. New York, Columbia University Press.

Grasse, J. G. T. 1971. In H. Plechl and G. Spitzbart, eds. *Orbis Latinus,* 4th ed. Braunschweig, Germany, Klinkhardt & Bierman.

Grassle, J. F. 1991. Deep-sea benthic biodiversity. *BioScience* 41: 464–469.

Gray, A. 1889. *Manual of the Botany of the Northern United States, Including the District East of the Mississippi and North of North Carolina and Tennessee.* 6th ed. New York, American Book Company.

Greene, E. L. 1983. *Landmarks of Botanical History.* Stanford, Calif., Stanford University Press.

Greuter, W. 1996. On a new BioCode, harmony, and expediency. *Taxon* 45: 291–294.

Greuter, W. and D. H. Nicolson. 1996. Introductory comments on the Draft BioCode, from a botanical point of view. *Taxon* 45: 343–348.

Greuter, W. et al. 1994. *International Code of Botanical Nomenclature (Tokyo Code) adopted by the Fifteenth International Botanical Congress, Yokohama. August–September 1993.* Regnum Vegetabile 131.

Grimaldi, D. 1987. Amber fossil Drosophilidae (Diptera), with particular reference to the Hispaniolan taxa. *American Museum Novitates* 2880: 1–23.

Grimaldi, D. A. 1990. Revision of *Zygothrica* (Diptera: Drosophilidae), Part II. The first African species, two new Indo-Pacific groups, and the *bilineata* and *samoaensis* species groups. *American Museum Novitates* 2964: 1–31.

Grimaldi, D. A. 1991. Mycetobiine woodgnats (Diptera: Anisopodidae) from the Oligo-Miocene amber of the Dominican Republic, and Old World affinities. *American Museum Novitates* 3014, 24 pp.

Groue, K. J. and L. J. Lester 1982. A morphological and genetic analysis of geographic variation among oysters in the Gulf of Mexico. *Veliger* 24: 331–335.

Grunshaw, J. P. 1995. The taxonomy of *Tylotropidius* Stål 1873 and related genera (Orthoptera Acrididae Eyprepocnemidinae). *Tropical Zoology* 8: 401–433.

Hageman, S. J. 1991. Approaches to systematic and evolutionary studies of perplexing groups: An example using fenestrate Bryozoa. *Journal of Paleontology* 65: 630–647.

Hairston, N. G., Jr. 1990. Problems with the perception of zooplankton research by colleagues outside aquatic sciences. *Limnology and Oceanography* 35: 1214–1216.

Hall, A. V. 1970. A computer-based system for forming identification keys. *Taxon* 19: 12–18.

Hanisak, M. D. and J. A. Kilar. 1990. Typification of *Sargassum filipendula* C. Agardh (Phaeophyceae, Fucales, Sargassaceae) and the names of two varieties. *Taxon* 39: 94–98.

Hartmann, H. T. and D. E. Kester. 1975. *Plant Propagation.* Englewood Cliffs, N.J., Prentice Hall.

Harvey, P. H. 1991. The state of systematics. *Trends in Ecology and Evolution* 6: 345–346.

Havel, J. E., P. D. N. Hebert, and L. D. Delorme. 1990. Genotypic diversity of asexual Ostracoda from a low arctic site. *Journal of Evolutionary Biology* 3: 391–410.

Hawksworth, D. L. 1974. *Mycologist's Handbook. An Introduction to the Principles of Taxonomy and Nomenclature in the Fungi and Lichens.* Kew, U.K., Commonwealth Mycological Institute.

Hawksworth, D. L. 1991. The fungal dimension of biodiversity: Magnitude, significance and conservation. *Mycological Research* 95: 641–655.

Hawksworth, D. L. 1995. Steps along the road to a harmonized bionomenclature. *Taxon* 44: 447–456.

Hayward, P. J. 1985. *Ctenostome Bryozoans.* London, E. J. Brill.

Hayward, P. J. and J. S. Ryland. 1979. *British Ascophoran Bryozoans.* London, Academic Press.

Heiser, C. N. 1965. Modern species concepts: Vascular plants. *Bryologist* 66: 120–124.

Heller, J. L. 1964. The early history of binomial nomenclature. *Huntia* 1: 33–70.

Hendler, G. 1995. New species of brittle stars from the Western Atlantic, *Ophionereis vittata, Amphioplus sepultus,* and *Ophiostigma siva,* and the designation of a neotype for *Ophiostigma isocanthum* (Say) (Echinodermata: Ophiuroidea). *Natural History Museum of Los Angeles County, Contributions in Science* 458: 1–19.

Hendler, G., J. E. Miller, D. L. Pawson, and P. M. Kier. 1995. *Sea Stars, Sea Urchins, and Allies: Echinoderms of Florida and the Caribbean.* Washington, D.C., Smithsonian Institution Press.

Hendricks, R. A. 1992. *Latin Made Simple,* rev. by L. Padol. New York, Doubleday.

Hennig, W. 1950. *Grundzüge einer Theorie der phylogenetischen Systematik.* Berlin, Deutscher Zentralverlag.

Hennig, W. 1966. *Phylogenetic Systematics.* Urbana, University of Illinois Press.

Hennig, W. 1981. *Insect Phylogeny.* New York, Wiley.

Henry, T. J. 1993. A striking new genus and species of bryocorine plant bug (Heteroptera: Miridae) from eastern North America. *Jeffersoniana* 2: 1–9.

Herman, S. G. 1980. *The Naturalist's Field Journal: A manual of instruction based on a system established by Joseph Grinnell.* Vermillion, S.D., Buteo Books.

Heron, C. 1992. The networks of botanical creation. *New Scientist* 133(1807): 39–44.

Hershkovitz, P. 1962. Evolution of Neotropical cricetine rodents (Muridae) with special reference to the phyllotine group. *Fieldiana, Zoology* 46: 1–524.

Hershler, R. 1996. Review of the North American aquatic snail genus *Probythinella* (Rissoidea: Hydrobiidae). *Invertebrate Biology* 115: 120–144.

Hewatt, W. G. 1937. Ecological studies on selected intertidal communities of Monterey Bay, California. *American Midland Naturalist* 18: 161–206.

Heywood, V. H. 1988. The structure of systematics, pp. 44–56 in D. L. Hawksworth, ed. *Prospects in Systematics.* Oxford, U.K., Clarendon Press.

Hillis, D. M., C. Moritz, and B. K. Mable. 1996. *Molecular Systematics,* 2d ed. Sunderland, Mass., Sinauer.

Hillman, R. E., S. E. Ford, and H. H. Haskin. 1990. *Minchinia teredinis* N. Sp. (Balanosporida, Haplosporidiidae), a parasite of teredinid shipworms. *Journal of Protozoology* 37: 364–368.

Hoare, R. D. and R. H. Mapes. 1995. Relationships of the Devonian *Strobilepis* and related Pennsylvanian problematica. *Acta Palaeontologica Polonica* 40: 111–128.

Hobbes, H. H., Jr. and M. Whiteman. 1987. A new, economically important crayfish (Decapoda: Cambaridae) from the Neches River Basin, Texas, with a key to the subgenus *Fallicambarus. Proceedings of the Biological Society of Washington* 100: 403–411.

Hoeksema, B. W. 1989. Taxonomy, phylogeny and biogeography of mushroom corals (Scleractinia: Fungiidae). *Zoologische Verhandelingen Uitgeveven door het Rijksmuseum van Natuurlijke Historie te Leiden* 254: 1–295.

Hoffman, R. L. 1965. The status of *Telonychopus meyeri* Verhoeff, and of the family name Telonychopidae. *Papeis Avulsos. Departmento Zoologia. Unversidade de São Paulo* 17: 243–253.

Hoffman, R. L. 1980. *Classification of the Diplopoda.* Geneva, Musée d'Histoire Naturelle, pp. 1–237.

Hoffman, R. L. 1992. On the taxonomy of the milliped genera *Pseudojulus* Bollman, 1887, and *Georgiulus,* gen. nov., of southeastern United States (Julida: Parajulidae). *Jeffersoniana* 1: 1–19.

Hoffman, R. L. 1996. Type locality and distribution of the crab spider *Xysticus emertoni* Keyserling (Aranaida: Thomisidae). *Banisteria* 8: 47–49.

Hoffman, R. L. 1998. An Appalachian species of *Rhysodesmus* (Polydesmida: Xystodesmidae: Rhydsodesmini). *Myriapodologica* 5: 77–83.

Hogans, W. E. and K. J. Sulak. 1992. *Diocus lycenchelus* new species (Copepoda: Chondracanthidae) parasitic on the eelpout *Lycenchelys verrillii* (Zoarcidae) from the Hatteras slope of the northwest Atlantic Ocean. *Bulletin of Marine Science* 51: 301–308.

Holmgren, P. K., N. H. Holmgren, and L. C. Banett, eds. 1990. *Index Herbariorum. Ed. 8, pt. 1. The Herbaria of the World.* New York, New York Botanical Garden.

Holroyd, P. A. and R. L. Ciochon. 1995. A new artiodactyl (Mammalia) from Eocene Pondaung sandstones, Burma. *Annals of Carnegie Museum* 64: 177–183.

Holsinger, J. R. 1989. Allocrangonyctidae and Pseudocrangonyctidae, two new families of holarctic subterranean amphipod crustaceans (Gammaridea), with comments on their phylogenetic and zoogeographic relationships. *Proceedings of the Biological Society of Washington* 102: 947–959.

Hopwood, A. T. 1959. The development of pre-Linnaean taxonomy. *Proceedings of the Linnaean Society (London)* 170: 230–234.

Houck, M. A., J. B. Clark, K. R. Petersen, and M. G. Kidwell. 1991. Possible horizontal transfer of Drosophila genes by the mite *Proctolaelaps regalis. Science* 253: 1125–1128.

Hubbell, T. H. 1954. The naming of geographically variant populations. *Systematic Zoology* 3: 113–121.

Hulsemann, K. 1991. *Calanus euxinus,* new name, a replacement name for *Calanus ponticus* Karavaev, 1894 (Copepoda: Calanoida). *Proceedings of the Biological Society of Washington* 104: 620–621.

Hunn, E. 1994. Place-names, population density, and the magic number 500. *Current Anthropology* 35: 81–85.

Huxley, J. S. 1940. *The New Systematics.* Oxford, U.K., Clarendon.

Huxley, J. 1958. Evolutionary processes and taxonomy with special reference to grades. *Uppsala Universiteits Arsskrift* 6: 21–39.

Imes, R. 1990. *The Practical Botanist.* New York, Simon & Schuster.

Ingram, G. and C. Corben. 1990. *Litoria electrica:* A new treefrog from western Queensland. *Memoirs of the Queensland Museum* 28: 475–478.

International Code of Zoological Nomenclature, 3d ed. 1985. London, International Trust for Zoological Nomenclature.

International Trust for Zoological Nomenclature 1997. *International Code of Zoological Nomenclature.* Web page, pp. 1–4. update of Oct. 14, 1997 (http://www.iczn.org/code.htm).

Isely, D. 1972. The disappearance. *Taxon* 21: 3–12.

Jaeger, E. C. 1955. *A Source-Book of Biological Names and Terms,* 3d ed. Springfield, Ill., Charles C. Thomas.

Jarne, P. and B. Delay. 1991. Population genetics of freshwater snails. *Trends in Ecology & Evolution* 6: 381–386.

Jeffrey, C. 1986. Some difference between the botanical and zoological codes, pp. 62–65 in W. D. L. Ride and T. Younès, eds. *Biological Nomenclature Today.* Oxford, U.K., IRL Press.

Jeffrey, C. 1989. *Biological Nomenclature,* 3d ed. London: Edward Arnold.

Jiménez-Guirado, D. 1990. Descripción de *Vanderlindia hispanica* sp. n. (Nematoda: Tylencholaimidae). *Graellsia* 46: 3–5.

Joly, L. J. and C. Bordon. 1996. Two new species of *Trochoideus* Buquet from Venezuela (Coleoptera, Endomychidae), with comments on the Neotropical species groups. *American Museum Novitates* 3169: 1–10.

Jussieu, A.-L. de. 1789. *Genera plantarum.* Paris, Hérissant and Barrois.

Karavaev, V. 1894. Contributions to the crustacean pelagic fauna of the Black Sea. *Kiev. Universitet. Obshchestvo estestvoispitalelei, Zapiski* 13: 35–61.

Karavaev, V. 1895. Contributions to the copepod fauna of the Black Sea. *Kiev. Universitet. Obshchestvo estestvoispitalelei, Zapiski* 14: 117–174.

Kelt, D. A. and R. E. Palma. 1992. A plea for the citation of authors of taxa in systematics studies. *Proceedings of the Biological Society of Washington* 105: 411–413.

Kennedy, R. S. 1987. New subspecies of *Dryocopus javensis* (Aves: Picidae) and *Ficedula hyperythra* (Aves: Muscicapidae) from the Philippines. *Proceedings of the Biological Society of Washington* 100: 40–43.

Kennedy R. S., P. A. Gonzales, and H. C. Miranda, Jr. 1997. New *Aethopyga* sunbirds (Aves: Nectariniidae) from the Island of Mindanao, Philippines. *The Auk* 114: 1–10.

Kennedy, R. S. and C. A. Ross. 1987. A new subspecies of *Rallina eurizonoides* (Aves: Rallidae) from the Batan Islands, Philippines. *Proceedings of the Biological Society of Washington* 100: 459–461.

Kennedy, W. J., W. A. Cobban, and N. H. Landman. 1996. New records of Acanthoceratid ammonoids from the Upper Cenomanian of South Dakota. *American Museum Novitates* 3161: 1–18.

King, G. M., C. Giray, and I. Kornfield. 1990. A new hemichordate, *Saccoglossus bromophenolosus* (Hemichordata: Enteropneusta: Harrimaniidae), from North America. *Proceedings of the Biological Society of Washington* 107: 383–390.

King, T. L., R. Ward, and E. G. Zimmerman. 1994. Population structure of eastern oysters (*Crassostrea virginica*) inhabiting the Laguna Madre, Texas, and adjacent bay systems. *Canadian Journal of Fisheries and Aquatic Sciences* 51 (suppl 1): 215–222.

Kirby, W. 1813. Strepsiptera, a new order of insects proposed; and the characters of the order, with those of its genera laid down. *Linnaean Society of London, Transactions* 11: 86–123.

Kirk, P. M. and P. F. Cannon. 1991. Problems facing indexers and the definition of effective publication, pp. 279–286 in D. L. Hawksworth, ed. *Improving the Stability of Names: Needs and Options.* Königstein, Germany, Koeltz Scientific Books.

Kirkman, J. 1992. *Good style: Writing for science and technology.* London, E & FN Spon.

Knowlton, N. and B. D. Keller. 1983. A new, sibling species of snapping shrimp associated with the Caribbean sea anemone *Bartholomea annulata. Bulletin of Marine Science* 33: 353–362.

Knowlton, N. and B. D. Keller. 1985. Two more sibling species of alpheid shrimps associated with the Caribbean sea anemones *Bartholomea annulata* and *Heteractis lucida. Bulletin of Marine Science* 37: 893–904.

Knowlton, N., E. Weil, L. A. Weigt, and H. M. Guzmán. 1992. Sibling species in *Montastraea annularis,* coral bleaching, and the fossil record. *Science* 255: 330–333.

Kohn, A. J. 1980. *Conus kahiko,* a new Pleistocene gastropod from Oahu, Hawaii. *Journal of Paleontology* 54: 534–541.

Koopman, K. F. 1989. Systematic notes on Liberian bats. *American Museum Novitates* 2946: 1–11.

Kott, P. 1990. The Australian Ascidiacea. Part 2, Aplousobranchia. *Memoirs of the Queensland Museum* 29(1): 1–298.

Krichagin, N. 1873. Contributions to the knowledge of the fauna of the Black Sea. Copepoda. *Kiev. Universitet. Obshchestvo estestvoispitalelei, Zapiski* 3: 370–429.

Kuris, A. M. and J. T. Carlton. 1977. Description of a new species, *Crangon handi,* and new genus *Lissocrangon,* of crangonid shrimps (Crustacea: Caridea) from the California coast, with notes on adaptation in body shape and coloration. *Biological Bulletin* 153: 540–559.

Lack, D. 1947. *Darwin's Finches: An Essay on the General Biological Theory of Evolution.* Cambridge, U.K., Cambridge University Press.

Lammers, T. G., T. J. Givnish, and K. J. Sytsma. 1993. Merger of the endemic Hawaiian genera *Cyanea* and *Rollandia* (Campanulaceae: Lobeliodeae) *Novon* 3: 437–441.

Lammers, T. G. and D. H. Lorence. 1993. A new species of *Cyanea* (Campanulaceae: Lobelioideae) from Kaua'i, and the resurrection of *C. remyi. Novon* 3: 431–436.

Langan-Cranford, K. M. and J. S. Pearse. 1995. Breeding experiments confirm species status of two morphologically similar gastropods (*Lacuna* spp.) in central California. *Journal of Experimental Marine Biology and Ecology* 186: 17–31.

Latreille, P. A. 1796. *Précis des caractères génériques des Insectes, disposés dans un Ordre naturel.* Paris, Boudeaux.

Lawrence, G. H. M., A. F. Günther Buchheim, G. S. Daniels, and H. Dolezal, eds. 1968. *Botanico-Periodicum-Huntianum.* Pittsburgh, Hunt Botanical Library.

Lawrence, M. A. and G. G. Musser. 1990. Mammal holotypes in the American Museum of Natural History: The lectotype of *Prionailurus alleni* Sody (1949). *American Museum Novitates* 2973: 1–11.

Lean, G., D. Hinrichsen, and A. Markham, eds. 1990. *Atlas of the Environment.* New York, Prentice Hall.

Lee, M. S. Y. 1997. Documenting present and past biodiversity: Conservation biology meets palaeontology. *Trends in Ecology & Evolution* 12: 132–133.

Lee, W. L. et al. 1978. Resources in invertebrate systematics, Part I. *American Zool.* 18: 167–185.

Leiggi, P., C. R. Schaff, and P. May. 1994. Field organization and specimen collecting, pp. 59–76 in P. Leiggi and P. May, eds. *Vertebrate Paleontological Techniques,* vol. 1. Cambridge, U.K., Cambridge University Press.

Lewis, W. H. 1959. The nature of plant species. *Journal of the Arizona Academy of Science* 1: 3–7.

Lillegraven, J. and M. C. McKenna. 1986. Fossil mammals from the "Mesaverde" Formation (Late Cretaceous, Judithian) of the Bighorn and Wind River basins, Wyoming, with definitions of Late Cretaceous North American land-mammal "Ages." *American Museum Novitates* 2840: 1–68.

Lindroth, S. 1983. The two faces of Linnaeus, pp. 1–62 in T. Frängsmyr, ed. *Linnaeus, the Man and His Work.* Berkeley, University of California Press.

Lindstrom, S. C. and K. M. Cole. 1990a. An evaluation of species relationships in the *Porphyra perforata* complex (Bangiales, Rhodophyta) using starch gel electrophoresis. *Hydrobiologia* 204/205: 179–1183.

Lindstrom, S. C. and K. M. Cole. 1990b. *Porphyra fallax,* a new species of Rhodophyta from British Columbia and Northern Washington. *Japanese Journal of Phycology* 38: 371–376.

Linnaeus, C. 1736. Methodus Juxta quam Physiologus accurate and feliciter concinnare potest Historiam cujuscunque Naturalis Subjecti, sequentibus hisce Paragraphis comprehensa. unpaged broadside bound with *Systema naturae,* 7th ed. Stockholm.

Linnaeus, C. 1749. Letter to Abraham Bäck, October 6, 1749, quoted in Lindroth 1983.

Linnaeus, C. 1751. *Philosophia Botanica.* Stockholm.

Linnaeus, C. 1753. *Species Plantarum,* 2 vols. Stockholm.

Linnaeus, C. 1758. *Systema naturae per regna tria naturae, secundum classes, ordines, genera, species cum characteribus, differentiis, synonymis, locis.* Editio decima, reformata, Tom I. Laurentii Salvii, Holmiae.

Linnaeus, C. 1764. *Genera plantarum,* 6th ed. Stockholm, Salvius.

Littler, D. S. and M. M. Littler. 1997. An illustrated marine flora of the Pelican Cays, Belize. *Bulletin of the Biological Society of Washington* 9: 1–149.

Littler, D. S., M. M. Littler, K. E. Bucher, and J. N. Norris. 1989. *Marine Plants of the Caribbean.* Washington, D.C., Smithsonian Institution Press.

Loeng, H. 1994. Application of geographical distribution maps in marine research. *Marine Pollution Bulletin* 29: 573–576.

Londt, J. G. H. 1995. An atlas of the Mecoptera of KwaZulu–Natal, South Africa. *Annals of the Natal Museum* 36: 169–187.

Lorini, M. L. and V. G. Persson. 1990. New species of *Leontopithecus* Lesson 1840 from Southern Brazil. Primates Callitrichidae. *Boletim Museu Nacional, Rio de Janeiro* 338: 1–14.

Lueders, E., ed. 1989. *Writing Natural History: Dialogues with Authors.* Salt Lake City, University of Utah Press.

Machado, M. M. and A. M. Costa. 1994. Enzymatic and morphological criteria for distinguishing between *Cardium edule* and *C. glaucum* of the Portuguese coast. *Marine Biology* 120: 535–544.

Maisey, J. G. 1989. *Hamiltonichthys mapesi,* g. & sp. nov. (Chondrichthyes; Elasmobranchii), from the Upper Pennsylvanian of Kansas. *American Museum Novitates* 2931: 1–42.

Manning, R. B. and D. K. Camp. 1993. Erythrosquilloidea, a new superfamily, and Tetrasquillidae, a new family of stomatopod crustaceans. *Proceedings of the Biological Society of Washington* 106: 85–91.

Marcus, L. F. 1993. The goals and methods of systematic biology, pp. 31–53 in

R. Fortuner, ed. *Advances in Computer Methods for Systematic Biology.* Baltimore, Johns Hopkins University Press.

Margulis, L. and K. V. Schwartz. 1988. *Five Kingdoms: An Illustrated Guide to the Phyla of Life on Earth,* 2d ed. New York, W.H. Freeman.

Masterson, J. 1994. Stomatal size in fossil plants: Evidence for polyploidy in majority of angiosperms. *Science* 264: 421–424.

Mathews, R. E. F. 1981. The classification and nomenclature of viruses: Summary of results of meetings of the International Committee on Taxonomy of Viruses in Strasbourg, 1981. *Intervirology* 16: 53–60.

Mathews, R. E. F., ed. 1983. *A Critical Appraisal of Virus Taxonomy.* Boca Raton, Fla., CRC Press.

Mathis, W. N. 1995. Shore flies of the Galápagos Islands (Diptera: Ephydridae). *Annals of the Entomological Society of America* 88: 627–640.

Mattingly, P. F. 1962. Towards a zoogeography of the mosquitoes. *Systematics Publication* 4: 17–16.

May, R. M. 1990. Taxonomy as destiny. *Nature* 347: 129–130.

May, R. M. 1991. A fondness for fungi. *Nature* 252: 475–476.

Mayr, E. 1942. *Systematics and the Origin of Species.* New York, Columbia University Press.

Mayr, E., ed. 1957. The species problem. *American Association for the Advancement of Science Publication* no. 50, 395 pp., Washington, D.C.

Mayr, E. 1963. *Animal Species and Evolution.* Cambridge, Mass., Harvard University Press.

Mayr, E. 1969. *Principles of Systematic Zoology.* New York, McGraw-Hill.

Mayr, E. 1982. *The Growth of Biological Thought.* New York, McGraw-Hill.

Mayr, E. 1988. Recent historical developments, pp. 31–43 in D. L. Hawksworth, ed. *Prospects in Systematics.* Oxford, U.K., Clarendon.

Mayr, E. and P. D. Ashlock. 1991. *Principles of Systematic Zoology,* 2d ed. New York, McGraw-Hill.

Mayr, E., E. G. Linsley, and R. L. Usinger. 1953. *Methods and Principles of Systematic Zoology.* New York, McGraw-Hill.

McClintock, D. 1969. *A Guide to the Naming of Plants, with Special Reference to Heathers.* n.p., The Heather Society.

McCloskey, J. M. and H. Spalding. 1989. A reconnaissance-level inventory of the amount of wilderness remaining in the world. *Ambio* 18: 221–227.

McGrady-Steed, J., P. M. Harris, and P. J. Morin. 1997. Biodiversity regulates ecosystem stability. *Nature* 390: 162–165.

McKinnon, A. D., W. J. Kimmerer, and J. A. H. Benzie. 1992. Sympatric sibling species within the genus *Acartia* (Copepoda: Calanoida): A case study from Westernport and Port Phillip Bays, Australia. *Journal of Crustacean Biology* 12: 239–259.

McNeill, J. and W. Greuter. 1986. Botanical nomenclature, pp. 3–25 in W. D. Ride and T. Younés, eds. *Biological Nomenclature Today.* Oxford, U.K., IRL Press.

Melville, R. V. 1986. Some aspects of zoological nomenclature, pp. 59–61 in W. D. L. Ride and T. Younès, eds. *Biological Nomenclature Today.* Oxford, U.K., IRL Press.

Messick, G. A. and E. B. Small. 1996. *Mesanophrys chesapeakensis* n. sp., a histophagous ciliate in the blue crab, *Callinectes sapidus,* and associated histopathology. *Invertebrate Biology* 115: 1–12.

Millar, R. H. and I. Goodbody. 1974. New species of ascidian from the West Indies. *Studies on the Fauna of Curaçao and Other Caribbean Islands* 15: 142–161.

Miller, D. R. and G. L. Miller. 1993. A new species of *Puto* and a preliminary analysis of the phylogenetic position of the *Puto* group within the Coccoidea (Homoptera: Pseudococcidae). *Jeffersoniana* 4: 1–35.

Miller, E. H. 1985. Museum collections and the study of animal social behaviour. *Occasional Papers of the British Columbia Provincial Museum* 25: 139–162.

Miller, W. V. 1994. New species of Heteroceridae (Coleoptera) from Canada and the United States. *Coleopterists Bulletin* 48: 11–18.

Minelli, A. 1993. *Biological Systematics: The State of the Art.* London, Chapman & Hall.

Mishler, B. D. and A. F. Budd. 1990. Species and evolution in clonal organisms: Introduction. *Systematic Botany* 15: 79–85.

Mitchell, J. C. and J. M. Anderson. 1994. Amphibians and reptiles of Assateague and Chincoteague Islands. *Virginia Museum of Natural History Special Publication* 2:1–120.

Moncrief, N. 1993. Geographic variation in fox squirrels (*Sciurus niger*) and gray squirrels *(S. carolinensis*) of the lower Mississippi River valley. *Journal of Mammalogy* 74: 547–576.

Moran, R. C. 1990. A new species of *Polypodium* (Polypodiaceae) and two new species of *Hypolepis* (Dennstaedtiaceae) from Mesoamerica. *Annals of the Missouri Botanical Garden* 77: 845–850.

Morin, N., R. D. Whetstone, D. Wilken, and K. L. Tomlinson, eds. 1989. Floristics for the 21st Century, Proceedings of the Workshop 4–7 May, 1988, Alexandria, Virginia. *Monographs in Systematic Botany from the Missouri Botanical Garden* 28, 163 pp.

Mortensen, T. 1932. Über den angeblichen Kieselschwamm *Microcordyla asteriae* Zirpolo. Ein Beitrag sur Geschichte der Pedicellarien. *Zoologischer Anzeiger* 97: 197–204.

Murant, A. F. 1985. Taxonomy and nomenclature of viruses. *Microbiological Sciences* 2: 218–220.

Muricy, G., N. Boury-Esnault, C. Bézac, and J. Vacelet. 1996. Cytological evidence for cryptic speciation in Mediterranean *Oscarella* species (Porifera, Homoscleromorpha). *Canadian Journal of Zoology* 74: 881–896.

Musser, G. G. 1982. Results of the Archbold Expeditions. No. 107. A new genus of arboreal rat from Luzon Island in the Philippines. *American Museum Novitates* 2730: 1–33.

Musser, G. G. 1982. Results of the Archbold Expeditions. No. 110. *Crunomys* and the small-bodied shrew-rats native to the Philippine Islands and Sulawesi (Celebes). *Bulletin of the American Museum of Natural History* 174: 1–95.

Myers, C. W. and M. A. Donnelly. 1991. The lizard genus *Sphenomorphus* (Scincidae) in Panama, with description of a new species. *American Museum Novitates* 3027: 1–12.

Naeem, S. and S. Li. 1997. Biodiversity enhances ecosystem reliability. *Nature* 390: 507–509.

National Research Council. 1980. *Research Priorities in Tropical Biology.* Washington, D.C., National Academy of Sciences.

Nichols, F. H. and M. M. Pamatmat. 1988. The ecology of the soft-bottom benthos of San Francisco Bay: A community profile. *United States Fish and Wildlife Service Biological Report* 85: 1–73.

Nicolson, D. H. 1991. A history of botanical nomenclature. *Annals of the Missouri Botanical Garden* 78: 33–56.

Nizinsky, M. S., B. B. Collette, and B. B. Washington. 1990. Separation of two species of sand lances, *Ammodytes americanus* and *A. dubius,* in the western North Atlantic. *Fishery Bulletin, U.S.* 88: 241–255.

Norris, J. N. and R. E. Bucher. 1989. *Rhodogorgon,* an anomalous new red algal genus from the Caribbean Sea. *Proceedings of the Biological Society of Washington* 102: 1050–1066.

Norse, E. A., ed. 1993. *Global Marine Biological Diversity: A Strategy for Building Conservation into Decision-Making.* Washington, D.C., Island Press.

Novacek, M. J. 1985. The Sespedectinae, a new subfamily of hedgehog-like insectivores. *American Museum Novitates* 2822: 1–24.

Novacek, M. J. 1992. The meaning of systematics and the biodiversity crisis, pp. 101–108 in N. Eldredge, ed. *Systematics, Ecology and the Biodiversity Crisis.* New York, Columbia University Press.

Novacek, M. J., I. Ferrusquía-Villafranca, J. J. Flynn, A. R. Wyss, and M. A. Norell. 1991. Wasatchian (early Eocene) mammals and other vertebrates from Baja California, Mexico: The Lomas las Tetas de Cabra fauna. *Bulletin of the American Museum of Natural History* 208, 88 pp.

Ockelmann, K. W. and B. Åkesson. 1990. *Ophryotrocha socialis* n. sp., a link between two groups of simultaneous hermaphrodites within the genus (Polychaeta, Dorvilleidae). *Ophelia* 31: 145–162.

O'Connor, M. 1991. *Writing Successfully in Science.* London, Chapman & Hall.

Office of Technology Assessment. 1987. *Technologies to Maintain Biological Diversity.* OTA-F-330. Washington, D.C., U.S. Government Printing Office.

O'Foighil, D. and C. Thiriot-Quiévreux. 1991. Ploidy and pronuclear interaction in Northeastern Pacific *Lasaea* clones (Mollusca: Bivalvia). *Biological Bulletin* 181: 222–231.

Ogilvie, R. T. 1985. Botanical collections in museums. *Occasional Papers of the British Columbia Provincial Museum* 25: 13–22.

Oken, L. 1815–1816. *Okens Lehrbuch der Naturgeschichte. 3. Theil: Zoologie,* 2 vols. Jena, Germany, A. Schmid.

Oliver, J. H. 1988. Crisis in biosystematics of arthropods. *Science* 240: 967.

Øresland, V. and F. Pleijel. 1991. An ectoparasitic typhoscolecid polychaete on the chaetognath *Eukrohnia hamata* from the Antarctic Peninsula. *Marine Biology* 108: 429–432.

Osburn, R. C. 1950. Bryozoa of the Pacific coast of North America. Part 1. Cheilostomata Anasca. *Allan Hancock Pacific Expeditions* 14: 1–269.

Osorio, D. and M. Vorobyev. 1997. *Sepia* tones, stomatopod signals and the uses of colour. *Trends in Ecology and Evolution* 12: 167–168.

Otte, D. and J. A. Endler, eds. 1989. *Speciation and its Consequences.* Sunderland, Mass., Sinauer.

Oug, E. 1990. Morphology, reproduction and development of a new species of *Ophryotrocha* (Polychaeta: Dorvilleidae) with strong sexual dimorphism. *Sarsia* 75: 191–201.

Ovtsharenko, V. and N. Platnick. 1995. On the Australasian ground spider genera *Anzacia* and *Adelphodrassus* (Araneae, Gnaphosidae). *American Museum Novitates* 3154: 1–16.

Pain, S. 1988. A touch of class in the field. *New Scientist* 15 Sept. 109: 49–55.

Palmer, M. W., G. L. Wade, and P. Neal. 1995. Standards for the writing of floras. *BioScience* 45: 339–345.

Palumbi, S. R. 1992. Marine speciation on a small planet. *Trends in Ecology and Evolution* 7: 114–118.

Palumbi, S. R. and E. C. Metz. 1991. Strong reproductive isolation between closely related tropical sea urchins (genus *Echinometra*). *Molecular Biology and Evolution* 8: 227–239.

Pankhurst, R. J. 1978. *Biological identification.* London, Edward Arnold.

Pereira, L. A. and R. L. Hoffman. 1993. The American species of *Escaryus,* a genus of holarctic centipeds (Gophilomorpha: Schendylidae). *Jeffersoniana* 3: 1–72.

Petersen, K. W. 1990. Evolution and taxonomy in capitate hydroids and medusae (Cnidaria: Hydrozoa). *Zoological Journal of the Linnaean Society* 100: 101–231.

Peterson, R. T. 1980. *A Field Guide to the Birds, a Completely New Guide to All the Birds of Eastern and Central North America,* 4th ed. Boston, Houghton Mifflin.

Peterson, R. T. and M. McKenney. 1968. *A Field Guide to Wildflowers, Northeastern and North-Central North America.* Boston, Houghton Mifflin.

Pettibone, M. H. 1990. New species and new records of scaled polychaetes (Polychaeta: Polynoidae) from the axial seamount caldera of the Juan de Fuca

Ridge in the northeast Pacific and the east Pacific Ocean off northern California. *Proceedings of the Biological Society of Washington* 103: 825–838.

Pettibone, M. H. 1995. New genera for two polychaetes of Lepidonotinae. *Proceedings of the Biological Society of Washington* 108: 577–582.

Pine, R. H. 1994. New mammals not so seldom. *Nature* 368: 593.

Pinker, S. 1995. *The Language Instinct.* New York, Harper Perennial.

Platnick, N. I. 1975. A revision of the South American spider genus *Trachylopachys* (Araneae, Clubionidae). *American Museum Novitates* 2589: 1–25.

Platnick, N. I. 1986. A review of the spider genus *Cyrioctea* (Araneae, Zodariidae). *American Museum Novitates* 2858, 9 pp.

Platnick, N. I. 1989. A revision of the spider genus *Segestrioides* (Araneae, Diguetidae). *American Museum Novitates* 2940: 1–9.

Platnick, N. I. and A. D. Brescovit. 1995. On *Unicorn,* a new genus of the spider family Oonopidae (Araneae, Dysderoidea). *American Museum Novitates* 3152: 1–12.

Platnick, N. I. and R. R. Forster. 1986. On *Teutoniella,* an American genus of the spider family Micropholcommatidae (Araneae, Palpimanoidea). *American Museum Novitates* 2854: 1–9.

Platnick, N. I. and M. U. Shadab. 1989. A review of the spider genus *Teminius* (Araneae, Miturgidae). *American Museum Novitates* 2963: 1–12.

Platnick, N. I. and D. X. Song. 1986. A review of the zelotine spiders (Araneae, Gnaphosidae) of China. *American Museum Novitates* 2848: 1–22.

Poirier, Y., D. E. Dennis, K. Klomparens, and C. Somerville. 1992. Polyhydroxybutyrate, a biodegradable thermoplastic, produced in transgenic plants. *Science* 256: 520–523.

Polloni, P. T., G. T. Rowe, and J. M. Teal. 1973. *Biremis blandi* (Polychaeta: Terebellidae), new genus, new species, caught by D.S.R.V. "Alvin" in the Tongue of the Ocean, New Providence, Bahamas. *Marine Biology* 20: 170–175.

Pomponi, S. A. 1988. Maximizing the potential of marine organism collections for both pharmacological and systematic studies. *Memoirs of the California Academy of Sciences* 13: 7–11.

Ptacek, M. B., H. C. Gerhardt, and R. D. Sage. 1994. Speciation by polyploidy in treefrogs: Multiple origins of the tetraploid, *Hyla versicolor. Evolution* 48: 898–908.

Putman, R. J. and S. D. Wratten. 1984. *Principles of Ecology.* Berkeley, University of California Press.

Quicke, D. L. J. 1993. *Principles and Techniques of Contemporary Taxonomy.* London, Blackie Academic and Professional.

Racheboeuf, P. R. and H. R. Feldman. 1990. Chonetacean brachiopods of the "Pink *Chonetes*" Zone, Onandaga Limestone (Devonian, Eifelian) central New York. *American Museum Novitates* 2982: 1–16.

Randall, J. E., T. H. Fraser, and E. A. Lachner. 1990. On the validity of the

Indo-Pacific cardinalfishes *Apogon aureus* (Lacepède) and *A. fleurieu* (Lacepède), with description of a related new species from the Red Sea. *Proceedings of the Biological Society of Washington* 103: 39–62.

Randall, J. E. and C. L. Smith. 1988. Two new species and a new genus of cardinalfishes (Perciformes: Apogonidae) from Rapa, South Pacific Ocean. *American Museum Novitates* 2926: 1–9.

Random-House Webster's Dictionary. 1993. New York, Ballantine.

Rauchenberger, M., K. D. Kallman, and D. C. Morizot. 1990. Monophyly and geography of the Río Pánuco basin swordtails (Genus *Xiphophorus*) with descriptions of four new species. *American Museum Novitates* 2975: 1–41.

Raven, P. H., B. Berlin, and D. E. Breedlove. 1971. The origins of taxonomy. *Science* 174: 1210–1213.

Règles interationales de la Nomenclature zoologique adoptées par les congrès internationaux de zoologie. 1905. Paris, F.R. de Rudeval.

Reimer, C. W. and J. J. Lee. 1988. New species of endosymbiotic diatoms (Bacillariophyceae) inhabiting larger Foraminifera in the Gulf of Elat (Red Sea), Israel. *Proceedings of the Academy of Natural Sciences of Philadelphia* 140: 339–351.

Reveal, J. L. and R. D. Hoogland. 1991. Protected plant family names: A new list for consideration, pp. 243–255 in D. L. Hawksworth, ed. *Improving the Stability of Names: Needs and Options.* Königstein, Koeltz Scientific Books.

Reyes, J. C., J. G. Meade, and K. V. Van Waerbeek. 1991. A new species of beaked-whale *Mesoplodon peruvianus* sp. n. (Cetacea: Ziphiidae). *Marine Mammal Science* 7: 1–24.

Reynolds, J. F. and D. J. Crawford. 1980. A quantitative study of variation in the *Chenopodium atrovirens–desiccatum–pratericola* complex. *American Journal of Botany* 67: 1380–1390.

Rice, M. E. 1978. Morphological and behavioral changes at metamorphosis in the Sipuncula, pp. 83–102 in F. S. Chia and M. E. Rice, eds. *Settlement and Metamorphosis of Marine Invertebrate Larvae.* New York, Elsevier North-Holland.

Rice, M. E. 1981. Larvae adrift: Patterns and problems in life histories of sipunculans. *American Zoologist* 21: 605–619.

Rice, M. E. 1988. Observations on development and metamorphosis of *Siphonosoma cumanense* with comparative remarks on *Sipunculus nudus* (Sipuncula, Sipunculidae). *Bulletin of Marine Science* 42: 1–15.

Rice, M. E. 1993. Two new species of *Phascolion* (Sipuncula: Phascolionidae) from tropical and subtropical waters of the central western Atlantic. *Proceedings of the Biological Society of Washington* 106: 591–601.

Ride, W. D. L. 1985. Preface, pp. x–xii in *International Code of Zoological Nomenclature,* 3d ed. Berkeley, University of California Press.

Ride, W. D. L. 1986. Zoological nomenclature, pp. 26–35 in W. D. L. Ride and T. Younès, eds. *Biological Nomenclature Today.* Oxford, U.K., IRL Press.

Rigby, J. K. and B. Senowbari-Daryan. 1996. Upper Permian inozoid, demospongid and hexactinellid sponges from Djebel Tebaga, Tunisia. *University of Kansas Paleontological Contributions, New Series* 7: 1–130.

Ritvo, H. 1997. *The Platypus and the Mermaid and Other Figments of the Classifying Imagination.* Cambridge, Mass., Harvard University Press.

Robbins, C. S., B. Bruun, and H. S. Zim. 1983. *Birds of North America.* New York, Golden Press.

Roberts, L. 1991. Ranking the rainforests. *Science* 251: 1559–1560.

Robilliard, G. A. and P. K. Dayton. 1972. A new species of platyctenean ctenophore, *Lyrocteis flavopallidus* sp. nov., from McMurdo Sound, Antarctica. *Canadian Journal of Zoology* 50: 47–52.

Robinson, H. 1988. Studies in the *Lepidaploa* complex (Vernonieae: Asteraceae) V. The new genus *Chrysolaena. Proceedings of the Biological Society of Washington* 101: 952–958.

Robinson, H. 1994. *Cololobus, Pseudopiptocarpha,* and *Trepadonia,* three new genera from South America (Vernonieae: Asteraceae). *Proceedings of the Biological Society of Washington* 107: 557–568.

Robson, N. K. B. 1990. Two new species and a new combination in *Vismia* (Guttiferae: Hypericoideae). *Annals of the Missouri Botanical Garden* 77: 410–411.

Roccataglia, D. and R. W. Heard. 1995. Two species of *Oxyurostylis* (Crustacea: Cumacea: Diastylidae), *O. smithi* Calman, 1912 and *O. lecroyae,* a new species from the Gulf of Mexico. *Proceedings of the Biological Society of Washington* 108: 596–612.

Rodríguez-Almaraz, G. A. and T. E. Bowman. 1995. *Sphaerolana karenae,* a new species of hyogean isopod crustacean from Nuevo Leon, Mexico. *Proceedings of the Biological Society of Washington* 108:207–211.

Röhner, M., R. Bastrop, and K. Jürss. 1996. Colonization of Europe by two American genetic types or species of the genus *Marenzellaria* (Polychaeta: Spionidae). An electrophoretic analysis of allozymes. *Marine Biology* 127: 277–287.

Rosell, D. and M.-J. Uriz. 1991. *Cliona viridis* (Schmidt, 1862) and *Cliona nigricans* (Schmidt, 1862) (Porifera, Hadromerida): Evidence which shows they are the same species. *Ophelia* 33: 45–53.

Ross, H. H. 1974. *Biological Systematics.* Reading, Mass., Addison-Wesley.

Roth, B. 1975. Description of a new terrestrial snail from San Nicolas Island, California (Gastropoda: Stylommatophora). *Bulletin of the Southern California Academy of Sciences* 74: 94–96.

Roux, J. P. 1994. Lectotypification of *Asplenium lucidum* Burm. f. (*Aspleniaceae*). *Taxon* 43: 641–642.

Rowe, F. W. E., A. N. Baker, and H. E. S. Clark. 1988. The morphology, development and taxonomic status of *Xyloplax* Baker, Rowe and Clark (1986) (Echinodermata:

Concentricycloidea), with description of a new species. *Proceedings of the Royal Society of London B* 233: 431–459.

Rozbaczylo, N. and J. I. Cañete. 1993. A new species of scale-worm, *Harmothoe commensalis* (Polychaeta: Polynoidae), from mantle cavities of two Chilean clams. *Proceedings of the Biological Society of Washington* 106: 666–672.

Rozen, J. G., Jr. 1989. Two new species and the redescription of another species of the cleptoparasitic bee genus *Triepeolus* with notes on their immature stages (Anthophoridae: Nomadinae). *American Museum Novitates* 2956, 18 pp.

Rozen, J. G., Jr. 1989. Morphology and systematic significance of first instars of the cleptoparasitic bee tribe Epeolini (Anthophoridae: Nomadinae). *American Museum Novitates* 2957: 1–19.

Rozen, J. G., Jr. and S. L. Buchmann. 1990. Nesting biology and immature stages of the bees *Centris caesalpiniae, C. pallida,* and the cleptoparasite *Ericrocis lata* (Hymenoptera: Apoidea: Anthophoridae). *American Museum Novitates* 2985: 1–30.

Ruppert, E. E. and R. S. Fox. 1988. *Seashore Animals of the Southeast.* Columbia, University of South Carolina Press.

Rützler, K. and S. Richardson. 1996. The Caribbean spicule tree: A sponge-imitating foraminifer (Astrorhizidae). *Bulletin de l'Institut royal des Sciences naturelles de Belgique. Biologie* 66 (suppl): 143–151.

Ryland, J. S. 1967. *Bugula* Oken, 1815, and *Scruparia* Oken, 1815 (Polyzoa): Proposed conservation under the plenary powers. Z.N.(S.) 1390. *Bulletin of Zoological Nomenclature* 24: 24–26.

Ryland, J. S. and P. J. Hayward. 1977. *British Anascans. Synopses of the British Fauna,* new series, no. 10. London, Academic Press.

Salman, S. D. and N. Jabbar. 1990. A new species of the genus *Cheiriphotis* Walker, from the north west Arabian Gulf, with a redescription of *C. megacheles* (Giles) (Amphipoda, Isaeidae). *Crustaceana* 58: 214–226.

Salmon, M. and S. P. Atsaides. 1968. Behavioral, morphological and ecological evidence for two new species of fiddler crabs (Genus *Uca*) from the Gulf Coast of the United States. *Proceedings of the Biological Society of Washington* 81: 275–289.

Salmon, M., S. D. Ferris, D. Johnston, G. Hyatt, and C. S. Whitt. 1979. Behavioral and biochemical evidence of species distinctiveness in the fiddler crabs, *Uca speciosa* and *U. spinicarpa. Evolution* 33: 182–191.

Salmon, M. and G. W. Hyatt. 1979. The development of acoustic display in the fiddler crab, *Uca pugilator,* and its hybrids with *Uca panacea. Marine Behaviour and Physiology* 6: 197–209.

Salmon, M., G. Hyatt, K. McCarthy, and J. D. Costlow, Jr. 1978. Display specificity and reproductive isolation in the fiddler crabs, *Uca panacea* and *U. pugilator. Zeitschrift für Tierpsychologie* 48: 251–276.

Salmon, M. and M. Kettler. 1987. The importance of behavioral and biochemical differences between fiddler crab taxa, with special reference to *Uca rapax* (Smith) and *U. virens* (Salmon and Atsaides). *Contributions in Marine Sciences* 30: 63–76.

Salmon, M., K. McCarthy, and J. D. Costlow, Jr. 1978. Display specificity and reproductive isolation in the fiddler crabs, *Uca panacea* and *U. pugilator. Zeitschrift für Tierpsychologie* 48: 251–276.

Saltzman, J. and T. E. Bowman. 1993. *Boreomysis oparva,* a new possum shrimp (Crustacea: Mysidacea) from an eastern tropical Pacific seamount. *Proceedings of the Biological Society of Washington* 106: 325–331.

Saunders, G. W. and G. T. Kraft. 1996. Small-subunit rRNA gene sequences from representatives of selected families of the Gigartinales and Rhodymeniales (Rhodophyta). 2. Recognition of the Halymeniales ord. nov. *Canadian Journal of Botany* 74: 694–707.

Savigny, J. C. ?1817. *Description de L'Égypt ou recueil des observations et des recherches qui ont été faites en Égypt pendant l'expédition de L'armée Française.* Histoire Naturelle, Planches, vol. 2. Paris, Publ. Gouvt.

Savory, T. 1962. *Naming the Living World.* New York: Wiley.

Schawaller, W., W. A. Shear, and P. M. Bonamo. 1991. The first Paleozoic pseudoscorpions (Arachnida, Pseudoscorpionida). *American Museum Novitates* 3009: 1–17.

Scheltema, R. S. 1996. Describing diversity. *Oceanus* 39: 16–18.

Schenk, E. T. and J. H. McMasters. 1936. *Procedure in Taxonomy.* Stanford, Calif., Stanford University Press.

Schenk, E. T. and J. H. McMasters. 1948. *Procedure in Taxonomy,* rev. ed., enlarged and in part rewritten by A. M. Keen and S. M. Muller. Stanford, Calif., Stanford University Press.

Schenk, E. T. and J. H. McMasters. 1956. *Procedure in Taxonomy,* 3d ed., enlarged and in part rewritten by A. M. Keen and S. M. Muller. Stanford, Calif., Stanford University Press.

Schluter, D. 1994. Experimental evidence that competition promotes divergence in adaptive radiation. *Science* 266: 798–801.

Schmidt, K. P. 1952. The "Methodus" of Linnaeus, 1736. *Journal of the Society for the Bibliography of Natural History* 2: 369–374.

Schopf, T. J. M. 1968. Ectoprocta, Entoprocta, and Bryozoa. *Systematic Zoology* 17: 470–472.

Schuh, R. T. 1989. Old World Pilophorini: Descriptions of nine new species with additional synonymic and taxonomic changes (Heteroptera: Miridae: Phylinae). *American Museum Novitates* 2945: 1–16.

Schultz, G. M. 1995. *Sinoniscus cavernicolus,* a new genus and species of terrestrial isopod crustacean from a cave in China. *Proceedings of the Biological Society of Washington* 108: 201–206.

Schultz, R. J. 1969. Hybridization, unisexuality and polyploidy in the teleost

Poeciliopsis (Poeciliidae) and other vertebrates. *American Naturalist* 103: 605–619.

Schultz, R. J. 1980. Role of polyploidy in the evolution of fishes, in W. H. Lewis, ed. *Polyploidy: Biological Relevance.* New York, Plenum.

Scudder, G. G. E. 1987. The next 25 years: Invertebrate systematics. *Canadian Journal of Zoology* 65: 786–787.

Shear, W. A. 1986. A cladistic analysis of the opilionid superfamily Ischyropsalidoidea, with descriptions of the new family Ceratolasmatidae, the new genus *Acuclavella,* and four new species. *American Museum Novitates* 2844: 1–29.

Shear, W. A. 1990. On the central and east Asian milliped family Diplomaragnidae (Diplopoda, Chordeumatida, Diplomaragnoidea). *American Museum Novitates* 2977: 1–40.

Siebert, D. J. and W. L. Minckley. 1986. two new catostomid fishes (Cypriniformes) from the northern Sierra Madre Occidental of Mexico. *American Museum Novitates* 2849: 1–17.

Silbermann, J. D., S. K. Sarver, and P. J. Walsh. 1994. Mitochondrial DNA variation and population structure in the spiny lobster *Panulirus argus. Marine Biology* 120: 601–608.

Simmons, N. B. 1996. A new species of *Micronycteris* (Chiroptera: Phyllostomidae) from northeastern Brazil, with comments on phylogenetic relationships. *American Museum Novitates* 3158: 1–34.

Simpson, C. T. 1901. Synopsis of the Naiades, or pearly freshwater mussels. *Proceedings of the National Museum* 22: 501–1044.

Simpson, G. G. 1945. The principles of classification and a classification of mammals. *Bulletin of the American Museum of Natural History* 85: 1–350.

Simpson, G. G. 1961. *Principles of Animal Taxonomy.* New York, Columbia University Press.

Sims, P. A. 1989. Some Cretaceous and Palaeocene species of *Coscinodiscus:* A micromorphological and systematic study. *Diatom Research* 4: 351–371.

Slobodchikoff, C. N., ed. 1976. *Concepts of Species.* Stroudsburg, Pa., Dowden, Hutchinson & Ross.

Smith, A. B. 1994. *Systematics and the Fossil Record: Documenting Evolutionary Patterns.* Oxford, U.K., Blackwell Scientific.

Smith, M. L. and R. R. Miller. 1986. Mexican goodeid fishes of the genus *Characodon,* with description of a new species. *American Museum Novitates* 2851: 1–14.

Smithe, F. B. 1974. *Naturalist's Color Guide Supplement.* New York, American Museum of Natural History.

Smithe, F. B. 1975. *Naturalist's Color Guide.* New York, American Museum of Natural History.

Smithe, F. B. 1981. *Naturalist's Color Guide Part III.* New York, American Museum of Natural History.

Sneath, P. H. A. 1986. Nomenclature of bacteria, pp. 36–47 in W. D. L. Ride and T. Younès, eds. *Biological Nomenclature Today.* Oxford, U.K., IRL Press.

Sneath, P. H. A. 1992. *International Code of Nomenclature of Bacteria,* 1990 rev., ed. by S. P. Lapage et al. Washington, D.C., Published for the International Union of Microbiological Societies by the American Society for Microbiology.

Sneath, P. H. A. and R. R. Sokal. 1973. *Numerical Taxonomy.* San Francisco, W.H. Freeman.

Sody, H. J. V. 1949. Notes on some primates, Carnivora, and the babirusa from the Indo-Malayan and Indo-Australian regions. *Treubia* 20: 121–190.

Sohn, I. G. 1994. Taxonomic synonymy, what is it and why? *Journal of Paleontology* 68: 669–670.

Sokal, R. R. 1973. The species problem reconsidered. *Systematic Zoology* 22: 360–374.

Sokal, R. R. and P. H. A. Sneath. 1963. *Principles of Numerical Taxonomy.* San Francisco, W.H. Freeman.

Solé-Cava, A. M. and J. P. Thorpe. 1986. Genetic differentiation between morphotypes of the marine sponge *Suberites ficus* (Demospongiae: Hadromerida). *Marine Biology* 93: 247–253.

Soule, D. F. and J. D. Soule. 1973. Morphology and speciation of Hawaiian and eastern Pacific Smittinidae (Bryozoa, Ectoprocta). *Bulletin of the American Museum of Natural History* 152: 365–440.

Souto Couri, M. and C. J. Einicker Lamas. 1994. A new species of *Hyperalonia* Rondani, 1863 (Insecta: Diptera: Bombyliidae: Exoprosopinae). *Proceedings of the Biological Society of Washington* 107: 119–121.

Stace, C. 1989. *Plant Taxonomy and Biosystematics,* 2d ed. London: Edward Arnold.

Stafleu, F. A. and R. S. Cowan. 1976. *Taxonomic Literature: A Selective Guide to Botanical Publications and Collections with Dates, Commentaries and Types,* 2d ed., 7 vols. The Netherlands, Bohn, Sheltema & Holkema.

Stap, D. 1990. *A Parrot Without A Name: The Search for the Last Unknown Birds on Earth.* New York, Knopf.

Stap, D. 1997. Finding the missing pieces. *Audubon* March–April: 82, 84–86.

Stearn, W. T. 1983. *Botanical Latin,* 3d rev. ed. London, David & Charles.

Stearn, W. T. 1992. *Botanical Latin: History, Grammar, Syntax, Terminology, and Vocabulary,* 4th rev. ed. Portland, Ore., Timber Press.

Steere, W. C. et al. 1971. *The Systematic Biology Collections of the United States: An Essential Resource: Part I: The Great Collections: Their Importance, Condition and Future,* New York, New York Botanical Garden.

Stepien, C. A. 1987. Color pattern and habitat difference between male, female and juvenile giant kelpfish (Blennioidei: Clinidae). *Bulletin of Marine Science* 41: 45–58.

Stepien, C. A. 1988. Regulation and significance of color patterns of the spotted kelpfish, *Gibbonsia elegans* Cooper, 1864 (Blennioidei: Clinidae). *Copeia* 1988: 7–15.

Stevens, P. F. 1994. *The Development of Biological Systematics: Antoine-Laurent de Jussieu, Nature and the Natural System.* New York, Columbia University Press.

Stiassny, M. L. J. 1990. Notes on the anatomy and relationships of the bedotiid fishes of Madagascar, with a taxonomic revision of the genus *Rheocles* (Atherinomorpha: Bedotiidae). *American Museum Novitates* 2979: 1–33.

Stiassny, M. L. J. 1991. Report on a small collection of fishes from the Wologizi Mountains of Liberia, West Africa, with a description of two new species of *Barbus* (Osteriophysi: Cyprinidae). *American Museum Novitates* 3015, 9 pp.

Stiassny, M. L. J. and P. N. Reinthal. 1992. Description of a new species of *Rheocles* (Atherinomorpha, Bedotiidae) from the Nosivolo Tributary, Mangoro River, eastern Malagasy Republic. *American Museum Novitates* 3031: 1–8.

Stobart, B. and J. A. H. Benzie 1994. Allozyme electrophoresis demonstrates that the scleractinian coral *Montipora digitata* is two species. *Marine Biology* 118: 183–190.

Stonedahl, G. M. and G. Cassis. 1991. Revision and cladistic analysis of the plant bug genus *Fingulus* Distant (Heteroptera: Miridae: Deraeocorinae). *American Museum Novitates* 3028: 1–55.

Stonedahl, G. M. and M. D. Schwartz. 1986. Revision of the plant bug genus *Pseudopsallus* Van Duzee (Heteroptera: Miridae). *American Museum Novitates* 2842: 1–58.

Strassman, J. E., C. R. Solís, J. M. Peters, and D. C. Queller. 1996. Strategies for finding and using highly polymorphic DNA microsatellite loci for studies of genetic relatedness and pedigrees, pp. 163–180 in J. D. Ferraris and S. R. Palumbi, eds. *Molecular Zoology: Advances, Strategies, and Protocols.* New York, Wiley.

Strickland, H. E. et al. 1842. Report of a committee appointed "to consider the rules by which the nomenclature of zoology may be established on a uniform and permanent basis." *Report of the Twelfth Meeting of the British Association for the Advancement of Science,* pp. 105–121.

Stuessy, T. F. 1990. *Plant Taxonomy: The Systematic Evaluation of Comparative Data.* New York: Columbia University Press.

Stuessy, T. F. and K. S. Thompson, eds. 1981. *Trends, Priorities, and Needs in Systematic Biology.* Lawrence, Kans., Association of Systematic Collections.

Sullivan, R. M. 1989. *Proglyptosaurus huerfanensis,* new genus, new species: Glyptosaurine Lizard (Squamata: Anguidae) from the Early Eocene of Colorado. *American Museum Novitates* 2949: 1–8.

Sylvester-Bradley, P. D. 1956. The species concept in palaeontology. *Systematics Association Publication* 2: 1–145.

Systematics Agenda 2000. 1994. *Systematics Agenda 2000: Charting the Biosphere.* Technical Report. United States, Systematics Agenda 2000.

Templeton, A. R. 1989. The meaning of species and speciation: a genetic perspective, pp. 3–27 in D. Otte and J. A. Endler, eds. *Speciation and Its Consequences.* Sunderland, Mass., Sinauer Associates.

Thomas, J. D. and M. Ortiz. 1995. *Leucothoe laurensi,* a new species of leucothoid amphipod from Cuban waters (Crustacea: Amphipoda: Leucothoidae). *Proceedings of the Biological Society of Washington* 108: 613–616.

Thomas, O. 1894. Descriptions of some new Neotropical Muridae. *Annals and Magazine of Natural History* 6(14): 346–366.

Thornton, I. W. B. 1991. Replacement name for *Aaroniella badonelli* Thornton (Psocoptera, Philotarsidae). *Bulletin de la Museum national de Histoire naturelle, Paris,* 4e sér., 13 Section A, no. 3–4: 483.

Thorpe, J. P., J. A. Beardmore, and J. S. Ryland. 1978. Genetic evidence for cryptic speciation in the marine bryozoan *Alyconidium gelatinosum. Marine Biology* 49: 27–32.

Thorpe, J. P. and J. S. Ryland. 1979. Cryptic speciation detected by ecological genetics in three ecologically important species of intertidal bryozoans. *Estuarine and Coastal Marine Science* 8: 395–398.

Thorpe, J. P., J. S. Ryland, and J. A. Beardmore. 1978. Genetic variation and biochemical systematics in the marine bryozoan *Alcyonidium mytili. Marine Biology* 49: 343–350.

Tilling, S. M. 1987. Education and taxonomy: The role of the Field Studies Council and AIDGAP. *Biological Journal of the Linnaean Society* 32: 87–96.

Tilman, D., C. L. Lehman, and K. T. Thomson. 1997. Plant diversity and ecosystem productivity: Theoretical consideration. *Proceedings of the National Academy of Science, U.S.A.* 94: 1857–1861.

Trehane, P., C. D. Brickell, B. R. Baum, W. L. A. Hetterscheid, A. C. Leslie, J. McNeill, S. A. Spongberg, and F. Vrugtman. 1995. International Code of Nomenclature for Cultivated Plants, 1995. *Regnum Vegetabile* 133.

Treiman, R., U. Goswami, and M. Bruck. 1990. Not all nonwords are alike: Implications for reading development and theory. *Memory & Cognition* 18: 559–567.

Tsuchiya, K. and T. Okutani. 1988. Subgenera of *Enoplueteuthis, Abralia* and *Abraliopsis* of the squid family Enoploteuthidae (Cephalopoda, Oegopsida). *Bulletin of the National Science Museum, Tokyo. Ser. A (Zoology)* 14:119–136.

Tucker, A. O., M. J. Maciarello, and S. S. Tucker. 1991. A survey of color charts for biological description. *Taxon* 40: 201–214.

Upchurch, G. R., P. R. Crane, and A. N. Drinnan. 1994. The megaflora from the Quantico Locality (Upper Albian), Lower Cretaceous Potomac Group of Virginia. *Virginia Museum of Natural History Memoir* 4: 1–64.

Vane-Wright, R. I., C. J. Humphries, and P. H. Williams. 1991. What to protect: Systematics and the agony of choice. *Biological Conservation* 55: 235–254.

Van Valen, L. 1971. Adaptive zones and orders of mammals. *Evolution* 25: 420–428.

Van Valen, L. 1976. Ecological species, multispecies and oaks. *Taxon* 25: 233–239.

Vari, R. P. 1992. Redescription of *Mesopristes elongatus* (Guichenot, 1866), an endemic Malagasy fish species (Pisces, Terapontidae). *American Museum Novitates* 3030: 1–7.

Veron J. E. N. 1990. New Scleractinia from Japan and other Indo-West Pacific countries. *Galaxea* 9: 95–173.

Verrill, A. E. 1869. The rules of zoological nomenclature. From the report of a committee appointed to report on the changes which they may consider desirable to make, if any, in the Rules of Zoological Nomenclature, drawn up by Mr. H. C. Strickland, at the instance of the British Association, at their meeting in Manchester in 1842. *American Journal of Arts and Sciences* 48: 92–110.

Virginia's Endangered Species: Proceedings of a Symposium. 1991. Coord. by K. Terwilliger. Blacksburg, Va., McDonald & Woodward.

Voss, E. G. 1952. The history of keys and phylogenetic trees in systematic biology. *Journal of Scientific Laboratories of Denison University* 43: 1–25.

Voss, R. S. 1991. An introduction to the Neotropical muroid rodent genus *Zygodontomys. Bulletin of the American Museum of Natural History* 210: 1–113.

Vrijenhoek, R. C., R. M. Dawley, C. J. Cole, and J. P. Bogart. 1989. A list of known unisexual vertebrates, pp 19–23 in R. Dawley and J. Bogart, eds. *Evolution and Ecology of Unisexual Vertebrates.* Albany, New York State Museum.

Vrijenhoek, R. C., S. J. Schutz, R. G. Gustafson, and R. A. Lutz. 1994. Cryptic species of deep-sea clams (Mollusca: Bivalvia: Vesicomyidae) from hydrothermal vent and cold-water seep environments. *Deep-Sea Research I* 41: 1171–1189.

Wallich, G. C. 1862. Note on the discovery of an extremely minute vertebrate lower jaw in mud dredged at St. Helena. *Annals and Magazine of Natural History* ser. 3, 10: 304.

Walters, S. M. 1961. The shaping of angiosperm taxonomy. *New Phytologist* 60: 74–84.

Ward, L. W. 1992. Molluscan biostratigraphy of the Miocene, Middle Atlantic coastal plain of North America. *Virginia Museum of Natural History Memoir* 2: 1–232.

Ward, R. D., and J. Andrew. 1995. Population genetics of the northern Pacific seastar *Asterias amurensis* (Echinodermata: Asteriidae): Allozyme differentiation among Japanese, Russian, and recently introduced Tasmanian populations. *Marine Biology* 124: 99–109.

Wasshausen, D. C. 1992. Three new species of *Justicia* (Acanthaceae) from Brazil. *Proceedings of the Biological Society of Washington* 105: 664–673.

Weber, A. 1991. *Globba unifolia* and *G. corneri* stat. et nom. nov. (Zingiberaceae; Peninsular Malaysia). *Plant Systematics and Evolution* 174: 1–4.

Weil, E. and N. Knowlton. 1994. A multi-character analysis of the Caribbean coral *Montastraea annularis* (Ellis and Solander, 1786) and its two sibling species, *M.*

faveolata (Ellis and Solander, 1786) and *M. franksi* (Gregory, 1895). *Bulletin of Marine Science* 55: 151–175.

Westfall, R. H., G. Dednam, N. Van Rooyen, and G. K. Theron. 1982. PHYTOTAB: A program package for Braun–Blanquet tables. *Vegetatio* 49: 35–37.

Westfall, R. H., H. F. Glen, and M. D. Panagos. 1986. A new identification aid combining features of a polyclave and an analytical key. *Botanical Journal of the Linnaean Society* 92: 65–73.

Wheelock, F. M. 1995. *Wheelock's Latin,* 5th ed., R. A. LaFleur, revision editor. New York, Harper Perennial.

White, M. J. D. 1978. *Modes of Speciation.* San Francisco, Freeman.

Wiley, E. O. 1978. The evolutionary species concept reconsidered. *Systematic Zoology* 27: 17–26.

Wiley, E. O. 1981. *Phylogenetics: The Theory and Practice of Phylogenetic Systematics.* New York, Wiley.

Williams, G. C. 1990. A new genus of dimorphic soft coral from the south-western fringe of the Indo-Pacific (Octocorallia: Alcyoniidae). *Journal of Zoology, London* 221: 21–35.

Willis, J. C. and G. U. Yule. 1922. Some statistics of evolution and geographic distribution in plants and animals, and their significance. *Nature* 109: 177–9.

Wilson, E. O. 1985. Time to revive systematics. *Science* 230: 1227.

Wilson, E. O. 1988. The current state of biological diversity, pp. 3–18 in E. O. Wilson, ed. *Biodiversity.* Washington, D.C., National Academy Press.

Wilson, E. O. 1989. The coming pluralization of biology and the stewardship of systematics. *BioScience* 39: 242–245.

Wilson, E. O. 1994. Biodiversity: Challenge, science, opportunity. *American Zoologist* 34: 5–11.

Wilson, E. O. 1995. *Naturalist.* New York, Warner Books, Inc.

Winchester, A. M. and H. E. Jacques. 1981. *How to Know the Living Things.* Dubuque, Iowa, Wm. C. Brown.

Winston, J. E. 1982. Marine bryozoans (Ectoprocta) of the Indian River area (Florida). *Bulletin of the American Museum of Natural History* 173: 99–176.

Winston, J. E. 1984. Shallow-water bryozoans of Carrie Bow Cay, Belize. *American Museum Novitates* 2799:1–38.

Winston, J. E. 1988. The systematists' perspective. *Memoirs of the California Academy of Sciences* 13: 1–6.

Winston, J. E. 1992. Systematics and Marine Conservation, pp. 144–168 in N. Eldredge, ed. *Systematics, Ecology and the Biodiversity Crisis.* New York: Columbia University Press.

Winston, J. E. In press. Libbie Hyman and the American Museum of Natural History. *American Museum Novitates.*

Winston, J. E. and E. Håkansson. 1986. The interstitial bryozoan fauna from Capron Shoal, Florida. *American Museum Novitates* 2865: 1–50.

Winston, J. E. and B. F. Heimberg. 1986. Bryozoans from Bali, Lombok, and Komodo. *American Museum Novitates* 2847: 1–49.

Winston, J. E. and K. S. Metzger. 1998. Trends in taxonomy revealed by published literature. *BioScience* 48: 125–128.

Witham, T. G. and C. N. Slobodchikoff. 1981. Evolution by individuals, plant-herbivore interactions, and mosaics of genetic variability, the adaptive significance of somatic mutation in plants. *Oecologia* 49: 287–292.

Wood, P. 1994. *Scientific Illustration.* New York, Van Nostrand Reinhold.

Woodford, P., ed. 1986. *Scientific Writing for Graduate Students.* Bethesda, Md., Council of Biology Editors.

Wynne, M. J. and J. N. Norris. 1976. The genus *Colpomenia* Derbès et Solier (Phaeophyta) in the Gulf of California. *Smithsonian Contributions to Botany* 35: 1–18.

Yancey, P. H. 1946. *Introduction to Biological Latin and Greek,* 3d rev. ed. Mt. Vernon, Iowa, F.G. Brooks.

Yoshida, T. 1978. Lectotypification of *Sargassum kjellmanianum* and *Sargassum miyabei* (Phaeophyta, Sargassaceae). *Japanese Journal of Phycology* 26: 121–124.

Younès, T. 1996. IUBS: The home union for biological nomenclature. *Taxon* 45: 295–300.

Zanoni, T. A. and R. G. Garcia G. 1995. Notes on the flora of Hispaniola. *Annals of Carnegie Museum* 64: 255–265.

Zhu, G. 1994. Lectotypification and Epitypification of *Dracontium gigas* (Seemann) Engler (Araceae). *Novon* 4: 404–407.

Zinsser, W. 1988. *On Writing Well,* 3d ed. New York, Harper & Row.

Zomlefer, W. B. 1994. *Guide to Flowering Plant Families.* Chapel Hill, University of North Carolina Press.

Zweifel, F. W. 1961. *Handbook of Biological Illustration.* Chicago, University of Chicago Press.

Zweifel, R. G. and F. Parker. 1989. New species of microhylid frogs from the Owen Stanley Mountains of Papua New Guinea and resurrection of the genus *Aphantophryne. American Museum Novitates* 2954: 1–20.

Subject Index

Numbers in boldface indicate definitions of terms

Author Index

Page numbers in brackets refer to authors cited as "et al." in text

Taxon Index

The taxonomic names included in this index are those used in examples (not sample titles) or mentioned in the text.